MODERN METHODS IN PLANT TAXONOMY

Other Related Books

Ayurvedic Medicinal Plants of India, Vol. 1-2 (Set)	*Bhutiya, R.K.*
Biodiversity and Environmental Biotechnology	*Dwivedi, P.*
Biodiversity Conservation	*Kumar, M.S.B.*
Biodiversity, Conservation, and Systematics	*Singh, P.*
Ethnic Aphrodisiac Plants	*Sood, S.K.*
Ethnic Indian Plants in Cure of Diabetes	*Sood, S.K.*
Ethnobotany and Medicinal Plants of India and Nepal, Vol. 3	*Singh, V.*
Ethnobotany of Bhil Tribe	*Singh, V.P.*
Flora of Andhra Pradesh (India) Vol.5 : Supplements	*Pullaiah, T.*
Herbaceous Plants as Natural Protective Food	*Rana, M.K.*
Herbal Technology: Recent Trends and Progress	*Daniel, M.*
Herbs of Tribal Land Jharkhand, India	*Jaipuriar, M.K.*
Identification of Common Indian Medicinal Plants	*Naik, V.N.*
Identification of Weeds and their Control Measures	*Singh, S.*
Indian Folk Medicines and other Plant-Based Products	*Singh, V*
Indigenous Knowledge: An Application	*Sahu, T.R.*
Medicinal Plants of India Vol. 1-3 (Set)	*Jadhav, D.*
Multipurpose Plants	*Srivastav, P.K.*
Non-Conventional Legumes and Useful Plants of North East India	*Kumar, S.*
Ornamental Plants for Gardening	*Chopra, V.L.*
Sacred and Magico-Religious Plants of India	*Sood, S.K.*
Taxonomy of Vascular Plants	*Lawrence, G.H.M*
The Identification of Trees and Shrubs 2nd Edition	*Makins, F.K.*
The Medicinal and Poisonous Plants of India	*Caius, J.F.*
Useful Herbs of Planet Earth	*Daniel, M.*

MODERN METHODS IN PLANT TAXONOMY

Edited by

V.H. HEYWOOD

Department of Botany, The University Reading, England

United Book Prints

1st Reprint in India

Published by:

United Book Prints
4806/24, Ansari Road,
Daryaganj,
New Delhi - 110 002
Tel.: 011-41511055
E-mail: unitedbookprints@gmail.com

Print: 2016

ISBN: 978-93-83692-33-0

Printed in India

List of Contributors

J. P. M. BRENAN, *Royal Botanic Gardens, Kew, Richmond, England.*

C. D. K. COOK, *The Hartley Botanical Laboratories, The University, Liverpool, England.**

A. CRONQUIST, *The New York Botanical Garden, Bronx, N.Y., U.S.A.*

J. CULLEN, *University of Liverpool Botanic Gardens, Ness, Neston, Cheshire, England.*

M. DALE, *Department of Botany, The University, Hull, England.*†

F. EHRENDORFER, *Botanisches Institut der Universität, Graz, Austria.*

D. E. FAIRBROTHERS, *Department of Botany, Rutgers–The State University, New Brunswick, New Jersey, U.S.A.*

F. J. F. FISHER, *Department of Biological Sciences, Simon Fraser University, Burnaby, British Columbia, Canada.*

V. H. HEYWOOD, *The Hartley Botanical Laboratories, The University, Liverpool, England.*‡

R. W. HOLM, *Department of Biological Sciences, Stanford University, Stanford, California, U.S.A.*

M. P. JOHNSON, *Department of Biological Sciences, Kent State University, Kent, Ohio, U.S.A.*

J. McNEILL, *The Hartley Botanical Laboratories, The University, Liverpool, England.*

C. R. METCALFE, *Jodrell Laboratory, Royal Botanic Gardens, Kew, Richmond, England.*

D. M. MOORE, *Department of Botany, The University, Leicester, England.*

O. T. SOLBRIG, *Botanical Gardens, University of Michigan, Ann Arbor, Michigan, U.S.A.*

W. T. STEARN, *Department of Botany, British Museum (Natural History), London, England.*

W. H. WAGNER, *Botanical Gardens, University of Michigan, Ann Arbor, Michigan, U.S.A.*

D. A. WILKINS, *Department of Botany, The University, Birmingham, England.*

Now at: **Institut für Systematische Botanik der Universität, Zurich, Switzerland.*
†*Division of Plant Industry, C.S.I.R.O., Canberra, Australia.*
‡*Department of Botany, The University, Reading, England.*

Preface

This volume is based on the papers presented at a Conference on Modern Methods in Plant Taxonomy, organized by the Botanical Society of the British Isles in association with the Linnean Society of London, held in the University of Liverpool on 11–12 September 1967. The conference was attended by over 200 botanists from sixteen different countries, including a very strong representation from North America.

A great deal of lively discussion was generated by the papers, one of the highlights being the exchange between the leading phylogenists Professor Armen Takhtajan of Leningrad and Dr. Arthur Cronquist from New York. It has not, however, proved possible to include these discussions, even in abridged form: although they were recorded on tape, their transcription and editing would have delayed publication of this book by several months and it was decided to follow the precept that symposium volumes ought to be published promptly or not at all.

Contributors were asked to prepare their contributions for publication in the form of a review wherever possible. The result is a major survey of the main fields of present day taxonomy with an emphasis on the application of newer techniques.

The papers are presented more or less in the sequence in which they were given and grouped under the major headings followed by the symposium. The sessions were chaired by the Presidents of the two Societies, Dr. J. G. Dony and Professor A. R. Clapham, F.R.S. respectively and by Professor J. G. Hawkes, and Professor Irene Manton, F.R.S. The main discussions were introduced by Professor A. Takhtajan, Dr. K. Jones, Professor V. H. Heywood and Dr. W. Ernst.

The Conference was preceded by a field excursion to Ainsdale dunes led by Miss Vera Gordon of the Liverpool Botanical Society. On the final afternoon a series of demonstrations illustrating most of the techniques discussed in the papers was mounted in the Hartley Botanical Laboratories.

The combined efforts of many people were responsible for the success of the meeting: in particular Dr. C. D. K. Cook and other members of staff and research students of the Hartley Botanical Laboratories. Thanks are due to them and also to the Officers of the University for permission to hold the meetings in the new Science Lecture Block and for the reception held in the Tate Hall of the University for the participants.

I should like to acknowledge the assistance given by Academic Press and the prompt and efficient way they have seen the book through the Press; the excellent work of the printers speaks for itself.

Reading and Madrid
July 1968

V. H. HEYWOOD

Contents

Introduction

1 | PLANT TAXONOMY TODAY

V. H. HEYWOOD

The Continuing Role of the Herbarium in Modern Taxonomic Research

2 | THE RELEVANCE OF THE NATIONAL HERBARIA TO MODERN TAXONOMIC RESEARCH IN THE UNITED STATES OF AMERICA

A. CRONQUIST

3 | THE RELEVANCE OF THE NATIONAL HERBARIA TO MODERN TAXONOMIC RESEARCH

J. P. M. BRENAN

4 | REGIONAL AND LOCAL HERBARIA

J. MCNEILL

5 | CURRENT DEVELOPMENTS IN SYSTEMATIC PLANT ANATOMY

C. R. METCALFE

The Role of Experimental Data

6 | THE KARYOTYPE IN TAXONOMY

D. M. MOORE

7 | FERTILITY, STERILITY AND THE SPECIES PROBLEM

O. T. SOLBRIG

8 | PHENOTYPIC PLASTICITY WITH PARTICULAR REFERENCE TO THREE AMPHIBIOUS PLANT SPECIES

C. D. K. COOK

9 | HYBRIDIZATION, TAXONOMY AND EVOLUTION

W. H. WAGNER, JR

Biochemistry, Computers and Taxonomy

10 | CHEMOSYSTEMATICS WITH EMPHASIS ON SYSTEMATIC SEROLOGY

D. E. FAIRBROTHERS

11 | BOTANICAL PROBLEMS OF NUMERICAL TAXONOMY

J. CULLEN

12 | ON PROPERTY STRUCTURE, NUMERICAL TAXONOMY AND DATA HANDLING

M. B. DALE

13 | NUMERICAL TAXONOMIC STUDIES IN THE GENUS *SARCOSTEMMA* R.BR. (ASCLEPIADACEAE)

M. P. JOHNSON AND R. W. HOLM

14 | OBSERVATIONS ON A COMPUTER-AIDED SURVEY OF THE JAMAICAN SPECIES OF *COLUMNEA* AND *ALLOPLECTUS*

W. T. STEARN

Geography and Ecology

15 | THE SCALE OF GENECOLOGICAL DIFFERENTIATION

D. A. WILKINS

16 | THE ROLE OF GEOGRAPHICAL AND ECOLOGICAL STUDIES IN TAXONOMY

F. J. F. FISHER

17 | GEOGRAPHICAL AND ECOLOGICAL ASPECTS OF INFRASPECIFIC DIFFERENTIATION

F. EHRENDORFER

Introduction

1 | Plant Taxonomy Today

V. H. HEYWOOD

*The Hartley Botanical Laboratories, The University, Liverpool 3, England**

INTRODUCTION

Nearly 20 years ago the Botanical Society of the British Isles held a conference on The Study of Critical British Groups which was edited by the late A. J. Wilmott and appeared as the first of a series of conference reports under the title *British Flowering Plants and Modern Systematic Methods* (1949). In the intervening years the Society has organized conferences covering a wide range of topics, from Local Floras and Mapping the British Flora to Reproductive Biology and a Darwin Centenary. On this occasion the Society wished to return to a consideration of the general field of systematics, and it was decided that it would be helpful and instructive to regard it as a progress report or reassessment of the past 20 years, taking the 1948 conference as a starting point.

It is significant, I feel, that this symposium is not restricted to British flowering plants as in 1949. Our horizons have widened considerably during the past 20 years and not only have we extended our investigations to continental Europe but we have found that the techniques developed and applied in other parts of the world, particularly in the United States, have a direct relevance to the understanding of the systematics and behaviour of British plants. Nor is the symposium restricted to problems at the specific and infraspecific levels where the newer techniques of the 1930's and 1940's had most application, but also with generic and familial problems. In other words there is renewed interest in synthetic classificatory problems such as the construction of classifications from various sources of evidence as opposed to concentrating on analytical problems such as specific delimitation. New sources of evidence and newer techniques such as those of numerical taxonomy are largely responsible for this shift in attitude. At the higher levels of classification activity is considerable if more measured than earlier in the century. Numerous "new" systems of classification of Dicotyledons and Mono-

* Now at Department of Botany, The University, Reading, England

cotyledons have been proposed but there is, fortunately, a tendency for a considerable measure of agreement between them to become apparent, largely I suspect because of two factors—the one is what might be called the application of the process of deweighting whereby undue emphasis on particular lines of evidence or types of character is gradually eliminated, the other being simply the consideration by the authors of the new systems of largely the same evidence, leading inevitably to similar results. There are still exaggerated cases of *a priori* weighting and consequently wide discrepancies between the new systems in some respects and little further progress is likely until some form of numerical method of assessment can be devised that is capable of sampling and processing the large range of material and data involved.

The history of the development of taxonomy is very largely a reflection of the history of development of techniques which have a taxonomic application. These techniques are seldom developed by taxonomists but by scientists working in related biological fields in the main, and more recently by workers in quite diverse fields. In the 1940's the newer techniques applied were those of cytology, ecology, genetics and combinations of these such as genecology and cytogenetics. This is very clearly seen in the programme of the 1949 meeting. Shortly afterwards Stebbins' classic *Variation and Evolution in Plants* was published summarizing much of the effect of the application of cytology, genetics, etc. to systematics although he was not directly concerned with classification. At the Montreal Botanical Congress in 1959 Stebbins summarized progress during the previous 10 years and I remember clearly gaining the impression that nothing strikingly new had happened in the interim, but rather that our knowledge of the mechanisms of variation and evolution had been confirmed and consolidated. As is clear from several of the papers in this symposium, this process has continued on an ever-increasing scale.

The picture since then is dramatically different, largely because of the application of new methods, in particular chemical techniques such as chromatography and electrophoresis which by rapid and efficient screening of chemical compounds and consequently their quick identification, led to the field known as biochemical systematics which has itself evolved at an astonishing pace during the past few years; and computer technology and electronic data processing which has had its major impact on systematics so far as Numerical Taxonomy or Taximetrics.

LIGHT AND ELECTRON MICROSCOPY

There have been considerable developments in microscopy which one must not overlook. These have had a marked effect on morphological, anatomical, cytological and palynological studies applied to systematics. At the level of the dissecting microscope with relatively low magnifications, improved instrument design and optics (including zoom lenses) has greatly eased the task of interpreting micromorphological features such as details of leaf, fruit and seed surfaces, indumentum types, etc. These are often used in the separation of polyploid races. The compound light microscope and accessories have shown many design and optical improvements which with improved photographic and staining techniques have also been applied to the study of micro-characters such as stomata and associated cells and other epidermal features in the grasses and other groups. Stace (1965*a*) has recently reviewed the value of cuticular studies in taxonomy and has also published studies on the epidermis in the Combretaceae (1965*b*) and on their general value in evolutionary studies (1966).

The debt of anatomists to improved microscopy is clear from Metcalfe's paper in this volume. In palynological studies light microscopy continues to play a major role although much attention is focused on the use of the electron microscope (see later). Various methods of microscopy are used for the study of pollen grains such as phase-contrast, polarizing or fluorescence, and when the fine detail of the intine and cytoplasm is required recourse is made to ultraviolet microscopy (see review by Erdtman, 1963). Karyological studies have been greatly assisted by improvements in microscope design and technique including the photomicroscope with various degrees of automation allowing rapid examination and recording of preparations.

The electron microscope has so far played a limited role in systematic studies of higher plants, the one notable exception being in palynology where the transmission electron microscope is widely used for the study of the ultrafine structure and stratification of the pollen grain walls (exines) by means of ultra-thin sections. This is a field which has developed over the past 15 years or so and examples of outstanding systematic papers are those of Rowley (1959) on Commelinaceae, Skvarla and Larson (1965) on Compositae and Skvarla and Turner (1966) again on Compositae. It should be emphasized that in all this work there has been an extensive background of technical development much of it undertaken by workers in fields other than botany.

More recently there have been a few attempts to employ ultrastructural details such as cell organelles in a taxonomic framework, and Turner (1967) suggests that the electron microscope properly belongs with the gadgetry of what he calls the Biochemical Period of systematics and cites papers which are beginning to tackle macromolecular problems by electron microscopy. The days when the nuclear sequences of the DNA strands might be read off by means of electron microscopy appear not to be far off bearing in mind the types of electron microscope now being developed.

The recently developed scanning electron microscope is likely to prove of value in a wider range of systematic studies than the transmission type.

Details of the scanning electron microscope are given by Oatley (1966) and its role in the future of electron microscopy is discussed by Coslett (1967). Briefly, its advantages over the conventional electron microscope are its enormous depth of focus and the small amount of preparation needed for the specimens to be studied. There is no need for sectioning or replicas since whole mounts can be examined and the picture built up by the electron beam, which scans the specimen and through various intermediate steps presents the image on a cathode ray display screen, is in fact in some ways better, as regards contrast formation, than that obtained by visual examination of a specimen (Oatley, 1966). The working distance between the object and the face of the final lens need never be less than 50 mm and when used at lower magnifications it can be much higher, thus allowing the examination of quite bulky specimens and the possibility of carrying out manipulations to the specimen while under observation. The resolution of the scanning electron microscope is comparatively low—about 250 Å—about 20 times as good as an optical microscope but again about 20 times lower than a conventional electron microscope for examining transparent sections. Fortunately, this is no great disadvantage since in many cases the value of the scanning electron microscope is in examining objects at relatively low magnifications where resolving power is unimportant. The range of magnification is from about 20 × to 30–40,000 and remarkable results have been obtained at the lower levels of magnification in studies of the surface topology of fruits (Heywood, 1967, 1968)) where the enormous depth of focus is shown to good advantage. In the optical microscope the depth of focus is limited at high magnifications; in the scanning electron microscope it is hundreds of times better. Very useful and equally striking results have been obtained in the examination of pollen grains (Echlin, 1968) where it is clear that many problems will be most usefully tackled by a joint approach using light microscopy, transmission and scanning electron microscopy in concert.

An important advantage of the scanning electron microscope, especially at lower magnifications, is the ease with which objects can be recorded by photography in almost three dimensions thus acting as a rapid means of obtaining and storing accurate information which would otherwise have to be built up by a series of separate observations into a composite and time-consuming drawing.

Recent work on human chromosome preparations suggest that the scanning electron microscope might eventually be of value in karyotype studies in plants permitting dimensions and structure to be observed in detail.

Another area where the scanning electron microscope has already demonstrated its potential value is in the study of surface wax excretions (Amelunxen *et al.*, 1967). This is already being studied by transmission electron microscopy using carbon replicas, and the chemical composition of the waxes by microchromatographic and microspectrometric analytical procedures (Eglinton and Hamilton, 1967). There is some doubt so far as to the taxonomic value of the various types of wax projections or extrusions but only a very small range of material has been studied so far.

TAXONOMISTS AND MACHINES

In the previous section I have made frequent mention of the use of new apparatus and machinery. This leads on to much more subtle and as yet little appreciated developments which will, I believe, come to occupy an increasingly important role in systematics. Perhaps I can introduce this best by a quotation:

> "Man's intelligent behaviour is due in part to his ability to select, classify and abstract significant information reaching him from his environment by way of his senses".

This, one might think, would make a fitting introduction to a text book of taxonomy. In fact, it was written by Dr C. A. Rosen of the Applied Physics Laboratory of the Stanford Research Institute, California, in a paper entitled "Pattern classification by adaptive machines". He goes on to say "The function, pattern recognition, has become a major focus of research by scientists working in the field of artificial intelligence. At the lowest level, pattern recognition is reduced to pattern classification, which consists of the separation, into desired classes of groups of objects, sounds, odors, events, properties, and the like".

If we remove, perhaps, the sounds, this sounds very like a description of the activities in which the taxonomist engages in the herbarium, the field and the laboratory. It is precisely this aspect of taxonomy which has received least attention and practically no significant development in techniques except, perhaps, through the attentions of numerical taxonomists with their necessary emphasis on questions of the selection and definitions of characters for machine processing. In other words, the basic activities of the taxonomist, the perception and handling of characters and character complexes, on which the fabric of routine identification and classification is based, has been almost entirely neglected by the taxonomist and is being tackled by physicists, psychologists, data processors and others. The last preserve of the taxonomist—the taxonomic eye, the so called intuition, is being probed and analysed into its components with a view to machine copying.

The salient feature of taxonomy today is the relationship between the taxonomist and the machine which in some cases has almost become as important as the relationship between him and the plants. I should like to develop this theme a little because in the long run it will, I think, become the most potent characteristic of many branches of systematics.

It has been suggested by Professor McLuhan that a new generation is growing up which is not so much visual as audio-tactile. Although I am not sure that I agree with his analysis or even understand it as he intends, nor do I wish to become involved in the controversy surrounding his ideas in general, there seems to be a valid point here for taxonomy. Traditional taxonomic practice is basically visual, man being primarily a visually gifted animal. In the herbarium and field character assessments and correlations are indeed made visually—the so-called taxonomy at a glance, although exactly how this is carried out is far from clear as discussion within a group of practising taxonomists soon makes clear. In higher plants at least, most of our classification is morphologically, i.e. visually, based. There are sound practical reasons for this, such as conveniences of handling and identification, but even if there were not, until the present day we have been forced to adopt such visual methods *faute de mieux*. According to McLuhan's thesis the new generations are being brought up on electronic media such as television and telephonic communication but it is not these (all-involving, as he terms them) media which affect taxonomy—though their effect is not insignificant—so much as instrumentation, much of it electronic-based, such as the electron microscope, the computer, the cathode-ray display tube, electrophoresis apparatus, spectrophotometers, etc.

An obvious, although not fully appreciated point about the use of machines in taxonomy, whether it be the microscope or gas chromatograph, is that the resultant data are essentially non-visual. Chromatograms are presented in a visible form and the patterns of spots can be compared by eye but more useful results derive from analysing the pattern of spots and listing the results in terms of presence or absence of particular compounds in various taxa compared. Likewise, a protein "fingerprint" can, of course, be seen and the different patterns for different species compared visually, but a proper interpretation has to be much more analytical—the presence or absence of spots in particular sectors of the "fingerprint" being counted and compared numerically.

What must be emphasized, however, is that this kind of information, however assessed, cannot be compared visually along with gross morphological features *at the same time* at a glance. It has to be added in afterwards if the traditional visual methods of classification are applied. The same applies to the results of microscopic observations such as anatomical or karyological features; of course, one sees the chromosomes but chromosome number, for example, has to be added in after the morphological assessment. Thus, all our classifications start off with an inbuilt bias due to primary reliance on morphological features. This is not entirely a new problem since, for example, details of flower structure cannot be seen adequately at the same time as a gross assessment of the rest of the plant and yet such features are extensively used, at higher levels of classification particularly. Here again I suspect that what we tend to do is to make our primary dispositions on overall similarities and then resort to these subvisual characters for confirmation and certainly for analysis, i.e. identification and keying of the results of the previous activity.

Until recently, this problem had not reached serious dimensions. We are now approaching a new and somewhat baffling situation where in many cases we will have available as raw material for classification more non-visual and subvisual data than visual. It is surely no longer satisfactory to say that we make a traditional morphologically based classification first and then add in piece-meal all the non-visual data that agree, quietly rejecting those that do not.

In other groups where this situation obtains, recourse has been made recently to numerical methods provided there is a sufficient number of characters available. We are all aware of the every-increasing application of numerical taxonomy to some groups of micro-organisms such as bacteria where previous classifications have often been highly artificial. In higher organisms some form of numerical classification will be necessary if non-visual information in quantity is to be used adequately and not just adorn the pages of journals. This, of course, brings us back to the machine—the most expensive one we have yet considered—the electronic computer where even the "software" costs as much as the most expensive type of optical microscope.

Numerical taxonomy is considered in detail elsewhere in this symposium. Out of the confused pattern of the results published so far, two trends seem to be emerging at least as regards higher plants (and animals):

(a) The use of numerical taxonomy for classificatory purposes in assisting to produce classifications where conventional methods have so far failed or have produced several alternatives none of which is apparently satisfactory.

(b) Use of classifications produced by numerical means as models containing information and hypotheses for later testing, i.e. regarding them not so much in classificatory terms but as biological models. From this use comes an answer to the frequently posed question, is it worth all the trouble and labour of using the large number of characters needed for numerical classifications when results may be only marginally better than conventionally produced classifications using a fraction of the characters (at least consciously) and taking a fraction of the time to produce? The answer of course is that numerical classifications contain vast amounts of information about character correlations which would not otherwise be evident thus raising a whole series of biological questions. Contrary to what is often adduced by way of criticism of numerical classifications, far from being static and mechanical (and by implication of little biological interest) they are full of interest and information to the biologist not directly interested in the classifications as such. As could have been predicted, numerical classifications have also been used as a starting point for statistical approaches to phylogenies' (e.g. Camin and Sokal, 1965) although not without falling into some of the customary pitfalls as Colless (1967) has recently pointed out and to which I have made reference in various publications.

The preparation of data for computer programming in turn presents a whole series of problems which we are only just beginning to appreciate fully. Anyone who has been involved in a numerical taxonomic programme will know all too clearly how exceedingly difficult it can be at times to decide what to treat as characters, how to assess them and convert them into suitable fodder for the machine. It is an exercise drawing upon all one's biological knowledge (and considerable ingenuity and commonsense) contrary to what critics of numerical taxonomy frequently allege.

Williams, at a symposium on phenetic and phylogenetic classification in Liverpool in 1964 (Heywood and McNeill, 1964) suggested that numerical taxonomists are not taxonomists at all and would probably be better not to be. By this he meant that the real taxonomist, as I have just outlined, decides what attributes to measure and prepares the raw material using his extensive knowledge and judgement while the numerical practitioner takes the information he is given, decides what coefficient is to be measured and what methods to use. In other words, he looks after the machine side whilst the taxonomist looks after the plant side.

If we accept this argument we will end up with a whole series of practitioners processing data provided by taxonomists who, in turn, will analyse and scrutinize the results. Fisher, later in this symposium, also proposes that it will be necessary to train a new class of taxonomic field technician to collect field data of the kind he considers necessary to express the multidimensional patterns of ecogeographical variation found in plants. Likewise, there are many aspects of conventional taxonomy, particularly in the herbarium, where no great demands are made on the intellectual capacity of the taxonomist (as opposed to writing a full-scale revision or monograph, for example) and which could perhaps be more economically undertaken by fairly low grade assistants of non-graduate status. One can go further and see that in some cases, as in comparative biochemistry, even the data which will be supplied to the numerical practitioners will be provided by highly specialist non-taxonomists. In the field of biochemical systematics there are probably ten times as many chemists or biochemists providing usable data as there are taxonomists. And it is likely with the high degree of specialization of techniques that this must be so.

Before long, if this trend continues, as it surely must, there will be more non-taxonomists than taxonomists operating in the field of systematics. And the field of taxonomy, because of this, will become by far the most expensive in the whole of biology. If one adds to this the application of machines to the field of information storage and retrieval which are vital to taxonomy but which have been surprisingly neglected by taxonomists in this computer age (see the penetrating discussion by Crovello, 1967), then perhaps McLuhan is nearer the truth than we are prepared to admit.

Certainly the problems of training taxonomists for this new situation will require careful study, and proposals even more wide-reaching than those recently put forward by Raven and Holm (1967) may well be needed. In its wider context the systematics of the immediate future will share the problems of the whole of biology and once again act as its focal point.

REFERENCES

AMELUNXEN, F., MORGENROTH, K. and PICKSAK, T. 1967. Untersuchungen an der Epidermis mit den Stereoscan-Electronmikroskop. *Z. Pflanzenphysiol.*, **57**: 79–95.

CAMIN, J. H. and SOKAL, R. R. 1965. A method for deducing branching sequences in phylogeny. *Evolution*, **19**: 311–326.

COLLESS, D. H. 1967. The phylogenetic fallacy. *Syst. Zool.*, **16**: 189–295.

COSLETT, V. E. 1967. The future of the electron microscope. *J. R. microsc. Soc.*, **87**: 53–76.

CROVELLO, T. J. 1967. Problems in the use of electronic data processing in biological collections. *Taxon*, **16**: 481–494.

ECHLIN, P. 1968. Pollen, *Scientific American*, **218** (4): 80–90.

EGLINTON, G. and HAMILTON, R. J. 1967. Leaf epicuticular waxes. *Science*, N.Y. **156**: 1322–1335.

ERDTMAN, G. 1963. Palynology. *In*: R. D. Preston (ed.). *Advances in Botanical Research*, **1**: 149–208.

HEYWOOD, V. H. 1967. *Plant Taxonomy*. Edward Arnold Ltd., London.

HEYWOOD, V. H. 1968. Scanning electron microscopy and micro-characters in the fruits of the Umbelliferae-Caucalideae. *Proc. Linn. Soc.*, **179**: pp. 287–289.

HEYWOOD, V. H. and MCNEILL, J. (eds). 1964. *Phenetic and Phylogenetic Classification. Syst. Ass. Publ.* **6**. London.

OATLEY, C. W. 1966. The Scanning Electron Microscope. *Sci. Progr.*, **54**: 483–495.

RAVEN, P. H. and HOLM, R. W. 1967. Systematics and the levels of organisation approach. *Syst. Zool.*, **16**: 1–5.

ROWLEY, J. R. 1959. The fine structure of the pollen wall in the Commelinaceae. *Grana Palynologica*, **2**: 3–30.

SKVARLA, J. J. and LARSON, D. A. 1965. An electron microscopic study of pollen morphology in the Compositae with special reference to the Ambrosieae. *Grana Palynologica*, **6**: 210–260.

SKVARLA, J. J. and TURNER, B. L. 1966. Systematic implications from electron microscopic studies of Compositae pollen—a review. *Ann. Missouri Bot. Gard.*, **53**: 220–256.

STACE, C. A. 1965*a*. Cuticular studies as an aid to plant taxonomy. *Bull. Br. Mus. nat. Hist. Bot.*, 4(1).

STACE, C. A. 1965*b*. The significance of the leaf epidermis in the taxonomy of the Combretaceae. I. A general review of tribal, generic and specific characters. *J. Linn. Soc.* (*Bot.*), 50: 229–252.

STACE, C. A. 1966. The use of epidermal characters in phylogenetic considerations. *New Phytol.*, 65: 304–318.

TURNER, B. L. 1967. Chemosystematics. Present and future applications. *In*: *Symposium on Newer Trends in Taxonomy*. 189–211. *Nat. Inst. Sci. India Bull.* No. 34.

The Continuing Role of the Herbarium in Modern Taxonomic Research

2 | The Relevance of the National Herbaria to Modern Taxonomic Research in the United States of America

ARTHUR CRONQUIST

The New York Botanical Garden, Bronx, N.Y., U.S.A.

INTRODUCTION

In discussing the relevance of the national herbaria to modern taxonomic research in the United States I have followed the definition of national herbaria suggested by Heywood (*in litt.*) "major herbaria where there is a large staff working on a wide range of taxonomic and floristic projects covering various areas and indeed countries, as opposed to small university or local herbaria whose programs are related primarily to local floras or a small range of activities". By this standard, if we are not too strict about the criterion of a large staff, there are perhaps seven or eight national herbaria in the United States.

In more or less geographic order these are, or are located at, the U.S. National Herbarium, the New York Botanical Garden, Harvard University (including the Gray, Arnold Arboretum and Farlow Herbaria, now all grouped under one roof), the University of Michigan, the Field Museum (also known as the Chicago Natural History Museum), the Missouri Botanical Garden, the University of California (Berkeley), and possibly Stanford University. A few other herbaria are as large as some of these, but do not have diversified programs extending beyond their own areas.

Much of what Brenan writes in chapter 3 (p. 23–32) of this book is relevant to the United States and only the examples would have to be changed. I will, therefore, approach the subject a little differently.

FUNCTIONS OF THE NATIONAL HERBARIA

The functions of these national herbaria in the United States might reasonably be grouped under four headings: (1) their own research programs; (2) service as repositories of type and other historical materials; (3) the loan of specimens for study at other institutions; and (4) the training of graduate students.

Much of the research carried on at these national herbaria would be difficult or impossible at institutions with lesser or different resources. It requires the examination of large numbers of specimens, of many species, often from widely differing geographic regions. It is of course possible to borrow specimens, or to travel to other herbaria, but it is still necessary to have one's principal tools at hand for daily work. Major Floras and broadscale studies of families, for example, can only be effectively based at the national herbaria.

Local and state Floras can and often do emanate from institutions that do not have national herbaria. We do not generally expect the writer of such a Flora to investigate for himself whether current taxonomic opinion is correct, or whether the name by which a plant has been generally known (botanically) in his area is indeed the right one. Instead we expect him to find out what grows in his area and hopefully the local distribution of each species, to establish the names as best he can from the literature, and to provide accurate keys for identification. All this is no small task, but it does not necessarily require a national herbarium.

Regional Floras emanating from national herbaria should be more authentic and reflect more original work in the definition of taxa and the establishment of names. Obviously a man who is writing a Flora cannot do a monograph of every genus, but if he is at one of the major herbaria he should be able to improve our knowledge of many groups, rather than merely recording what is known or believed.

Floristic work on tropical and subtropical America is largely concentrated in the national herbaria of the United States. An increasing amount of work is being done in the countries directly concerned, however, and the end point of the change has not yet been reached. The situation might be roughly compared to that in the early part of the nineteenth century, when most of the floristic work on the present United States of America was being done in Europe. This sort of taxonomic ontogeny is probably inevitable, but it does create problems for the future. American botanists must now go to Europe to see type specimens of many of our species, particularly from the eastern states, and our friends from south of the border are going to have to consult North American herbaria in the future to establish the identities of their plants.

Broadscale studies of families, in which some attention is paid to generic limits and the relationships among genera, are carried on chiefly at the national herbaria. Staff members at the larger institutions are apt to list their interests as whole families or large regional floras, in contrast to the genera and state or local Floras more often listed by people in smaller herbaria. There are of course plenty of exceptions and borderline cases. The outstanding American student of the Crassulaceae operates at a university herbarium, admittedly a rather large one, which does not really qualify as one of the national herbaria. Even so, his work is mostly biosystematic rather than broadscale. The generic flora of the southeastern United States now being prepared at Harvard is an interesting combination of floristic and broadscale revisionary work. The principal function of this generic flora, in my view, is to provide guidance to botanists at smaller institutions who are interested in the southeastern flora but do not have the facilities to determine proper generic limits.

Biosystematic studies are also carried out at the national herbaria, especially those which are at the same time university herbaria. Harvard and Michigan have active programs in biosystematics as well as in those fields which are by nature largely restricted to the national herbaria, and the current activities at both Berkeley and Stanford are mainly biosystematic. One of the staff members at Michigan is also undertaking the preparation of a state Flora which will reflect taxonomic and nomenclatural studies of the sort to be expected in regional Floras, as well as a consideration of detailed local distribution.

Most general systems of angiosperms have emanated from major herbaria in one or another part of the world. Interestingly enough, however, a major herbarium may be less essential for this sort of work than for work at the level of genera within a family. At the level of the system of families and orders, much of the information must be taken from the literature rather than directly from the specimens. C. A. Bessey, the spiritual godfather of all modern angiosperm phylogenists, was never associated with a major herbarium after his student days. Another American botanist who has never been associated with a major herbarium also has a system of angiosperms nearly completed. Still, it is clear enough that it is useful to have the resources of a major herbarium at one's disposal when doing this sort of thing. The most serious weaknesses of Bessey's system are not in the principles but in the execution, and this is precisely where he might have been helped by working in a larger herbarium.

As all staff members of such institutions know, a major botanical library is a necessary adjunct to a major herbarium. We must not only study the specimens, but we must also read what our predecessors and colleagues have said about them. Taxonomy is unique among the natural sciences in that we must continually refer to publications of past eras, in order to make sure that we know what we are talking about.

The importance of large herbaria as repositories of types and historical materials needs no documentation. The scattered smaller herbaria could not perform this function nearly so well even if their continuity and safekeeping could always be guaranteed.

The major American herbaria lend great numbers of specimens for taxonomic research at other institutions. Last year the New York Botanical Garden sent out more than 20,000 specimens on loan. These represented more than 200 separate loans, involving nearly a hundred different institutions. Other major herbaria lend comparable numbers. A very large proportion of these loans are used by graduate students in the preparation of doctoral theses. Many institutions, even those with only small herbaria, undertake to give doctoral training in taxonomy. The necessary field, garden, and cytological studies for a biosystematic thesis can be based at a small institution just as well as at a large one, but if the job is to be properly completed, the specimens from a number of different herbaria must be assembled for comparative study. There is no effective substitute for the herbarium in determining the geographic and morphological limits of a species. No one person can hope to see in the field any large number of species, in each of its various guises and throughout its geographic range, but in the herbarium these very things can quickly be done for many species, at least those of temperate regions. In the herbarium one can travel from Maine to Alaska and California and back again in five minutes.

All of the national herbaria in the United States are associated in one way or another with universities which provide graduate training in taxonomy. The Harvard, Michigan, Stanford and University of California herbaria are integral parts of their own universities. The New York Botanical Garden and the Missouri Botanical have longstanding agreements with Columbia University and Washington University, respectively, designed to facilitate cooperation in student training. The United States National Herbarium and the Field Museum have less close and perhaps less satisfactory relationships with nearby universities. It is no accident that these two institutions, among the national herbaria, train the fewest students.

Having considered the general aspects of the relevance of the national herbaria to taxonomic research in the United States, I shall now interpret my assigned title liberally and outline some of my own recent work at the New York Botanical Garden Herbarium.

A PROPOSED PHYLOGENY OF THE MONOCOTYLEDONS

I have been interested for some time in the general system of angiosperms. Many of my ideas are similar to those of Armen Takhtajan, although there are still a number of unresolved differences. In the monocotyledons I have put some old ideas together with some new ones to come up with a scheme which I think is a real improvement over past ones. This I shall now outline.

The dicotyledons which gave rise to the monocotyledons must have had apocarpous flowers with a fairly ordinary (not highly specialized) perianth, and with uniaperturate pollen. They must also have been herbs which did not have a very active cambium, and they presumably had laminar placentation. The only order of dicotyledons which fits these specifications is the

Nymphaeales. I suggest that the premonocotyledonous dicotyledons were indeed something like the modern Nymphaeales. It is noteworthy also that the Nymphaeales are aquatic, that they lack vessels, and that they show tendencies toward the union of two cotyledons into one.

I believe that the monocotyledons had an aquatic ancestry, and furthermore that this aquatic ancestry has had a profound influence on the subsequent evolutionary history of the group. Loss of cambium, loss of vessels (except for some vestiges in the roots), and reduction of the leaf to a phyllode are all probably associated with the aquatic habitat. When some members of the group returned to dry land, their further evolutionary progress was hindered by the absence of secondary thickening, the virtual absence of vessels, and the lack of a normal leaf blade. It is always easier to lose something than to get it back, in evolution as well as in human activities, and some things, once lost, are never regained. Although some of the more advanced monocotyledons have managed to develop a functioning vessel system and broad, net-veined leaves, none of them has regained a typical cambium. All secondary thickening in the monocotyledons is "anomalous". The fact that habit differences are often taxonomically more important among the monocotyledons than they usually are in the dicotyledons reflects the evolutionary difficulties under which the monocotyledons labor.

It is possible that loss of the cambium by the ancestral premonocotyledons preceded their entry into the water, rather than resulting from it. That would not make a lot of difference to the evolutionary history of the monocotyledons *per se*, and perhaps we do not need to resolve that question. Still, it would be a little easier if we figured that the herbaceous, terrestrial premonocotyledons, before they took the fateful plunge into the water, had some cambial activity, and that vessels were largely confined to the secondary tissues. Then the loss of cambium, in association with the aquatic habitat, would also get rid of most or all of the vessels.

I do not accept Cheadle's argument (1953) that vessels originated entirely independently in the monocotyledons and dicotyledons, and that the evolutionary divergence of the two groups must have preceded the origin of vessels in either, because vessels appear first (phyletically) in the stem in dicotyledons and in the roots in monocotyledons. If the vessels in the early monocotyledons or premonocotyledons are vestigial, then they could just as well have survived in the roots as anywhere else, especially inasmuch as roots seem to be phyletically rather conservative anyway. Nor do I see any reason why there might not have been a phyletic regression in the degree of specialization of vessels, under conditions in which they had no survival value. The concept that evolutionary advances in the xylem, and specifically in vessel structure, are irreversible is based on land plants, in which these progressive changes can be presumed to have survival value. Remove the selective force and you have a different situation. The deletion of a single enzyme, by mutation, can break a biosynthetic chain and push a complicated structure back to an earlier evolutionary stage, as is well shown by the peloric flower forms of the cultivated snapdragon.

However, if those who insist that the monocotyledons have a primitively vesselless ancestry will also admit that the Nymphaeales are primitively vesselless, I won't make a big issue of it. Takhtajan (1959) does in fact consider the Nymphaeales to be primitively vesselless, but in one of his papers on vessels in monocotyledons Cheadle (1953) indicates that he thinks the Nymphaeales have lost their vessels, rather than never having had them.

In any case, I believe it is necessary to consider that the monocotyledons do have an aquatic ancestry, that these aquatic ancestors either lost or did not have a functional cambium to begin with, that they either lost or did not have a functional vessel system to begin with, and furthermore that these ancestors had much in common with the modern Nymphaeales.

The typical parallel-veined leaf of monocotyledons is a modified, bladeless petiole. All stages in the change from an ordinary leaf with petiole and net-veined blade to a bladeless, parallel-veined petiole can be seen in a single population of a single species of *Sagittaria*, according to the depth of the water, as I have observed in Minnesota, and as De Candolle pointed out as long ago as 1827. The blade of *Sagittaria* is probably only a secondarily expanded petiole-tip, not strictly homologous with the blade of dicotyledons, but the present structure is nonetheless that of a petiolate leaf with a well-defined, palmately net-veined blade. *Sagittaria* provides us with the equivalent of a see-it-now television production. It may not be the real thing, but it helps us to understand what the real thing was like.

The sheath-blade structure of monocotyledon leaves—here I am calling that flattened petiole-tip a blade—reflects the cotyledonary structure. Drawing on some still existing parallels and intermediates in the Nymphaeales, I suggest that the single cotyledon of monocotyledons is derived by lateral fusion of two cotyledons, followed by the reduction and loss of one of the lobes. Regardless of the morphological nature of the single cotyledon, it appears that throughout the monocotyledons it is basically the same organ. There is no need to assume a polyphyletic origin of the group. Monocotyledonous dicotyledons are something else again. At least some of these have merely lost or suppressed one of the two cotyledons.

The adventitious, fibrous root system of monocotyledons reflects the lack of a cambium. Having no adequate means of secondary thickening, the individual roots cannot persist, enlarge, and ramify.

The unique septal nectary of many monocotyledons helps to unify the class and also to strengthen the concept that the Alismatidae are near-basal (Brown, 1937). Septal nectaries probably take their origin in the mostly apocarpous subclass Alismatidae. *Sagittaria* and other Alismataceae have nectaries between the petals and staminodes and between and around the staminodes and lower carpels. *Alisma* itself, with a single whorl of carpels, has a nectary at the base of the split between any two adjacent carpels. The palms, which range from apocarpous to syncarpous, have correspondingly alismatoid to septal nectaries. Presumably a similar change has occurred in the lines leading to the Liliidae and Commelinidae.

I recognize four subclasses of monocotyledons: Alismatidae, Arecidae, Commelinidae and Liliidae. These compare closely with the subclasses proposed and named by Takhtajan (1964, 1966), except that the Bromeliales, Zingiberales, Juncales, Typhales and Cyperaceae belong (in my opinion) to the Commelinidae instead of the Liliidae.

Each of the four subclasses of monocotyledons tends to exploit a different ecological niche or set of niches, although with much overlapping. The Alismatidae are chiefly aquatic, whereas the other subclasses are chiefly terrestrial. I should point out that the progressive adaptation from semi-aquatic to fully aquatic to finally marine habitats in the Alismatidae probably reflects another return to the water after a more or less terrestrial interlude, rather than being the original invasion of the water by monocotyledons The Arecidae typically have large, often petiolate leaves and are usually either arborescent or have the flowers crowded into a spadix. Except for the Arales, they have a well developed vessel system. The palms are the principal culmination of the Arecidae. The Commelinidae have intensively exploited the avenue of floral reduction and wind pollination, culminating in the grasses and sedges. The Liliidae have intensively exploited insect pollination, and they also have bulbs, tubers, or corms much more commonly than do the other subclasses. The orchids are the culmination of the Liliidae.

It may be of some interest to follow the partly divergent and partly parallel series of advances in the Commelinidae and Liliidae. The hypothetical common ancestor of these two groups may be assumed to have been a narrow-leaved herb with the vessels confined to the roots, and with several subsidiary cells associated with each stomate, and with the sepals and petals well differentiated from each other, with a superior ovary, septal nectaries, and a starchy endosperm. Such a plant would probably be referred to the Commelinidae if we had it, but no such plant exists today.

Following the Commelinid line, we find first an early divergence of the Zingiberales, leading to broad, veiny leaves and epigynous flowers, although there is no obvious reason why these two features should be linked. No other group appears to be derived from the Zingiberales, which are so distinctive that they might with some reason be treated as a separate subclass. Returning to the main Commelinid line, we next see the development of a vessel system in all vegetative organs. When this development was nearing completion the

Bromeliales diverged into their specialized ecologic niche as xerophytes and then epiphytes, with an inferior ovary. The Bromeliales are a terminal group, which has given rise to nothing else. The main Commelinid line then loses its nectaries, and we reach the order Commelinales. Several lines of floral reduction then lead from the Commelinales to the terminal groups Eriocaulales, Restionales, Juncales, Cyperales (including the Gramineae) and Typhales. We do not know why the Commelinales lost the ability to produce nectar, but the loss evidently set the stage for floral reduction associated with wind pollination. In at least some of the Commelinales insect visitors to the flowers collect pollen instead of nectar. It is easy to suppose that the need to produce more pollen to offset the losses to marauding insects might also prepare the way for anemophily.

Picking up the Liliid line at its pre-Commelinid base, we find an immediate critical change in the perianth, so that both the sepals and the petals are petaloid. Concomitantly with this change, so far as present information shows, the number of subsidiary cells associated with each stomate was reduced to two. At this stage the Pontederiaceae and probably the Philydraceae diverged. The Pontederiaceae, like the Alismatidae, may well be only secondarily aquatic. In any case the remainder of the Liliidae are basically land plants (or epiphytes). The next step in the main Liliid line is the substitution of other food reserves (especially cellulose or oils) for starch in the endosperm, followed or accompanied by the loss of all subsidiary cells from the stomates. We are now in the midst of the order Liliales. Three of the characters which have been attained by the Liliid line at this stage are nearly or quite without parallel in the Commelinidae: (1) petaloid sepals; (2) nonstarchy endosperm; (3) stomates without subsidiary cells. Some of the Liliaceae now develop vessels in the stem and leaves, usually in association with an expansion of the leaf surface, and some of these vesseliferous lines develop an inferior ovary—e.g. the Dioscoreaceae. The larger group of typical Liliidae, however, moves separately into epigyny and thence into mycotrophy and the development of highly complex pollinating mechanisms, as in the Orchidaceae.

REFERENCES

Brown, W. H. 1937. The bearing of nectaries on the phylogeny of flowering plants. *Proc. Am. Phil. Soc.*, 79: 549–595.

Cheadle, V. I. 1953. Independent origin of vessels in the monocotyledons and dicotyledons. *Phytomorphology*, 3: 23–44.

De Candolle, A. P. 1827. *Organographie Végétale*. Paris.

Takhtajan, A. L. 1959. *Die Evolution der Angiospermen*. Gustav Fischer Verlag, Jena.

Takhtajan, A. L. 1964. The taxa of plants above the rank of order. *Taxon*, 13: 160–164.

Takhtajan, A. L. 1966. *Systema et Phylogenia Magnoliophytorum*. Soviet publishing institution "Nauka", Moscow and Leningrad.

3 | The Relevance of the National Herbaria to Modern Taxonomic Research

J. P. M. BRENAN
Royal Botanic Gardens, Kew, Richmond, England

INTRODUCTION

I have noticed an occasional tendency among less well-informed propounders of other disciplines of botany to believe that, because plant classification has a lengthy history extending back over several centuries, it must naturally in the course of time have become decrepit, static and generally out-dated. Here, as elsewhere, there is always a danger of being too anthropomorphic! The tendency to regard systematic botany, particularly when carried out in herbaria, in this way has perhaps been encouraged by the fact that its methods are not always easy to put across, that its practice has sometimes seemed more an art than a science, and perhaps that intuition born of skill and experience is more necessary than it sometimes is elsewhere. I believe that this condescending view is less prevalent now than it was—perhaps it helped when systematics acquired its less easily understood but more erudite-sounding title of taxonomy!

It is not my intention to present an apologia for systematic botany. At the same time I must make it clear that the outdated image suggested above does not, of course, apply to the national herbaria in Great Britain.

PURPOSE AND FUNCTIONS OF THE NATIONAL HERBARIA

Before one can consider the relevance of the national herbaria to modern taxonomic methods, it is important to know something of what these herbaria are and, even more important, to consider their purpose and functions. The three great national herbaria in Britain are, of course, located at the Royal Botanic Garden, Edinburgh, the British Museum (Natural History) and the Royal Botanic Gardens, Kew. Although these three great institutions are each of them conducting research in the same field of knowledge, the co-operation between them, both official and unofficial, is so close that the obvious danger of their overlapping in work is reduced as much as possible. Indeed, a few years ago an agreement was reached under the aegis of the Treasury between Kew and the British Museum which to some extent delimited the official spheres of work of these two herbaria. Thus, for instance, under this agreement, official research on British and European plants as well as on those of North America is now carried out exclusively by the British Museum and not by Kew. There are other such delimitations which I need not go into here. I shall use mainly Kew in my discussion, although realizing that the factors and problems are in large measure common to all three places.

As I said previously, it is essential to know something of the functions of the national herbaria before assessing their relevance to modern taxonomy, and it is this aspect that I wish to deal with first.

The national herbaria were brought into existence and have subsequently continued primarily in order to fulfil public needs, as is indeed only reasonable since they are supported by public funds. This service to public interests must therefore take a high place among their functions, and must inevitably have a considerable effect on their staffing and the sort of research carried out. These public interests embrace a wide range of people and bodies—ordinary private members of the public, amateurs of botany, those merely interested in plants, schools, universities and training colleges, and research institutes and Government departments both at home and abroad.

The general purpose of the national herbaria is the accurate identification of plants and plant material, and the making of this information available. The latter is done in numerous ways. At one end of the scale, letters arrive almost daily from members of the public containing requests for the right names of unknown plants found in gardens or in the wild. These must be answered as helpfully and courteously as possible, and without undue delay. At the other end of the scale, there are requests for the identification of large collections made in other countries or the preparation of Floras covering extensive areas of the globe. Some of these collections may be very important for economic purposes. Thus, large collections from Ethiopia and Kenya have been identified for the U.S. Department of Agriculture in order that the samples of the plants collected might be tested chemically for their efficacy in the treatment of cancer. At present, Kew has partial or complete responsibility for Floras of Tropical East Africa, West Tropical Africa, the Zambesi region, Cyprus and Iraq, besides providing help on a smaller scale to a number of other Floras of other regions. These tasks may, of course, be very time-consuming and require the cooperation of a number of botanists. In order that the identification of plants may be as accurate and efficient as possible, it is necessary, in addition, for active basic research to be undertaken so that important, neglected or difficult groups of plants may be taxonomically revised. If a genus has not been monographed or revised recently, then identification may be often difficult or impossible. It is clear that, if the national herbaria are to fulfil their service to the public, fundamental and applied research are both relevant and play their part, although the boundary between them may be sometimes hard to draw. It also happens that the demands of the applied side may be in opposition to those of fundamental research, particularly where there is competition for use of limited time and limited staff. The "Floras *versus* monographs" controversy is an old one and has been much argued. The balance is not always easy to hold but the requirements of both these aspects of research must as far as possible be met and balanced if the national herbaria are to continue to fulfil their useful public role.

The title of this chapter compels me to answer the question of what relevance the national herbaria have to modern taxonomic methods. One is tempted to follow this with the question "What are the modern taxonomic methods?" We at Kew have tried to follow the principle that as much evidence as possible from as many different lines of research as possible should be taken into account in making taxonomic decisions. We thus believe that modern taxonomic research has not only great relevance to Kew, but that it is, in fact, what we do there. This was, I believe, the thought behind the foundation of the Jodrell Laboratory at Kew as long ago as 1876 and certainly behind its recent expansion, rebuilding and re-opening in 1965. In this Laboratory active research is carried out in the fields of cytology, physiology and anatomy, and there is no reason why its activities should not expand into other fields should the need be shown. I can perhaps best indicate the relevance of these newer methods of research by giving a few examples of their practical use in taxonomic research at Kew.

ANATOMY

The value of anatomy to plant taxonomy has long been recognized at Kew and a tangible result of this is the two massive volumes on the Systematic Anatomy of the Dicotyledons by Metcalfe and Chalk (1950). Two additional volumes on the Monocotyledons appeared in 1960 and 1961 and others are in course of preparation.

Let me give also one or two examples of smaller pieces of research. Cutler (1965) conducted an anatomical investigation into the strange South American genus *Thurnia* Hook. f., previously although uneasily assigned to the families Juncaceae or Rapateaceae, or else placed as a family of its own. As a result, he found various remarkable anatomical features, for example, inverted bundles of an unique type in the leaf, and thus provided further positive evidence for the view that *Thurnia* should be assigned to a family of its own, the Thurniaceae, separate from the Juncaceae and Rapateaceae. Here is an instance of anatomy giving a positive taxonomic lead. It is noteworthy that this investigation was made entirely on dried herbarium material, revived in boiling water, cooled, and fixed with formalin acetic alcohol. Forman (1966) using anatomical and pollen characters (the latter inconclusive) concluded that the family Pandaceae, previously generally considered monotypic and limited to one species in West Africa, should be expanded by the inclusion of the genera *Galearia* Zoll. and Mor. and *Microdesmis* Hook. f., both previously placed in *Euphorbiaceae*. As Metcalfe stated (1966:319) "Taking all of these facts into account the anatomy [of the leaf and stem] appears to provide strong evidence that *Panda* and *Galearia* are quite closely allied to one another."

A further example of the way in which vegetative anatomy, together with concurrent studies of pollen morphology, can give a positive lead in suggesting the correct affinities of genera and species whose family position had been hitherto doubtful is provided by the joint research of Erdtman and Metcalfe (1963). But these are just a few examples taken from among many of fruitful cooperation between anatomy and taxonomy.

There can be no doubt that the great collections of dried plants in the national herbaria provide an immense potential wealth of material for anatomical study. Although the objection may be made that material of this sort, dried and pressed flat, is unsuitable for this sort of research on account of the distortion of tissues consequent on drying and pressing, recent investigation into ways of "reviving" such specimens has demonstrated that their value as research material to the anatomist may be great. It is true that, particularly where tissues are exceptionally thick or fleshy or where organs are very fragile or evanescent, no treatment will effect satisfactory reconstitution. Yet it is among plants where this difficulty is most acute that the greatest wealth of spirit material of the national herbaria lies. Also, these groups of plants are a comparatively small proportion of the whole, and the difficulty imposed by fleshy or fragile tissue often resides in a single organ. Dr W. R. Ernst of the Smithsonian Institute in Washington, who has recently been studying the vascular anatomy of the receptacle in Papaveraceae in conjunction with his more general work on this family informs me that herbarium material, suitably pretreated, has provided very satisfactory (and sometimes the only available) material for his studies.

CYTOLOGY

The days are, I am glad to say, past when it might have been necessary to justify to a predominantly taxonomic audience the potential value of cytology in taxonomic research. Some of us may have little understanding of the chromosomes and their behaviour or may even be disposed to regard them as morphologically similar to and of the same incomprehensible genus as "those damn dots", as Lord Randolph Churchill once referred to decimal points! Their immense value in elucidating difficult taxonomic questions, particularly at the level of species and below, has been most conclusively demonstrated in a multitude of problems for this line of research to be treated with anything other than the greatest respect by the taxonomist. The value of cytology has been openly recognized at Kew since the Cytological Department was established as part of the Jodrell Laboratory in 1960 under Dr Keith Jones. Since then there has been a fruitful history of cooperation at Kew between the taxonomists and cytologists. In my own studies on Commelinaceae I have received most valuable help from Dr Jones and his colleagues, whose work has provided unexpected but most relevant evidence about the evolution of the family. A further example of fruitful joint work is given by the work on the fern genus *Cheilanthes* of Manton *et al.* (1966).

Kew, where a rich Herbarium and a great Botanic Garden are adjacent to one another, is particularly well placed for this sort of cooperation in research. The facilities of the gardens provide a fine opportunity of growing living material for cytological research, though there are, of course, serious limitations which must be recognized. Much of the taxonomic research in the national herbaria is carried out on the plants of tropical countries, partly because there tend to be more unsolved taxonomic problems in the tropics, and partly in response to the need of developing countries to assess their vegetation and flora. Where plants for study need to have the protection of glass in order to be grown satisfactorily in this country, the number of individuals and species which can be grown, with only a small area of glasshouses available, is thus automatically limited. With large shrubs and trees the competition for limited space becomes, of course, especially acute, and only a token representation of woody tropical species can therefore be grown. Also, many tropical trees cannot be grown satisfactorily in this country because they do not reach the stage of flowering and fruiting until they have attained a greater height than the dimensions of normal greenhouses allow, and cytotaxonomic research on them is thus hampered. There is no doubt, however, that, in spite of these limitations, cytology has a valuable role to play as far as the national herbaria are concerned, particularly where herbaceous plants are concerned and where the cultural demands made by them are not too great.

The national herbaria are also relevant to cytological research as providing a place where authentically documented specimens of plants which have been investigated may be permanently deposited. A number of earlier cytological results remain doubtful because voucher specimens were not kept. A naïve trust was sometimes placed in the names of plants and seed-packets from botanic gardens, with the result that it is now sometimes impossible to verify the precise identity of the plants investigated. Voucher specimens for cytological research preserved in this manner are often the only means whereby subsequent workers can test the basis of previous research, and the national herbaria have their part to play in housing them.

PALYNOLOGY

Within the last 20 years the study of pollen has expanded to such an extent that it now has its extensive separate literature and is virtually a discipline of its own. Pollen grains are comparatively little altered by the processes of preparing specimens for the herbarium, and as a result herbaria have already been extensively used as a source of reference material for comparative studies in palynology. There can be little doubt that great herbaria with world-wide coverage, such as our own national ones, provide an unrivalled wealth of palynological material, although naturally there is a tendency, which must be guarded against, to pillage specimens too often of their pollen, and this is particularly relevant when the specimens of a genus or species are very few.

It is a real pleasure here to pay tribute to the friendly cooperation which the Royal Botanic Gardens, Kew, has received over the years from Professor G. Erdtman and his staff in Stockholm. We have often asked for his help in examining the pollen of plants either new or of doubtful taxonomic position, and never in vain. Furthermore, he has been generous in giving to Kew duplicates of the pollen preparations which he has made, with the result that there is a very useful nucleus of a pollen slide collection.

The joint work of Erdtman and Metcalfe (1963) has already been mentioned as an example of the way in which palynology and anatomy can cooperate in solving taxonomic problems, though this relates only to two small genera and one species of doubtful position.

An additional noteworthy example is given by the work of my colleague, Mr C. Jeffrey, on the family Cucurbitaceae, in which, having put forward a new system of classification with the family based on gross morphology, he enlisted the help of Professor Erdtman in putting it to the test by means of a comparison with the results of an extensive study of the pollen morphology of the family (Jeffrey, 1964). To quote Jeffrey's words (p. 473):

"The completion of a set of pollen-slides of Cucurbitaceae, kindly prepared and donated to Kew by Prof. G. Erdtman of Stockholm-Solna, has enabled the classification previously outlined by me [see *Kew Bull.*, 15: 338–346 (1962) and 16: 483 (1963)] to be tested against the natural groupings within the family as indicated by pollen morphology. The degree of correspondence between the groups of the classification (based on general morphology) and the pollen morphological ones may be taken as an indication of the success of the classification in its attempt to define as exactly as possible the natural groupings of genera within the family. It is very satisfactory to be able to report that the degree of correspondence is high, much more so than with any previous scheme of classification."

He goes on to say (p. 474):

"Two important facts emerge: one, that the natural limits of subfamilies, tribes and subtribes within the family have now been definitely delimited, the other, that no future worker on the taxonomy of Cucurbitaceae can afford to ignore pollen morphology."

There can be little doubt that there is still great scope for the employment of palynological evidence in tackling other taxonomic problems, and this is certainly a field of research to which the national herbaria are very relevant and which is itself important for the work of the herbaria.

PHYSIOLOGY

The institution of a Physiological section at the Royal Botanic Gardens, Kew, is a recent event, dating only from the opening of the rebuilt Jodrell Laboratory in 1965. It is thus still early to forecast how closely and relatively physiological research may be integrated into the taxonomic work at Kew. There is no doubt at all that physiological data, particularly in the phytochemical field, can be important in making taxonomic decisions. Differences in the chemical constitution of species can be as reliable and significant characters as those of gross morphology and may also, to a much greater extent than cytological evidence, give valuable pointers towards the correct delimitation of families and genera. Interesting work has been recently carried out on the comparative study of proteins in leguminous seeds as shown by paper chromatography tests. Specimens from the Kew Herbarium have been employed and results have shown that seeds preserved as herbarium specimens can be satisfactorily employed as material for this sort of research.

Recent research by Mr T. Reynolds and Miss Susan Mills, as yet unpublished, involving a chromotographic survey of anthocyanin types in the genus *Rhododendron*, has provided very interesting taxonomic evidence. So far the research is in its early stages and is confined to red and purple-flowered members of 22 series in this vast genus. The analysis of these results will be lengthy but it is clear that they are full of taxonomic implications.

NUMERICAL TAXONOMY

It is, of course, obvious that numerical taxonomy carries out a valuable function in that it may express in mathematical and hence theoretically verifiable terms the bases on which classifications are made. The taxonomist's intuition or "hunch" now becomes susceptible to closer and more reasoned criticism and he must surely welcome this. Classical taxonomy has always been closely concerned with measuring organs, and numerical methods, used in an unsophisticated sort of way perhaps, are nothing new or strange.

The suggestion is sometimes made that it might be desirable to computerize the whole herbarium. It seems easy to overestimate the potentiality of the computer and I am far from convinced that a good well-managed herbarium does not incorporate in its own arrangement its own index, as easy and efficient to use as any provided by mechanical means.

I have mentioned before the world-wide coverage in the national herbaria. It is not always enough appreciated that this wealth of material often provides a literally unrivalled record of geographical distribution and variation of plant species. In the Kew Herbarium alone there are between four and five million specimens and although, naturally, coverage is unequal, i.e. some parts of the world are, as one might expect, very much better represented than others, nevertheless it is possible to gain a remarkably accurate picture of the general distribution of most species and also by close study of the specimens to analyse and assess their variation. Where the coverage is relatively good and there is a good proportion of modern, well-annotated specimens, as for most of the continent of Africa for example, I have repeatedly found that a close study of the material of a given species will reveal whether there is any recognizable pattern of variation and will often show whether that pattern is correlated with geography, climate or ecology. It may be objected that the specimens are not random samples, and this may be true, though I think that the criticism misses the point in that, although it may be possible to make random samples within a population and that this method may give more accurate results as far as that population is concerned, yet the difficulties of sampling numerous scattered populations of a widely distributed species may in practice be formidably great, particularly in terms of time and money. Numerous specimens from many parts of the total range of a species may be a fairly close approximation to random sampling for the total species-area.

In well-known countries such as our own, where collections are often made by those familiar with the identity and appearance of the plants they see, it may happen that an undue proportion of collections may be abnormal or untypical, the variant or sort of a species tending to be more frequently collected than the common typical plant; but I do not believe that this tendency is so marked in the tropics where collectors tend to be much less familiar with the species they meet with. These collections in the national herbaria will, I am sure, be increasingly recognized as an invaluable repository of geographical variation and I believe that it will be possible, should there be need, for much of this to be analysed in numerical terms.

CONCLUSIONS

Although I have mentioned some of the more recent developments in taxonomy and tried to show that they are relevant to the work of the national herbaria, and *vice versa*, none of these techniques has really shown that it is more than a new line of attack on the general problems of classification. There has been no sign that classical taxonomic methods with their reliance on gross morphology are in danger of being toppled from their pedestal or indeed that they are seriously unstable. Classical taxonomy has made many mistakes, due sometimes to lack of evidence or to poor work, and some of these have been rectified with the assistance of new techniques. There are also some problems where the orthodox taxonomist finds himself baffled, particularly where hybrid complexes are concerned, or excessive variation or apomixis, and there is no doubt that new methods, particularly cytology, have shed new and revealing light. As Constance (1964) has well said:

"The relevance to classification is that every few years some new approach or technique is proclaimed which this time is going to be successfully exploited to get the stone—that is, taxonomy—over the crest of the ridge dividing intuitive art from exact science. Anatomy, paleobotany, embryology, palynology, cytology, and genetics, to name a few, have all had their try; chemistry and mathematics are chafing eagerly in the wings to have their day upon the stage. One may be reasonably sure that other actors lurk behind these, although their features are not as yet quite discernible."

He went on to quote:

" 'Up to the present,' according to Sharp, 'our science has been, in part, an art, intuitive and descriptive, and of necessity, in part must continue to be so . . . the artist and the computer must both be retained.' "

Taxonomists in the national herbaria will not immediately or necessarily make use of all these new lines of research, for obvious enough reasons, but they should be aware of them and take them into account, in affirmation of the principle that the wider the evidence on which taxonomy is based the more firmly it is likely to stand.

REFERENCES

CUTLER, D. F. 1965. Vegetative anatomy of *Thurniaceae*. *Kew Bull.*, **19**: 431–441.

CONSTANCE, L. 1964. Systematic botany—an unending synthesis. *Taxon*, **13**: 257–273.

ERDTMAN, G. and METCALFE, C. R. 1963. Affinities of certain genera *incertae sedis* suggested by pollen morphology and vegetative anatomy. *Kew Bull.*, **17**: 249–256.

FORMAN, L. L. 1966. The reinstatement of *Galearia* Zoll. & Mor. and *Microdesmis* Hook. f. in the *Pandaceae*. *Kew Bull.*, **20**: 309–318.

JEFFREY, C. 1964. A note on pollen morphology in *Cucurbitaceae*. *Kew Bull.*, **17**: 473–477.

MANTON, I., ROY, S. K. and JARRETT, F. M. 1966. The cytotaxonomy of some members of the *Cheilanthes farinosa* complex in Africa and India. *Kew Bull.*, **18**: 553–565.

METCALFE, C. R. 1966. Notes on the anatomy of leaf and stem of *Panda oleosa* and *Galearia celebica*. *Kew Bull.*, 20: 318–319.

METCALFE, C. R. and CHALK, L. 1950. *Anatomy of the Dicotyledons*. Vols. 1 and 2. Clarendon Press, Oxford.

4 | Regional and Local Herbaria

J. McNEILL

The Hartley Botanical Laboratories, The University, Liverpool, England.

INTRODUCTION

The importance of the great national herbaria, or *major* herbaria, as Davis and Heywood (1963: 259) more accurately term them, is not often, if at all, seriously called into question and the two preceding chapters in this book ably demonstrate the enormous contribution that they are making and will continue to make not only to taxonomic research but to the whole of systematics. The same cannot be said of what in the British Isles we usually think of as university and local herbaria. At the turn of the century in a University Botany Department the maintenance and development of a herbarium was regarded as axiomatic; it was, too, a period of flourishing local natural history or botanical societies often possessing their own herbarium or else actively associated with the collections maintained by a local museum. Today the whole pattern of university teaching has changed, the scope of botanical research has broadened immeasurably, and the naturalist is rightly more interested in conservation than in a stamp-collecting approach. One might ask if, with Clapham, Tutin and Warburg (Clapham *et al.*, 1959, 1962), Ross-Craig (1948–) and even Keble-Martin (1965) to identify one's plants, the BSBI Atlas (Perring and Walters, 1962) or perhaps a Dony (1953, 1967) or a Travis (Savidge *et al.*, 1963) to find where they grow, there is any need for a herbarium. Should one not, perhaps, destroy the specimens, dispose of the cupboards and release the space for a conservation exhibit or a molecular biology laboratory? Or, being less ruthless, one could donate the specimens to a national institute such as the British Museum (Natural History) who could certainly be able to keep them safely although exigencies of space might mean that they were not readily accessible. Certainly some of the comments in Kent's (1958) index of *British Herbaria* lend force to the argument for centralization. "———'s Herbarium formerly here, but apparently disposed of"; "deteriorated by damp during the war"; "destroyed as worthless *c.* 1949"; "society defunct, herbarium not traced" . . . and so it goes on. Even more important are accounts such as that given a few years ago in the *Flora Malesiana Bulletin* (den Hartog, 1965) of the bad condition of taxonomically very important early collections lying neglected in a French University Herbarium. Coming nearer home the ravages of beetle which caused the destruction of much of the Liverpool University Herbarium after the First World War (Heywood, 1956) is another reminder of the vulnerability of university and local herbaria.

Before examining these arguments, and, I hope, demonstrating that a less facile solution is required, it is worth reminding ourselves of the actual situation with regard to university and local herbaria. Of course, many of the major herbaria of the world are themselves university foundations—one need only mention the Harvard University herbaria, or the herbarium of the Lund Museum or of Botanical Institutes such as those in Florence or Prague. Although this situation does not apply in Britain or in Ireland, I would prefer to use and develop a little, Davis and Heywood's (1963: 259–260) "classification" of herbaria. In addition to major herbaria, Davis and Heywood speak of *regional* and *local* herbaria in which the collections concentrate on one region or on a more local area within the region and also of *working* herbaria which are essentially identification collections for a particular area or for a particular taxonomic group. These terms are, of course, a little imprecise and I shall be highly subjective in my usage of them. They are preferable, however, to the size criterion adopted, for example, by Beaman (1965*a*) in his account of university herbaria in the U.S.A. Not only are there large herbaria (i.e. with at least half a million specimens) which I would not regard as major herbaria, but the size character itself is not as objective as it might appear; for example, the Kew Herbarium has long been recorded as having more than six million specimens (Lanjouw and Stafleu, 1952; Davis and Heywood, 1963: 261) yet as a result of a recent careful statistical assessment there appear to be only between four and five million (*cf.* chapter 3 of this volume, p. 30). The position with Leningrad is similar where it has been established that the herbarium has three and a half to four million specimens (Takhtajan, pers. comm.) and from personal experience estimation is almost as difficult, even if one is curating only around 150,000 sheets.

Size, of course, does enter into it. I cannot envisage a *major* herbarium today with much under half a million specimens, even if it were exclusively devoted to vascular plants. So far as the terms regional and local are concerned, a herbarium which concentrated on, say, N.W. Europe I would rate as *regional*, though perhaps only just, but if it devoted itself exclusively to the British Isles then I would call it *local*, even if it had half a million sheets.

Index Herbariorum, part I (Lanjouw and Stafleu, 1964), is, of course, the standard reference work on the herbaria of the world; the latest (5th edition) details some 850 institutes of which no more than 50 can be regarded as possessing a major herbarium. This leaves me with rather a large slice of the cake and also a rather crumbled one. And, of course there are many collections of dried plants which do not aspire to a full entry in *Index Herbariorum.* This is evidenced by various national accounts such as that of Kent (1958) for the United Kingdom and the Republic of Ireland, and that for Hungary of Priszter (1962–1966). I trust, therefore, that I will be excused if I focus most sharply on the situation in Britain and Ireland.

REGIONAL AND LOCAL HERBARIA IN THE BRITISH ISLES

Kent's (1958) account of British herbaria, although nearly 10 years old still provides a useful basis from which to derive statistics on vascular plant collections in these islands. He gives details of more than 270 herbaria or collections, about 60 of which had no current information to supply or else could not be traced. Of the remainder, at least 20 claimed to have less than 1000 specimens and at best are simply working herbaria, while a further 60 or so comprising the donated specimens of a single collector were what I suggest should be called *idle* herbaria! The remainder, representing about 120 institutes, were either general herbaria (size rarely stated) or else had at least two individual collections, and were sufficiently active taxonomically either to reply to Kent's questionnaire or at least to let him in to see the collections: not very demanding criteria. Of these 120, it may be of interest to note that excluding the three major herbaria (Kew, British Museum (Natural History) and Edinburgh), 61% belonged to national or local museums or to local Natural History Societies; 26% to universities or other institutes of further education; 7% to English Public Schools; and 6% to Research Institutes (including Botanic Gardens) or to professional societies, such as the Linnean Society. When one starts, however, to look at the sort of herbaria involved a different picture emerges. Although some curators may dissent, it appears that there are only 14* herbaria of vascular plants in the British Isles to which the term regional, as I have defined it, should be applied; of these, nine belong to universities and five are the property of national, municipal or private museums. The remaining 200 and more comprise *local* herbaria, some such as that of the Royal Albert Museum, Exeter being of considerable size, *working* herbaria and probably a good number of *idle* herbaria, but without a greater personal knowledge of the collections and institutes concerned than I possess, the proportions in each category cannot easily be estimated, but there must be at least 20 reasonably active local herbaria in the British Isles in addition to the 14 regional herbaria. There are, I am sure, also a number of *special* herbaria with regional or greater

* ABD, BIRM, CGE, NMW, TCD, DBN, LEIC, LIV, LIVU, SLBI, MANCH, NOT, FHO, OXF (using the standard *Index Herbariorum* abbreviations of Lanjouw and Stafleu, 1964).

coverage of particular groups: the Southampton University Herbarium, for example, may represent this situation (*cf.* Watson, 1965; Watson *et al.*, 1967). This statistical digest of herbaria presents the background against which the modern role of regional and local herbaria must of necessity be examined.

FUNCTIONS OF A HERBARIUM

There is not the space here to review thoroughly the uses of a herbarium; it has been done often enough in text books and elsewhere in more or less detail and with varying emphases (*cf.* Lawrence, 1951; Benson, 1957, 1962; Kruckeberg, 1960; Davis and Heywood, 1963; Fosberg and Sachet, 1965: 17–23; Beaman, 1965*b*; Rollins, 1965; A. H. Smith, 1965; A. C. Smith, 1966). Suffice to say that there are two primary functions served by the herbarium, namely accurate identification and α-taxonomic research, both monographic and floristic. By this, I mean taxonomic research which, although it will take account of the fact that apomixis, say, is known in a particular group or that a certain character exhibits single-gene inheritance, does not seek to do more than provide an α-taxonomy (Turrill, 1938)—a basis or framework for other biological studies and notably for biosystematic and genecological research, and for all comparative work whether cytological, phytochemical, palynological or whatever. But these are only the *primary* functions of the herbarium; there are *secondary* functions in which there is a much closer interaction between the student of general systematics (following the Simpsonian usage of systematics—*cf.* Simpson, 1961: 6–9) or one might even say between the biologist and the herbarium, with the plants it contains.

This starts from the simple use of the herbarium as an adjunct to population studies with their inevitable limitations of geographical coverage: the herbarium collections are the material upon which one seeks to justify or otherwise the extrapolation being made from the sample populations. I. Hedberg's (1961) well known biometric work on herbarium material of *Anthoxanthum odoratum* and of the plants termed *A. alpinum* is a good example of this. Despite considerable correlation between ploidy and the characters purporting to distinguish the diploid "*A. alpinum*" ($2n = 10$) from the tetraploid *A. odoratum sensu stricto*, she found from sampling herbarium material that without chromosome counts it was quite impossible in many cases to determine the cytodeme to which a particular plant belonged. She even arranged for a colleague to select the specimens and obscure the label so that she would not be biased in her measuring.

Whether or not one agrees with her conclusions, the work demonstrates the vital and necessary use of the herbarium in thorough biosystematic or cytotaxonomic work. Other contemporary herbarium functions are more unexpected. For herbarium taxonomists it was, perhaps, the one sad aspect of the cytotaxonomic bandwagon when it began to roll along 10 to 15 years ago; that once a plant was on a herbarium sheet its chromosomes could no longer be counted—unless indirectly if you could raise seedlings or sporelings; certainly the Linnean type was uncountable (save perhaps in *Nelumbo* —*cf.* Ødum, 1965: 10). This, of course, is not so with other micro-characters, such as pollen-grain or stomatal size (*cf.* Aalders and Hall's (1962) identification as between two segregate species of the sterile type of *Vaccinium angustifolium* Aiton)—or seed morphology as shown by Ball and Heywood's (1962, 1964) discrimination of diploids and tetraploids in *Petrorhagia prolifera* (L.) Ball and Heywood *sensu lato*. Moreover, herbarium material can usually be resuscitated sufficiently for anatomical studies, for example with ethylene glycol (Schwabe, 1961) and many examples of the application of such research are given in chapter 5 of this volume. But by far the most exciting development in this direction is the relatively recent discovery that many chemical constituents of plants remain detectable in herbarium material often as well as in freshly collected plants and even in specimens of relatively great age.

The results of the first attempts to examine this thoroughly were published by Bate-Smith in 1965. He compared fresh and herbarium material of three species of the Rosaceae tribe Quillajeae and found that so far as the phenolic constituents studied were concerned there was no significant difference between the fresh and the dried, although one of the herbarium specimens was more than 100 years old. Harborne (1967: 181) refers to similar success with herbarium and living specimens of genera of the Umbelliferae in his recent book on phenolic constituents of plants and he has also found (Harborne, pers. comm.) on sampling 12 herbarium specimens of *Genista pilosa* with collection dates ranging from 1832 to 1964 that all clearly contained a particularly striking flavone (luteolin 5-glucoside) and that the only one showing the spot slightly less brightly under the ultraviolet light was not from the oldest collection but from a more recent scrappy specimen with scarcely any leaves to sample. The phytochemical bandwagon, if it be one, will certainly use the herbarium but we cannot expect to foist our useless fragments on to the chemist. That this use of herbarium specimens is not confined to phenolic compounds is shown by Nooteboom's (1966) recording of alkaloids from fresh and herbarium specimens of a few species of the Simaroubaceae, although in this case the alkaloids were not identified and the herbarium specimens were all relatively recent, being collected between 1948 and 1958. One of my students here in Liverpool is currently examining the effects of length of storage on seed protein in Caryophyllaceae using acrylamide gel disc electrophoresis with small seed samples such as are occasionally available from herbarium material.

These secondary functions of the herbarium, and there are others (*cf.* Rollins, 1965), could in theory at least be served by the major herbaria. In practice, I doubt if this would be so, for while few curators object to occasional dissection or to giving one leaf once from historically and nomenclaturally unimportant specimens, continual foraging expeditions by ravenous phytochemists or biometric analyses of floral minutiae on a herculean scale are quite another matter. A diversity of herbaria would at least spread the load.

INTERACTION WITH OTHER BIOLOGICAL FIELDS

There is, however, a more subtle role which a herbarium and particularly a university herbarium can play in contemporary biological research. This is the enabling of a point of contact, a cross fertilization of ideas, and sometimes even a cooperation in effort to be developed between taxonomists with their general knowledge of the pattern of variation and range of diversity of different groups, and cytologists, phytochemists, electron microscopists or computer addicts each with their specialist skills. Too often one finds specialist studies designed to resolve the taxonomy or more often "the phylogenetic relationships" of problematical groups utilizing only those species or genera (usually those readily available in cultivation) which in themselves present no taxonomic problems. A not very serious example of this, in that it was incidental to the main purpose of the study, was Böcher and Larsen's (1958) comment, after examining the cytology of *Moehringia trinervia* (L.) Clairv. and *Arenaria serpyllifolia* L. that the differences they found helped to justify the maintenance of *Moehringia* as a separate genus. Even before Favarger (1962) discovered just how wide a range of base number and chromosome morphology was to be found within *Arenaria*, as a taxonomist working on the group it was evident to me that Böcher and Larsen's was not the critical comparison: that between *Moehringia trinervia* and the Asiatic and N. American *Arenaria lanuginosa* (Michaux) Rohrb. would have been much more enlightening; a comparison incidentally yet to be made (*cf.* McNeill, 1962: 93).

However much he may use anatomical, cytological, phytochemical, ultramicroscopic or computational facilities, the taxonomist requires an adequate herbarium and associated library, for the herbarium remains the one place where a reasonable sample of the variation and diversity of the groups with which he is dealing can be brought together in the same place, at the same time and in the same condition, and as I think I have already shown in a condition which retains greater information content than many non-taxonomists suppose.

But what do I mean by an adequate herbarium? Certainly not all the 270 in Kent's (1958) list nor even the 120 institutional herbaria which make some effort to curate their collectors. It requires to be a *regional* herbarium or at least a *special* herbarium with world-wide scope in particular taxonomic groups. But that is not all; there must be active addition to the collection along planned lines of development; there must be a utilization of developing techniques in curating practice and in making available the information potential of the collections; there must always be active cooperation not only with other herbaria but also with other types of organism-orientated research; in short if I may use the phrase it must be a *living* herbarium.

THE FUTURE OF REGIONAL HERBARIA IN THE BRITISH ISLES

If we project this onto the British (and Irish) scene what picture do we get? The 14 regional herbaria could potentially all serve this function though I suspect that only one half to two-thirds do so. Expensive though it is, the formation of new herbaria or the extension of small local ones should not be entirely ruled out; for example the Leicester University Herbarium was effectively founded only in 1948 and it is reckoned that there are now about 70,000 specimens (Tutin, pers. comm.); it is one in my "regional" category. Personally, I would question if there is a need for more regional herbaria in these islands; the development of a special herbarium, which treated a limited field in depth seems to me the wisest course for those institutes engaged in systematic research, yet at present lacking taxonomic facilities.

On the other hand it seems tragic that fuller use is not being made of the regional herbaria that do exist. They represent a considerable capital investment, in some cases priceless, yet in several instances there is practically no return. The answer I am certain is not to donate indiscriminately to the major herbaria, although the transfer of historically important specimens from other regions than that in which the institute specializes may be desirable. There is certainly a case for some merging or regrouping of regional herbaria (*cf.* Pavlov, 1960), possibly on the "permanent loan" basis, as in the transfer of Glasgow University's non-British Herbarium to the Royal Botanic Garden, Edinburgh (*cf.* Stafleu, 1966), but in most situations there is a need for active development of the herbarium. This is particularly true of those regional herbaria which are also university herbaria for if taxonomy is to serve the needs of biology, post-graduate schools of taxonomy must be maintained and developed in a number of universities (Royal Society, 1963; Systematics Association, 1964).

The running costs of an active regional herbarium such as I have outlined, unlike the capital costs, are not exorbitant. I have analysed the expenditure on maintaining the Liverpool University Herbarium over the past three years, a moderately expansive period in which accessions have averaged around

8000 specimens per year, but on the other hand one in which no very sophisticated curating techniques have been employed and in which most of the cryptogamic herbarium has been quiescent. The running costs for basic herbarium materials were between £250 and £300 per annum of which the largest item is undoubtedly mounting paper; this is only 4% of that proportion of the Department of Botany's recurrent grant not spent on replacing equipment—i.e. 4% of the material running costs of the Department. It includes ancillary equipment such as drying paper, and renewing presses, but excludes all capital expenditure, e.g. on the cupboards themselves, on drying rooms or fumigation equipment. Metal cupboards such as are currently being installed at Kew give a costing of £30 per 1000 specimens accommodated but facilities for drying and fumigation are not difficult to manufacture and if suitable rooms or outhouses exist, conversion should not cost more than about £100, for the needs of most regional herbaria. The most important exclusion however is also the most costly, namely the labour necessary to maintain the herbarium. Because at Liverpool this work is shared with other duties by some junior technicians and there is no record of the time members of the academic staff spend on herbarium matters, it is difficult to give a precise figure, but I suppose £1000 p.a. on technical staff salaries and perhaps a moiety of £500 for scientific staff would be reasonably near our actual costs, but there is in fact a very considerable backlog of work requiring supervision by scientific staff. Even allowing another £500 for this, the cost of herbarium labour at around £2000 p.a. also represents a very small proportion (<4%) of the department's total salaries bill.

THE ROLE OF LOCAL HERBARIA

So far I have dealt exclusively with regional herbaria; local herbaria have an active role for example in identification and in distributional studies, but by their nature they will not be involved in taxonomic research as such and their scale of operations will be altogether much smaller. However, it would be as a great a mistake to think of local herbaria as complete closed collections with no active role as it would be to regard regional herbaria in this light. The flora of these islands and of every single county or district is continually changing. A controlled but deliberate accession policy is essential for any local herbarium that seeks to serve as anything more than a historical item illuminating the land-use pattern and the social mores of the late nineteenth and early twentieth centuries. I cannot accurately estimate the expense involved but assuming the collection is currently in good condition and technically well-curated, perhaps a dangerous assumption, the running cost in terms of materials would be trifling compared with that outlined for an active regional herbaria.

NEW DEVELOPMENTS

In concluding, I would like to draw attention to one field in which both local and regional herbaria may have an important role to play which is sometimes denied to the major herbaria. This is that of experimentation in curating techniques and in the application of new technological developments to herbarium practice. The major herbaria are inevitably conservative; their collections are irreplaceable and so they cannot afford to make a mistake in a curatorial decision (the same is partly true of some regional herbaria). Moreover their scale of operation is so vast that the advantages of any more time-consuming new method of operation must be well-proven before its adoption can be considered. Working continuously day and night, every day of the year in the Kew Herbarium it would take eight to ten man-years to perform so trifling an operation as to involve one minute's work per specimen. Local and regional herbaria seem therefore particularly suited for practical experimentation with new technological developments.

Manuals such as Savile (1962) and Fosberg and Sachet (1965) give a wealth of information on curatorial techniques applicable in all parts of the world, but as the latter point out (Fosberg and Sachet 1965: 5) this is a continually developing field and new methods of preparing herbarium specimens, preserving them from damage and making them more readily accessible are continually being proposed (*cf.* Ber, 1959, 1960, 1963, 1968; Cody, 1966; Eusebio and Stern, 1964; Fosberg, 1965; Franks, 1965; Fry, 1965; Popov, 1964, 1965; Porter, 1967: 42–54; Simon, 1962; Whitmore, 1965).

There is one technological development, however, which merits particular mention. This is the application of electronic data-processing equipment in herbarium practice. Sokal and Sneath (1966) have reviewed its opportunities and scope in taxonomy as a whole and make fascinating predictions of "1984" taxonomic and herbarium practice, while Crovello (1967) has analysed its application to biological collections. It is a vitally important field of trial for regional and local herbaria. The reports of Soper's work in Toronto (Soper and Perring, 1967) and Perring's (1967) discussion of developments in this country are encouraging but the field is still only opening up and the economics of it have yet to be proved at least in herbarium practice. Zoologists appear to be more enthusiastic and the Museum of Natural History at the Smithsonian Institute, a major repository certainly, have adopted an EDP scheme for the documentation of their new collections (Squires, 1966).

I have outlined the continuing role of the regional and local herbarium in biological studies, but this is another rather different role for which some of them at least are well suited, namely technological research designed ultimately to increase the efficiency of taxonomy and of its service to systematics and to biology in general.

SUMMARY

A distinction is drawn between major, regional, local and working herbaria and an analysis made of the present distribution in each category, principally with reference to the herbaria in the British Isles. Regional and local herbaria are sometimes thought to be valueless and are not always adequately preserved from damage or destruction. The diverse and unique contributions of the herbarium not only to contemporary systematics but to the whole of biology are outlined and the role of regional herbaria in biological research is emphasized. An analysis is made of the costs of maintaining an active regional herbarium and consideration is given to the continuing role of local herbaria in ecological and distributional studies. The importance of regional and local herbaria in experimenting with new techniques in curatorial practice and in particular in the development of the use of electronic data processing (EDP) equipment is stressed. A comprehensive bibliography of recent publications on herbarium policy and curatorial practice is included.

REFERENCES

Aalders, L. E. and Hall, I. V. 1962. New evidence on the cytotaxonomy of *Vaccinium* species as revealed by stomatal measurements from herbarium specimens. *Nature, Lond.*, **196**: 694.

Ball, P. W. and Heywood, V. H. 1962. The taxonomic separation of cytological races of *Kohlrauschia prolifera* (L.) Kunth sensu lato. *Watsonia*, **5**: 113–116.

Ball, P. W. and Heywood, V. H. 1964. A revision of the genus *Petrorhagia*. *Bull. Br. Mus. nat. Hist.* (*Bot.*), **3**: 121–172.

Bate-Smith, E. C. 1965. Investigation of the chemistry and taxonomy of sub-tribe Quillajeae of the Rosaceae using comparisons of fresh and herbarium material. *Phytochemistry*, **4**: 535–539.

Beaman, J. H. 1965*a*. The present status and operational aspects of university herbaria. *Taxon*, **14**: 127–133.

Beaman, J. H. 1965*b*. The herbarium in the modern university: a symposium. Introduction. *Taxon*, **14**: 113.

Ber, V. G. 1959. Zashchita botanicheskikh kollektsiĭ ot brediteleĭ. *Bot. Zh. SSSR*, **44**: 1260–1270.

Ber, V. G. 1960. *Methodology of Botany: Protection of botanical collections from insects.* OTS: 60–41, 378; JPRS: 5566. Office of Technical Services, Washington (translation of Ber, 1959).

Ber, V. G. 1963. Zashchita botanicheskikh kollektsiĭ ot brediteleĭ (Soobshchenie vtoroe). *Bot. Zh. SSSR*, **48**: 384–395.

Ber, V. G. 1968. *Protection of botanical collections from pests.* RTS 4455. National Lending Library for Science and Technology, Boston Spa (translation of Ber, 1963).

Benson, L. D. 1957. *Plant classification.* D. C. Heath and Co., Boston.

Benson, L. D. 1962. *Plant taxonomy: methods and principles.* Ronald Press, New York.

Böcher, T. W. and Larsen, K. 1958. Experimental and cytological studies on plant species. IV. Further studies on short-lived herbs. *Biol. Skr.*, **10** (2): 1–24.

CLAPHAM, A. R., TUTIN, T. G. and WARBURG, E. F. 1959. *Excursion Flora of the British Isles*. University Press, Cambridge.

CLAPHAM, A. R., TUTIN, T. G. and WARBURG, E. F. 1962. *Flora of the British Isles*, ed. 2. University Press, Cambridge.

CODY, W. J. 1966. *New herbarium equipment: mobile shelves and drying rack for mounted specimens*. Plant Research Institute, Research Branch, Canada Department of Agriculture, Ottawa.

CROVELLO, T. J. 1967. Problems in the use of electronic data processing in biological collections. *Taxon*, **16**: 481–494.

DAVIS, P. H. and HEYWOOD, V. H. 1963. *Principles of angiosperm taxonomy*. Oliver and Boyd. Edinburgh and London.

DONY, J. G. 1953. *Flora of Bedfordshire*. Corporation of Luton Museum and Art Gallery, Luton.

DONY, J. G. 1967. *Flora of Hertfordshire*. Hitchin Urban District Council, Hitchin.

EUSEBIO, M. A. and STERN, W. L. 1964. Preservation of herbarium specimens in the humid tropics. *Philipp. Agric.*, **48**: 16–20.

FAVARGER, C. 1962. Contributions a l'étude cytologiques des genres *Minuartia* et *Arenaria*. *Bull. Soc. neuchâtel. Sci. nat.*, **85**: 53–81.

FOSBERG, F. R. 1965. Lauryl pentachlorphenate protecting herbarium specimens: application with brushes. *Taxon*, **14**: 165–166.

FOSBERG, F. R. and SACHET, M. H. 1965. *Manual for tropical herbaria* (*Regnum veg.*, **39**). International Bureau for Plant Taxonomy and Nomenclature, Utrecht.

FRANKS, J. W. 1965. *A guide to herbarium practice*. The Museum Association Handbook for Museum Curators, part E. Section 3. The Museum Association, London.

FRY, W. G. 1965. Methods in taxonomy. *Nature, Lond.*, **207**: 245–246.

HARBORNE, J. B. 1967. *Comparative biochemistry of the flavonoids*. Academic Press, London and New York.

HARTOG, C. DEN. 1965. The herbarium of the Institut Botanique de la Faculté des Sciences, Caen. *Flora Malesiana Bull.*, **20**: 1270–1271.

HEDBERG, I. 1961. Cytotaxonomic studies in *Anthoxanthum odoratum* s. lat. 1. Morphologic analysis of herbarium specimens. *Svensk bot. Tidskr.*, **55**: 118–128.

HEYWOOD, V. H. 1956. The University Herbarium, Liverpool. *Taxon*, **5**: 10–11.

KEBLE-MARTIN, W. 1965. *The Concise British Flora in Colour*. Ebury Press and Michael Joseph [London].

KENT, D. H. 1958. *British Herbaria: an index to the location of herbaria of British vascular plants with biographical references to their collectors*. The Botanical Society of the British Isles, London.

KRUCKEBERG, A. R. 1960. Location and housing of institutional herbaria in the United States and Canada. *Brittonia*, **12**: 295–297.

LANJOUW, J. and STAFLEU, F. A. 1952. *Index herbariorum, part 1. The herbaria of the world* (*Regnum veg.*, **2**). International Bureau for Plant Taxonomy and Nomenclature, Utrecht.

LANJOUW, J. and STAFLEU, F. A. 1964. *Index herbariorum, part 1. The herbaria of the world*, ed. 5 (*Regnum veg.*, **31**). International Bureau for Plant Taxonomy and Nomenclature, Utrecht.

LAWRENCE, G. H. M. 1951. *Taxonomy of vascular plants*. The Macmillan Company, New York.

MCNEILL, J. 1962. Taxonomic studies in the Alsinoideae: I. Generic and infrageneric groups. *Notes R. bot. Gdn Edinb.*, **24**: 79–155.

NOOTEBOOM, H. P. 1966. Flavonols, leuco-anthocyanins, cinnamic acids, and alkaloids in dried leaves of some Asiatic and Malesian Simaroubaceae. *Blumea*, **14**: 309–315.

ØDUM, S. 1965. Germination of ancient seeds: floristical observations and experiments with archaeologically dated soil samples. *Dansk bot. Ark.* 24 (2): 1–70.

PAVLOV, V. N. 1960. O spravochnoi literature i o roli gerbariev v nauchnoi rabote (V poryadke obsuzhdeniya). *Bot. Zh. SSSR*, 45: 1834–1835.

PERRING, F. H. 1967. Information retrieval techniques for museums. *Classific. Soc. Bull.*, 1 (3): 29–30.

PERRING, F. H. and WALTERS, S. M. (eds) 1962. *Atlas of the British Flora.* Thomas Nelson and Sons Ltd., London and Edinburgh.

POPOV, K. P. 1964. Do pitannya pro zakhist gerbariїv vid shkidnikiv. *Ukr. bot. Zh.*, 21: 102–104.

POPOV, K. P. 1965. K voprosu o predotvraschenii povrezhdeniĭ gerbarnykh kollektsii nacekomymi. *Bot. Zh. SSSR*, 50: 368–370.

PORTER, C. L. 1967. *Taxonomy of flowering plants*, ed. 2. W. H. Freeman & Co., San Francisco and London.

PRISZTER, S. 1962–66. Magyar herbáriumok. *Bot. Közl.*, 48: 109–115, 300–306; 49: 122–123, 331–333; 52: 157–159.

ROLLINS, R. C. 1965. The role of the university herbarium in research and teaching. *Taxon*, 14: 115–120.

ROSS-CRAIG, S. 1948– . *Drawings of British plants* (parts I–XXIV. 1948–1967). G. Bell and Sons Ltd., London.

ROYAL SOCIETY, 1963. *Taxonomy: report of the committee appointed by the council of the Royal Society.* Royal Society, London.

SAVIDGE, J. P., GORDON, V. and HEYWOOD, V. H. (eds) 1963. *Travis's Flora of South Lancashire.* Liverpool Botanical Society, Liverpool.

SAVILE, D. B. O. 1962. *Collection and care of botanical specimens* (Publ. No. 1113). Research Branch, Canada Department of Agriculture, Ottawa.

SCHWABE, H. 1961. Glicol de etilene: un nuevo método en histologia para ablander material de herbario. *Boln Soc. argent. Bot.*, 9: 383–394.

SIMON, C. 1962. Erfahrungen mit wenig bekannten Methoden der Herbartechnik. *Bauhinia*, 2: 63–69.

SIMPSON, G. G. 1961. *Principles of animal taxonomy.* Columbia University Press, New York, Oxford University Press, London.

SMITH, A. C. 1966. Advice to administrators of systematic collections. *Taxon*, 15: 201–205.

SMITH, A. H. 1965. The role of the herbarium in cryptogamic botany. *Taxon*, 14: 121–126.

SOKAL, R. R. and SNEATH, P. H. A. 1966. Efficiency in taxonomy. *Taxon*, 15: 1–21.

SOPER, J. H. and PERRING, F. H. 1967. Data processing in the herbarium and museum. *Taxon*, 16: 13–19.

SQUIRES, D. F. 1966. Data processing and museum collections: a problem for the present. *Curator*, 9: 216–227.

STAFLEU, F. A. (ed.) 1966. News and notes: Glasgow University Herbarium. *Taxon*, 15: 46.

SYSTEMATICS ASSOCIATION, 1964. Development and support of systematics in Britain. *Nature, Lond.*, 203: 358–359.

TURRILL, W. B. 1938. The expansion of taxonomy with special reference to Spermatophyta. *Biol. Rev.*, 13: 342–373.

WATSON, L. 1965. Taxonomic significance of certain anatomical variations among Ericaceae. *J. Linn. Soc. (Bot.)*, 59: 111–125.

WATSON, L., WILLIAMS, W. T. and LANCE, G. N. 1967. A mixed-data numerical approach to angiosperm taxonomy: the classification of Ericales. *Proc. Linn. Soc. Lond.*, 178: 25–35.

WHITMORE, T. C. 1965. Lauryl pentachlorphenate protecting herbarium specimens: the Honiara technique. *Taxon*, 14: 164–165.

5 | Current Developments in Systematic Plant Anatomy

C. R. METCALFE

Jodrell Laboratory, Royal Botanic Gardens, Kew, Richmond, England.

INTRODUCTION

It is assumed that most taxonomists will agree that data concerning the histological structure of the vegetative organs of flowering plants can usefully be employed for the following purposes: (1) The identification of fragmentary material, which may often be of economic importance. (2) The preliminary identification of herbarium specimens. (3) As an aid towards establishing the interrelationships of taxa at and above the species level. Below the species level other methods of attack are generally more rewarding.

If there are any who still doubt that anatomical evidence can be employed in these ways, numerous concrete examples could be given to illustrate how histological data have in fact been used with great success.

In spite of these achievements of the anatomical method it is equally important to realize that this method of approach to the taxonomy of the higher plants has its own limitations and difficulties, to *some* of which attention will now be drawn.

DIFFICULTIES AND LIMITATIONS OF THE ANATOMICAL METHOD

1. Inadequate Factual Data

It cannot be too strongly emphasized that we have to face taxonomic problems today at a time when our knowledge of the histological characters of the plants to be classified is still very meagre. The systematic anatomist is at a grave disadvantage compared with the herbarium botanists who are far more numerous than systematic anatomists and who have been accumulating descriptive data and accurately named specimens for several hundred years. The fact that the reference collection at the Jodrell Laboratory contains getting on for 40,000 microscope slides means very little when we realize that it covers only a small fraction of the world's flowering plants. If we turn to the literature we are still not much better off, for, although the number of species that have been examined is much greater than those represented in the Kew slide collection it still represents only a very small part of those that exist.

In these circumstances it is scarcely surprising that we are not always able to answer the questions that are put to us by our herbarium colleagues. When, for example, we are asked to suggest possible affinities for an unidentified, sterile, or otherwise incomplete herbarium specimen, how do we set about it? If the plant is a dicotyledon, we turn to the tables of diagnostic characters at the end of "Anatomy of the Dicotyledons" and try to discover which families, if any, have a *combination* of histological characters that correspond to those of the specimen to be identified. Sometimes we get an answer quite readily, and by turning to the description of the family that has been suggested by the tables we may get a little further. If we are very lucky we then find microscope slides in the reference collection that "fit" exactly with the specimen for name. But more often than not we can get no further than suggesting possible relationships for the specimen in question because the available data are inadequate. We can then go back to our herbarium colleagues with suggestions that enable them to narrow their search, and, by our joint endeavours, success can sometimes be attained. But there are occasions when our data are insufficient to enable us to make any suggestion at all; disappointment ensues and the belief is engendered that the anatomical method "is no good". It would be just as reasonable to cast aspersions on the herbarium method if herbarium botanists failed to identify plants because they had access to inadequate material for comparison. When confronted with questions such as whether a particular specimen should be placed in family A or family B the systematic anatomist is again sometimes defeated by lack of data concerning the families in question. And so one might go on, but the moral to be drawn from all this is that the anatomical method is at present more frequently being defeated by the lack of adequate data on which to form an opinion than from any inherent weakness in the anatomical method itself.

2. *Dangers of Drawing Conclusions from Mechanically "Processed" Data*

It is becoming increasingly common to make use of mechanical or electronic devices of one kind or another to sort out and interpret the vast masses of data that are involved in taxonomy. There is no objection to doing this, and indeed we should be foolish not to make use of modern inventions to aid us in our work. There are, however, certain dangers in relying too much on these machines, the first of which is that they are mechanical robots and not equivalent to human intelligence. Many of us will have heard about the surprise of Mrs Bloggs when she found three large vans of doughnuts being delivered at the small village shop of which she was the proprietor. This was because the computer at the bakery had not been sufficiently "intelligent" to spot that Mrs Bloggs, in her innocence, had put a tick in the wrong column on the form when she placed her order. This sort of thing can and does happen in systematic botany too, but the errors, being less obvious, may be more difficult to detect.

There is also the danger that arises because it is so much easier and quicker to "process" data in a mechanical device than it is to obtain the data to be processed by examining plants. In consequence those who wish to establish generalizations quickly may be tempted to feed inadequate or dubious data into the machines to keep them going. This applies particularly with histological data, which often take longer to obtain than those based on exomorphic characters, because of the techniques involved.

3. *Conclusions Based on Facts that are not Significant*

Those whose experience of systematic anatomy is relatively limited are apt to think that every histological difference that they encounter is taxonomically significant. Those who are at the beginning of their botanical careers often find themselves exposed to the influences of the "publish or perish" philosophy, which are sometimes held by grant-awarding bodies. In consequence articles that are overloaded with trivial details of no taxonomic significance come to be printed. If one is dealing with a relatively large taxonomic group such as the Cyperaceae it is desirable to obtain a bird's eye view of the structure of a wide range of genera and species within the family before launching out into detailed descriptions of any particular species. This type of fault is sometimes made with the result that important characters are inadequately treated whilst others of relative insignificance are over emphasized.

4. *Consequences of Inadequate Knowledge of Previous Publications*

The literature of plant anatomy is very imperfectly known because this branch of botany is relatively unfashionable. Consequently there is a tendency for workers in systematic anatomy, more perhaps than in many other branches of biology, to embark on investigations that have already been done, or to publish articles that are less satisfactory than others which already exist. In one recent example that came to my notice an author with inadequate library facilities was about to publish an article on certain aspects of the anatomy of the Dioscoreaceae without having seen several very important articles on the same subject. It was possible to prevent the publication of misleading information by drawing attention to these earlier investigations.

5. Difficulties Imposed by the Constitution of the Plants Themselves

Herbarium botanists are fully aware that characters which are of considerable taxonomic significance for one family are often quite useless for another. The same principle applies with systematic anatomy. The investigator has to take the plants on which he is working as he finds them and to make the best use of the characters which they exhibit. Provided the plants under investigation have at least one, or preferably more, of the following attributes there are reasonable prospects that the systematic anatomist will be able to make something of their taxonomy. With plants that do not fulfil these conditions he is less likely to be successful. The desirable attributes are as follows: (a) well developed secondary xylem; (b) distinctive trichomes or other dermal appendages; (c) a characteristic distribution pattern of sclerenchyma; (d) ergastic substances such as crystals and siliceous bodies which are deposited in the plant body in distinctive morphological forms. The categories (a)–(d) will now be discussed in greater detail.

(a) *Secondary Xylem* It is obvious that the diagnostic characters of the secondary xylem, which can be so easily investigated in arborescent and arboreal dicotyledons, are of less significance if we are dealing with herbs. It is also evident that since secondary xylem is lacking from the monocotyledons we are unable to study their wood structure in the same way as we can for dicotyledons, nor can we make any direct comparison between the two groups so far as secondary xylem is concerned. It is a fact well known to wood anatomists that the wood structure to be found in certain families is so highly distinctive in its cellular organization that the wood of such a family can be readily recognized. This applies, for example, to the Ebenaceae and Proteaceae. On the other hand there are families where there is nothing very constant or diagnostic for the wood structure of the family as a whole; Euphorbiaceae and Flacourtiaceae can be cited as examples. Then again there are generalized types of wood structure that may occur in a number of distinct and unrelated families.

Facts such as these, which tend to make wood anatomy easier to use taxonomically in some families than in others, have their counterparts amongst the characters used by herbarium botanists. It is generally easy, for example, to recognize a plant as being a member of the Papilionaceae from its flower, but in the Rosaceae we find a much greater range of floral form.

(b) *Trichomes and epidermal appendages* Trichomes and other epidermal appendages are of great *diagnostic* value, but clearly this can apply only to plants in which these appendages exist. The diagnostic rather than the taxonomic value has been emphasized, however, because similar types of these structures occur in families that are clearly unrelated to one another.

It is therefore evident that similar appendages are likely to have arisen more than once in the course of evolution. To take an extreme example, some of the complex types of trichomes and scales that occur in certain Rhododendrons have their nearest counterpart in the Bromeliaceae.

(c) *Distribution pattern of sclerenchyma* The distribution pattern of sclerenchyma can be of great value, but only in plants where this tissue is well developed. It is of particular value in monocotyledons. This is well known for the Gramineae where, for example, the distribution pattern of sclerenchyma in the leaves of *Festuca* enables species that are otherwise difficult to identify to be distinguished from one another.

It is of very great value in the Cyperaceae, as my own current work shows very clearly, e.g. for separating species of *Carex*, and it is also of value at the genus level.

Similarly, E. Ayensu's work at the Smithsonian Institution, Washington D.C., U.S.A., on Velloziaceae shows that *Vellozia* and *Barbacenia* can be distinguished from one another by the form of the sclerenchyma in the leaf. Dr Lyman Smith, a leading expert on the taxonomy of the Velloziaceae, with whom Ayensu was collaborating, agrees with the results obtained by the anatomical method. Of special interest is one taxon which was described as a member of the Velloziaceae on exomorphic evidence alone. When the leaf anatomy of this plant had been investigated it was found to exhibit a sclerenchyma distribution pattern that is so different from that of either *Vellozia* or *Barbacenia* that Dr Ayensu concluded that it had been wrongly placed in the Velloziaceae. When consulted about this, Dr Lyman Smith at first thought that Ayensu must be wrong. However, after a lapse of time and further investigation he agreed that Ayensu was right, and a fresh publication on the subject is being prepared.

(d) *Ergastic substances* The occurrence and distribution of these substances can be most useful and informative. This applies particularly to crystals and silica-bodies. Silica-bodies provide characters that have been used by grass taxonomists for many years. They are also of value in other monocotyledonous families such as orchids and palms, not to mention the family with which I am currently concerned, the Cyperaceae. I shall return to the Cyperaceae later, but I would indicate at this stage that the Cyperaceae, unlike the Gramineae, have one dominant type of silica-body which is conical, but it has not always been appreciated that the variations of this conical type are of diagnostic value within the family. Furthermore, contrary to general belief, there are other types of silica-body within the Cyperaceae but these are of much more restricted distribution in the taxonomic sense.

If the structure of the plants is such that data of the kinds mentioned in (a)–(d) above cannot be obtained it may well turn out that other disciplines than systematic anatomy will be of greater taxonomic significance. This applies, for example, to the Marantaceae, where Tomlinson's work has shown that a great diversity of external form conceals a remarkable uniformity of histological detail. This applies also to some of the aquatic monocotyledons such as those which Dr Stant has been investigating, e.g. those belonging to the Alismataceae and Butomaceae and other families at the beginning of the Monocotyledons in Hutchinson's system. One of the significant points that emerges from her work is the contrast that exists between the fleshy, aerenchymatous structure of the families on which she has been working in contradistinction to aquatic grasses, sedges and members of the Typhaceae and Juncaceae which retain the histological imprints of their respective families and are totally different in structure to the Butomaceae and Alismataceae, particularly because they have a well developed system of sclerenchyma.

CURRENT POSITION OF HISTOLOGICAL INVESTIGATIONS CENTRED ON THE JODRELL LABORATORY

There are some who seem to think that our work consists of sheer compilation from the literature. This I most strenuously deny. We naturally use the existing body of knowledge as a starting point and indeed it is a matter of scientific etiquette that we should do so. But our researches have taken us far beyond the confines of previously published data and we are attempting to assess the taxonomic implications of our investigations as well as to record the data on which our opinions are based.

It would be an understatement to say that we are overwhelmed by the number of problems that need investigation. At the same time even to organize and present the information that we already possess in a form in which it can be readily understood and used by others is in itself a major task.

One of the principal difficulties is that it is impossible to complete even a single volume of a reference book quickly, especially if the volume is one that runs to some 700 pages. We already have much information that has been awaiting publication for a disappointingly long time. Our publishers have, however, now agreed to produce shorter, more frequent volumes. This change of attitude has already meant that the proofs of Vol. III of "Anatomy of the Monocotyledons" have begun to come in and other volumes are at an advanced state of preparation. Volume III by P. B. Tomlinson (now at the Fairchild Tropical Garden, Miami, Florida, U.S.A.) covers the families which comprise the Commelinales, Xyridales, Eriocaulales, Bromeliales and Zingiberales in Hutchinson's "The Families of Flowering Plants". This volume consists wholly of Tomlinson's own work apart from the article on Rapateaceae by Sherwin Carlquist of the Rancho Santa Ana Botanic Garden, Claremont, California, U.S.A. My colleague D. F. Cutler and I hope shortly to complete our survey of the Glumiflorae in two volumes. It will be recalled that the Gramineae have already been covered in Vol. I. Besides this, E. Ayensu (now at the Smithsonian Institution, Washington, D.C., U.S.A.) is writing a very interesting account of the Dioscoreales. And then there is to be a volume on the aquatic monocotyledons by Dr Margaret Stant.

1. Some Interesting Points from Current Work

(a) *Commelinaceae–Marantaceae: P. B. Tomlinson* Dr Tomlinson's contribution is particularly valuable because he has been able to study many of the plants which he has investigated in the field as well as in the laboratory and he has done so in Singapore, Ghana, parts of tropical S. America and the West Indies as well as in the U.S.A. What he has to say is likely to provoke much thought and discussion amongst taxonomists. Here are just a few points that emerge from his work as a foretaste of what is to come.

Commelinaceae. There is ample justification for excluding *Cartonema* and treating it as a separate family, the Cartonemataceae, as has already been done by Hutchinson. In Tomlinson's opinion *Triceratella* should be placed in a distinct subfamily. With these genera removed the rest of the family is anatomically relatively homogeneous. Nevertheless there are groups of genera that "hang together" on histological characters, but these groups cut across the subdivisions which Mr J. P. M. Brenan feels to be more important on the basis of exomorphic characters.

Flagellariaceae. Smithson's previous opinion is confirmed that *Hanguana* is so unlike *Flagellaria* and *Joinvillea*, i.e. the other two genera which with *Hanguana* constitute the family, that it is doubtful if *Hanguana* should be placed here at all. Since these anatomical observations were made the family Hanguanaceae has been erected by H. K. A. Shaw (1965). *Flagellaria* and *Joinvillea* bear some resemblance to the Gramineae in their anatomical structure.

Mayacaceae. Anatomical evidence suggests that *Mayaca*, the sole genus in this family, probably originated from the same stock that has produced Eriocaulaceae, Commelinaceae and Xyridaceae.

Xyridaceae. This is anatomically a very heterogeneous family. The genera are individually distinctive in their anatomy and there are also characters of diagnostic value at the species level.

Rapateaceae. Dr Carlquist's work on this family is partly based on newly discovered species generously placed at his disposal by Dr Bassett Maguire of the New York Botanical Garden. There is close agreement between the anatomy and taxonomy of the family as recently revised by Bassett Maguire.

The family resembles the Xyridaceae in some respects but differs in possessing silica-bodies, tannin cells, in having another type of chlorenchyma cell, and the root structure is very different.

Eriocaulaceae. The form and structure of the trichomes, especially those on the reproductive organs, are very valuable taxonomically and sometimes critical for the separation of species. The genera are not otherwise circumscribed anatomically and there is a structural trend from xeromorphic to hydromorphic forms. The range of root structure in the family is remarkable. It is interesting to note that the anatomy of *Aphyllanthes* (a genus of uncertain taxonomic position generally ascribed to the Liliaceae) is very unlike that of the Eriocaulaceae. This is in contradistinction to the morphology of the pollen grains which is said to be similar.

Zingiberales. Tomlinson's work has shown that the structure of the Zingiberales is, on the whole, less diverse than that in the other orders that are covered in his volume, although the Cannaceae and Marantaceae are thought to occupy rather isolated positions within the order. Tomlinson supports the view of Nakai that *Heliconia* should be given the status of a distinct family. Indeed Tomlinson reached this opinion before his attention had been drawn to Nakai's publication. *Ravenala*, *Strelitzia* and *Phenakospermum* are thought to constitute a natural taxonomic unit. It is recommended that the Costaceae should be separated from the Zingiberaceae and there is anatomical support for Hutchinson's raising of *Orchidantha* to the rank of a family, the Lowiaceae.

(b) *Juncales: D. F. Cutler* Dr Cutler has shown that it is often easier to identify members of the Restionaceae by examining transverse sections of the stems than it is from exomorphic characters. He has also, with collaboration from H. K. Airy Shaw, provided evidence that the two genera *Anarthria* and *Ecdeiocolea* are so different from other members of the family that it is logical to transfer them to new families, Anarthriaceae and Ecdeiocoleaceae, allied to but distinct from the Restionaceae (Cutler and Shaw, 1965).

(c) *Cyperales: C. R. Metcalfe* The Cyperaceae, which constitute the sole family in Hutchinson's Cyperales, are taxonomically intricate. The anatomical data collected from the rather considerable literature by my colleague Miss M. Gregory, as well as from my own observation on a very wide range of genera, including some that have not previously been examined, will probably be enough to promote discussion amongst cyperologists for some time to come. No claim is made that all of the taxonomic problems that abound in this family have been solved. The material on which this investigation is based consists partly of British specimens collected by my late colleague Mr E. Nelmes and myself and which were identified by Mr

Nelmes. Other specimens I have myself collected in Jamaica, Trinidad and Florida, besides which I am indebted to specialists on Cyperaceae in various parts of the world who have sent material. Their contribution is greatly appreciated but I trust they will forgive me if they are not mentioned by name at this stage. In addition, Professor Vernon I. Cheadle very kindly placed at my disposal fixed material which he himself had collected in Australia and S. Africa. Besides this wealth of specially collected material it has been possible to examine specimens of many genera from the Kew Herbarium. In this connection it is a great pleasure to acknowledge the assistance of Miss S. Hooper who specializes in the study of Cyperaceae at Kew.

It is not possible to anticipate more than a very small fraction of the data and conclusions that will appear in my forthcoming volume on Cyperaceae, but it is hoped that the following comments may be of interest.

(1) The family is much more structurally diverse than is commonly realized. This lack of understanding has arisen because many botanists have concentrated mainly on genera such as *Carex* that are common in the northern hemisphere. This enormous genus is relatively homogeneous anatomically when considered in relation to the family as a whole.

(2) I believe the family is a natural group with the possible exception of a few genera that might be more appropriately placed in or near the Restionaceae.

(3) In spite of the superficial similarity in habit I do not think that the Cyperaceae are very closely related to the Gramineae. They have some histological characters in common, due perhaps to convergence, but there are notable histological differences. One example is afforded by the silica-bodies. There are fewer types than in the Gramineae and they are all different from those of the Gramineae. Other differences are in the organization of the mesophyll.

If the Cyperaceae can be pictured without any silica-bodies the structure is much more like that of the Juncaceae than of the Gramineae. The Juncaceae are often like Cyperaceae without silica-bodies.

(4) The tribes as at present constituted are not very homogeneous anatomically.

(5) Some of the larger genera are not homogeneous, e.g. *Cyperus*. It seems desirable to subdivide these.

(6) It is often possible, on histological grounds, to decide when two or more genera are related to each other.

(7) Histological characters of the leaf can be found which are of diagnostic value at the species level. As in the Gramineae these largely depend on variations in outline in transverse sections taken at a standard position in the leaf and also in the distribution pattern of sclerenchyma. Some details of the

silica-body distribution as seen in surface view preparations of the leaf are also of specific diagnostic value.

(8) There are numerous histological characters that are of considerable diagnostic value because of their restricted occurrence, but they do not appear to afford any evidence of closer relationship. An example is the occurrence of septate hairs in the four genera *Costularia*, *Everardia*, *Fuirena* and *Scleria*.

(9) Silica-bodies. The following are examples of types with a restricted taxonomic distribution:

Wedge-shaped bodies are restricted to *Mesomelaena*, *Neesenbeckia*, *Ptilanthelium* (conical bodies present as well), *Scirpodendron*, *Thoracostachyum*.

Bridge-shaped bodies have been noted in *Mapania* and *Thoracostachyum*.

Spherical and hemispherical warty bodies have been noted in *Acriulus*, *Bisboeckelera*, *Scleria*.

Spherical and hemispherical echinulate bodies have been seen in certain species of *Rhynchospora* and *Scleria*.

Smooth hemispherical bodies (accompanied by other types) occur in *Lophoschoenus*.

(10) Chlorenchyma. The cells of the chlorenchyma in most Cyperaceae are polyhedral, spherical or columnar (palisade-like) and very frequently lobed. By contrast the cells are axially elongated in *Elynanthus*, *Gahnia* (cells moniliform), *Tetraria*, *Thoracostachyum*, *Tricostularia*.

(11) Bundle sheaths. The types of bundle sheath that have been recognized and the genera in which they have been seen are as follows:

Two-layered, with the inner sheath fibrous and outer sheath parenchymatous. Present in most Cyperaceae.

Three-layered, with the inner and outer sheaths parenchymatous and the middle sheath fibrous. This type has been noted in *Abildgaardia*, *Bulbostylis*, *Carpha*, *Crosslandia*, *Cryptangium* (special type), *Fimbristylis*, *Mapania* (tendency), *Nelmesia*, *Scirpus* (some species only).

Two-layered, with the inner sheath parenchymatous and the outer sheath fibrous. This type has been noted in *Cyperus* (species belonging to one subdivision of the genus only), *Juncellus*, *Kyllinga*, *Lipocarpha*, *Mariscus*, *Oreobolus*, *Pycreus*, *Remirea*, *Torulinium*.

COMPARISON WITH DATA OBTAINED FROM OTHER DISCIPLINES

1. *Chemotaxonomy*

It is clearly evident that it is only a short step from the study of the micromorphology of ergastic substances to chemotaxonomy. Like histology, chemotaxonomy will no doubt have its own limitations and difficulties.

Nevertheless it is abundantly clear from many discussions with Dr E. C. Bate-Smith and from a rapid skim through such monumental works as Hegnauer's "Chemotaxonomie" that no *major* conflict between conclusions based on chemotaxonomy and systematic anatomy respectively seems likely to arise. There may be minor points of difference but it seems that the significance of these may well be capable of resolution.

Although there are more recent examples where chemotaxonomic evidence and the evidence from systematic anatomy point to similar or identical taxonomic conclusions, I still find that the results recorded in a joint paper by Dr Bate-Smith and myself (Bate-Smith and Metcalfe, 1957) concerning the occurrence of leuco-anthocyanins in numerous dicotyledonous families afford a good example of what can be achieved by the joint endeavours of anatomists and chemotaxonomists. A particular point of interest at the time when this work was undertaken was the fact that by looking rapidly through our reference collection of microscope slides I was able to predict in which families Dr Bate-Smith would detect leuco-anthocyanins *before he had applied his tests*. When the tests had been applied a very high proportion, indeed nearly all, of these predictions were confirmed. It was possible to make these predictions because the sections in our microscope slides were stained in such a way that certain cells, which I had up till then referred to as tanniniferous cells, showed up very clearly. These cells in fact contained leuco-anthocyanins, so by observing the cells one could predict that the presence of leuco-anthocyanins would be confirmed by chemical tests.

2. Palynology

In many discussions and exchanges of ideas with Professor G. Erdtman it has been encouraging to find that the facts of palynology very frequently lead to similar taxonomic conclusions to those obtained by anatomical comparisons. A good example is afforded by our recent joint paper (Erdtman and Metcalfe, 1963). In this article we record how our independent investigations on pollen morphology and vegetative anatomy respectively have provided strong evidence concerning the affinities of certain genera of which the relationships were previously uncertain. Thus *Kania eugenioides* Schltr. was shown to have affinities with the Myrtaceae rather than with the Guttiferae, Saxifragaceae or Philadelphaceae. This conclusion was subsequently confirmed by Weberling (1966), who demonstrated the occurrence of stipules in *Kania* which had previously been overlooked.

Reverting to the joint article by Erdtman and Metcalfe, we also demonstrated that *Tristania merguensis* Griff. (*Thorelia deglupta* Hance) is typically myrtaceous and that it is unlike Lythraceae. We also demonstrated the campanulaceous affinity of *Berenice arguta* Tulasne, a species originally referred to the Saxifragaceae-Escallonioideae.

Results such as these are most encouraging. Unfortunately, however, the evidence of pollen morphology and vegetative anatomy does not always point to the same taxonomic conclusions. This has already been made clear by a few of the observations mentioned earlier in this article. Another example is afforded by *Ceratostigma* of the Plumbaginaceae which on anatomical evidence seems to be rightly placed, but which on the evidence of pollen morphology seems to Professor Erdtman to belong more rightly to the Linaceae. The evidence of comparative anatomy does not, however, appear to support this view.

Then again there is the genus *Emblingia*, the affinities of which have always been and still are doubtful. Here the palynological evidence is suggestive of affinities with the Polygalaceae whereas the anatomy of the vegetative organs seems rather to support Engler's original view that the plant is allied to the Goodeniaceae to which it also bears a strong morphological resemblance. Both palynological and comparative anatomical evidence agree in pointing to a lack of affinity between *Emblingia* and the Capparidaceae, although an affinity with this family had previously been suggested.

These examples where the evidence provided by different disciplines does not lead to a single taxonomic conclusion are presented here not in any spirit of controversy but with a genuine desire to find out why this conflict in conclusions should have arisen. It is clearly evident that palynology and comparative anatomy are both valuable in deciding between alternative taxonomic affinities. Why then should the evidence obtained by these two disciplines agree for some plants and not for others? I do not pretend to know the final answer, but it seems that both disciplines are alike in having inadequate factual data on which to base conclusions. There are also difficulties about interpreting what one sees under the microscope, especially where critical structures are under consideration. These points of difference should not, however, cause despondency but should serve rather as a stimulus to further investigation.

CONCLUSION

I hope this survey will serve to show some of the ways and directions in which our current histological studies of the vegetative organs of flowering plants can serve as an integral part of taxonomy. As we have seen, like all approaches to taxonomy, the anatomical method has its limitations. Nevertheless as more histological data are accumulated and their taxonomic significance is analysed, it is to be expected that the anatomical approach will be more widely appreciated and fully understood. The great need seems to be for a greater number of fully trained anatomists who are prepared to contribute to these studies. When we remember how many plants there are in the world that await histological examination we are presented with a challenge that should serve as a stimulus to all who are interested in this field of enquiry.

REFERENCES

BATE-SMITH, E. C. and METCALFE, C. R. 1957. Leuco-anthocyanins 3. The nature and systematic distribution of tannins in dicotyledonous plants. *J. Linn. Soc.* (*Bot.*), **55**: 669–705.

CUTLER, D. F. and SHAW, H. K. A. 1965. Anarthriaceae and Ecdeiocoleaceae; two new monocotyledonous families separated from the Restionaceae. *Kew Bull.*, **19**: 489–99.

ERDTMAN, G. and METCALFE, C. R. 1963. Affinities of certain genera *incertae sedis* suggested by pollen morphology and vegetative anatomy. *Kew Bull.*, **17**: 49–56.

SHAW, H. K. A. 1965. Diagnoses of new families, new names etc. for the seventh edition of Willis's 'Dictionary'. *Kew Bull.*, **18**: 249–73.

WEBERLING, F. 1966. Additional notes on the myrtaceous affinity of *Kania eugenioïdes* Schltr. *Kew Bull.*, **20**: 517–20.

The Role of Experimental Data

6 | The Karyotype in Taxonomy

D. M. MOORE

Department of Botany, The University, Leicester, England

INTRODUCTION

There is little doubt that, of the genetical tools made available to the taxonomist during the last six or seven decades, the data provided by the chromosome cytologist have gained the widest currency and there is now an extensive literature pertaining to the taxonomic role of chromosome studies. The great volume of this literature precludes any comprehensive review in the space available, and anyhow a convenient summary and access to previous work has recently been provided by Ehrendorfer (1964). I wish to summarize briefly some of the points to be considered when assessing the present taxonomic role of chromosome studies. Only the number, size and gross structure of the mitotic chromosomes, the karyotype as defined by Levitsky (1924), will be dealt with here and, although data on chromosome number and structure are frequently obtained from studies of meiosis, information that can be derived more conveniently from the meiotic chromosomes is covered in Chapter 7 of this volume.

COMPONENTS OF THE KARYOTYPE

1. Chromosome Number

Angiosperms and pteridophytes show a wide range of chromosome numbers, varying from four in somatic cells of *Haplopappus gracilis* (Nutt.) Gray to 265 in those of *Poa litorosa* Cheeseman and 500–520 in *Ophioglossum vulgare* L. The great diversity of chromosome numbers, their frequent correlation with taxonomic groupings, their general constancy within populations and species and the relative ease in their determination when compared with obtaining information on the other components of the karyotype means that most discussions concerning the karyotype and taxonomy revolve around chromosome numbers.

The commonest variation in chromosome number involves polyploidy, which may well have occurred in at least 47% of angiosperm species (Grant, 1963). Groups which contain diploid and polyploid species are unified by the *basic number*, x, which is the gametic number of the diploid species. Thus, in a diploid species the zygotic chromosome number ($2n$) $= 2x$, in a tetraploid $2n = 4x$, in a hexaploid $2n = 6x$, and so on. Difficulties arise in groups without diploid species so that the basic number must be inferred; the accuracy of the inference obviously depends upon the amount of chromosome data available for the group. The basic number is very frequently constant for a genus or higher taxon and has proved useful in supraspecific studies. In many instances more than one basic number can be present in a group, either by the gain or loss of centromeres at diploid or polyploid levels, or by amphidiploid combination. In initial studies of a group such numbers, whether original or changed, as fall within the prescribed limit of basic numbers at the diploid level, i.e. $x = 2$–13 (Grant, 1963), may conveniently be referred to as *primary basic numbers*, whilst the remainder are termed *secondary basic numbers*, x_2. Additional information often reveals that x_2 may be less than 13 in particular groups where polyploids are derived from species with basic numbers below 7. Darlington (1963) has further modified the nomenclature by using the terms dibasic or tribasic for numbers derived through amphidiploidy involving two or three primary numbers. It is clear that there are various degrees of inference when using the observed gametic numbers to determine the different types of basic number and detailed cytogenetic studies are often necessary to approach a full understanding of the situation. Only in a relatively few well-studied groups (e.g. Fig. 1) is the necessary information available to make full taxonomic use of basic numbers and it is important that the level of inference involved in their derivation for a particular group should be made quite clear since this is directly relevant to their taxonomic value.

Other than faulty observation, the principal difficulties in using chromosome numbers arise when supernumerary chromosomes or satellites (trabants) are present. Satellites can usually be seen to be attached to the main body of the chromosome to which they belong but sometimes, particularly early in mitosis, they are widely separated and may be incorrectly assigned to the chromosome complement as, for example, in *Drimys lanceolata* (Poir.) Baill. (Raven and Kyhos, 1965). Supernumerary chromosomes differ from the standard complement of A chromosomes by their inconstant behaviour and inheritance. They may vary from the so-called B-chromosomes, which can be fairly readily recognized by their smaller size and largely or entirely heterochromatic structure, to extra chromosomes indistinguishable from the standard complement without detailed analysis (e.g. *Clarkia unguiculata* Lindl.: Lewis, 1951; *Narcissus bulbocodium* L.: Fernandes, 1949)

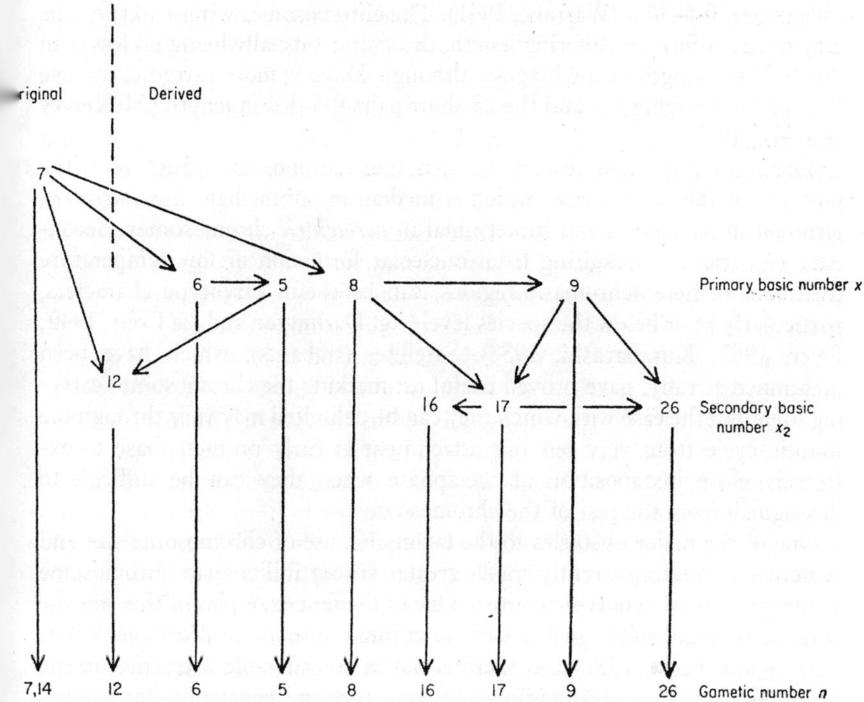

FIG. 1. Diagram illustrating the interrelationships of the various kinds of basic chromosome number and the gametic chromosome number by reference to the genus *Clarkia*. Data from Lewis and Lewis (1955) with supplementary information from Mosquin (1961) and Moore and Lewis (1966). $x_2 = 12,17$ and $x_2 = 26$ are, respectively, dibasic and tribasic numbers. Note the separate bases of $n = 8$ and $n = 16$ and of $n = 6$ and $n = 12$.

and the latter type may well be responsible for much of the unspecified intraspecific variation in chromosome numbers which has been recorded (Darlington, 1963: p. 30). Although variable frequencies of supernumerary chromosomes within species can sometimes be correlated with, for example, geographical distribution (*Centaurea scabiosa* L.: Fröst, 1958), ecological preference (*Festuca pratensis* Huds.: Bosemark, 1956; *Clarkia unguiculata* Lindl.: Mooring, 1960) or breeding system (*Plantago coronopus* L.: Paliwal and Hyde, 1959), they have not been used in any formal systematic studies.

2. Chromosome Size and Structure

Although chromosome volume, expressed as a function of DNA content (Rees, 1963), can occasionally be used to characterize the size of the karyotype, it is customary to use chromosome length, which in most plants is in the range *c.* 0·5–30 μ (Warmke, 1941). The chromosomes within a karyotype may be of uniform or differing length, the shortest usually being no less than 0·5–0·33 the length of the longest, although *Yucca* is more extreme, the five long pairs averaging 6 μ and the 25 short pairs 0·5–1 μ in length (McKelvey and Sax, 1933).

The most important feature for denoting chromosome structure is the position of the centromere, which is median or submedian in *metacentric*, terminal in *telocentric*, and subterminal in *acrocentric* chromosomes. Secondary constrictions, resulting from nucleolar formation or low temperature treatment of heterochromatic regions, can be useful karyotype characters, particularly at or below the species level (e.g. Darlington and La Cour, 1940; Dyer, 1963; Kurabayashi, 1958). Satellites (trabants), which have been mentioned already, have proved useful for marking the chromosomes carrying them but the ease with which they can be delimited may vary through the mitotic cycle from very tenuous attachment at early prometaphase to extremely close juxtaposition at metaphase when they can be difficult to distinguish from the rest of the chromosome.

One of the major obstacles to the taxonomic use of chromosome size and structure is their apparently much greater susceptibility than chromosome number to casual genotypic control. One of the first examples of this was the demonstration of single gene control of chromosome size in *Matthiola* R. Br. (Lesley and Frost, 1927) and there is now a considerable literature on the genotypic control of chromosome size, structure and behaviour (for reviews see: Rees, 1961; Lewis and John, 1963: Chapter 2). These features of the karyotype are also subject to environmental modification (e.g. Swanson, 1957: 153), a point to remember when comparing data, such as idiograms obtained at different times.

USE OF THE KARYOTYPE

1. General Considerations

In a recent clarification of the principles of modern taxonomy Davis and Heywood (1963: 2) have specified two main approaches, the empirical and the interpretative, which may overlap and intermingle but which nevertheless provide convenient headings for summarizing the taxonomic contribution made by karyotype studies.

(a) *Empirical or Phenetic Studies* One of the principal reasons why karyotype studies, particularly those involving chromosome number, have played a prominent part in taxonomy must be that they generally reinforce, or at least are not contrary to other sorts of data, usually morphological and geographical. Chromosome number, whether the gametic number at the species level or the basic number in higher categories, tends to exhibit a greater constancy than many other characters more immediately subject to environmental modifications. In addition, of course, even though the chromosomes, like other characters, are ultimately under the control of genes, they do give rise directly to other chromosomes, while sepals do not give rise to sepals nor hairs to hairs (*cf.* Lewis and John, 1963: 421). There has been much discussion about the importance of the chromosomes relative to other characters and Davis and Heywood (1963: 208) have recently provided valid grounds for not weighting chromosome data. It is perhaps not entirely flippant to suggest that the attention paid by taxonomists to chromosome numbers is in part due to their brevity and convenience, which encourages their citation in Floras and monographs. Formulae for the size and structure of the chromosomes are much less crisp (for their use see, e.g. Fernandes, 1966) while data on fertility, crossability etc. are altogether too cumbersome for the usual taxonomic presentation.

(b) *Interpretative or Phylogenetic Studies* The chromosomes provide a tangible link with the genetical mechanism underlying the evolutionary processes that have formed the conceptual framework for taxonomic work during the last 100 years or so. They have a reasonably predictable behaviour and occurrence which encourages their use in phylogenetic speculation and, although it is easy to overstate their value, most taxonomists would in general agree with Darlington (1963: 119) that "chromosome studies . . . point not only to the past but to the future". Experience has given a number of guide-lines when using chromosome studies to interpret relationships, thus:

(i) Polyploids are derived from diploids.

(ii) Small chromosomes are frequently encountered in more derived plants than are large chromosomes.

(iii) Karyotypes containing metacentric chromosomes of uniform size are likely to be found in less derived plants than those with acro- or telocentric chromosomes of unequal size.

(iv) Original primary basic numbers give rise to derived primary basic numbers and one or both are precursors of secondary basic numbers.

It is obviously necessary to use good sense in the application of these "rules" since there are exceptions to (ii) and (iii). Just as the cytologist should heed the warning of Smith-White (1959: 277) that "the thesis that primitive morphological type may be used to infer primitive chromosome genome is unsafe, and must often lead to error", so should the cytotaxonomist be aware of the converse proposition.

2. *Karyotype Studies and the Taxonomic Hierarchy*

Attempts have been made to utilize karyotype data in taxonomic studies at all levels up to that of the family and even above, but most involve species and groups of species. It is impossible even to begin to review here the large amount of available information, but it may be worthwhile exemplifying the use of various sorts of karyotype data. For convenience, the examples will be divided into those above and those at or below the genus-level.

(a) *Above the Genus-Level* There are few examples of the use of karyotype studies for differentiating families. Darlington (1963) has provided a diagram illustrating the potential use of basic numbers for suggesting relationships among the families of woody angiosperms, in which other features of the karyotype are difficult to use because of their generally small chromosomes. At the present state of knowledge the taxonomic value of speculation on this scale must be small and we are necessarily cautioned that "the names of the families are taken to represent directly, not their external structure, but their chromosome numbers; and the relation between the two is just the problem to which we cannot give a uniform solution" (Darlington, 1963: 125). Warburg (1938) concluded that, while the Geraniaceae, Oxalidaceae and Tropaeolaceae were cytologically close to each other, the Limnanthaceae differed markedly in chromosome size, number and behaviour. This relationship is reflected in the present separation of the sub-order Limnanthineae, with a single family, from the Geraniineae, which contains all the other families (Melchior, 1964). However, the Balsaminaceae, which Warburg found to be cytologically intermediate between the above two groups, are now separated in the Sapindales; clearly, as already noted, the chromosomes provide a modest reflection of other data at this level of taxonomic enquiry. Similarly, in a recent study of chromosome numbers in the Annonales (Raven and Kyhos, 1965) karyotype affinities can be derived for the closely related Magnoliaceae, Himantandraceae and Degeneriaceae and also for the Schizandraceae and Illiciaceae, which are not thought to be closely related, while there is a marked difference in chromosome numbers of the Lauraceae ($x = 12$) and the closely allied Hernandiaceae ($n = 20, 40$).

Although karyotype studies are of no value within chromosomally uniform families such as Pinaceae, Juglandaceae etc., there are many examples of their use for differentiating groups within families. The classical case of the Gramineae, in which the number and size of chromosomes provided valuable confirmatory evidence for the delimitation of tribal boundaries (Avdulov, 1931; Hubbard, 1948), has recently been brought into relief by the new chromosome data for *Spartina* Schreb. (Marchant, 1963), which can now be safely assigned to the Chlorideae on cytological as well as morphological grounds. In the Compositae, chromosome numbers have proved valuable in assigning to tribes various genera of uncertain affinities, and indeed the tribes appear to have different original basic numbers, while it is also interesting that chromosome numbers coincide with morphological variability in indicating that Helenieae are a diverse group of various affinities (Raven *et al.*, 1960; Raven and Kyhos, 1961; Ornduff *et al.*, 1963; 1967; Payne *et al.*, 1964; Solbrig *et al.*, 1964). Using data on chromosome number and structure, Hair (1963) has proposed a cytological classification of the Podocarpaceae "as a basis for a re-examination of the family since no fully satisfactory systematic treatment of the podocarps has yet been devised", while the survey of mitotic cycles in the Onagraceae (Kurabayashi *et al.*, 1962) should be noted because it is a rare example of the use of such detailed studies at the family level. The Onagraceae comprises six tribes containing 22 genera (Raven, 1964; Carlquist and Raven, 1966), of which 18 are known cytologically. Kurabayashi *et al.* (1962) recognized three modally distinct groups of tribes: (i) Fuchsieae, Lopeziaeae, Circaeae; (ii) Onagreae; (iii) Jussiaeae, Epilobieae, which differ in karyotype uniformity, chromosome size and intrachromosomal contraction cycles between interphase and mitotic metaphase. Not only is this work interesting because of the correlation between these data and tribal groupings but also because all species characterized by translocation systems of any sort, a potent force in the evolution of the diversity, fall into group (ii), in which the chromosomes are subequal, uniformly differentiated throughout the mitotic cycle, metacentric and of medium size.

(b) *The Genus-Level and Below* The taxonomic value of karyotype studies is generally greater at these levels than at those already considered, no doubt partly because in any particular investigation it is easier to amass more chromosomal data in relation to the size of the group and partly because causal relationships are usually more evident as a result of the shorter evolutionary time scale than at higher levels of the taxonomic hierarchy. The karyotype is frequently of no assistance in generic delimitation, as in *Lolium* L. and *Festuca* L. (Peto, 1933), which can have chromosome complements indistinguishable even by very refined analysis or, at a cruder level, *Luxuriaga* Ruiz & Pav. and *Enargea* Banks, respectively New Zealand and South American representatives of the small family Philesiaceae (Moore, 1967), but chromosomes have been useful in many instances. The classical case is the use of chromosome number and size to delimit *Crepis* L. from such related genera as *Cymboseris* Boiss. and *Youngia* Cass. and to include in it the former genera *Phaecasium* Cass. and *Pterotheca* Cass., but it is worth keeping in perspective the role of the chromosome data. Babcock (1947: p. 4) clearly states "The systematic treatment of *Crepis* . . . rests primarily on comparative morphology. . . . It has even been stated that our phylogenetic conclusions were based on chromosome number alone; nothing could be further from the truth". The cytological data suggested a re-examination of morphological characters which soon revealed serious discrepancies in their use and led to the selection of others which permitted a treatment more in accord with the chromosomes. The frequency with which differences in chromosome number, particularly basic number, can be used to distinguish genera and groups of related genera makes it unnecessary to belabour the point with examples and, where sufficient data are available, it is common experience that "Because basic numbers are usually stable within genera (of the Onagraceae) they are good indicators of relationships when considered together with the degree of phenotypic similarity" (Lewis and Raven, 1961: 1468). An interesting side-effect of using chromosome data in assessing generic limits is provided by Raven and Lewis (1960) when confirming the separation of *Hauya* Moç. & Sessé ex DC. and *Xylonagra* Donn. Sm. & Rose (Onagraceae). The two genera were found to have different basic numbers, respectively 10 and 7; furthermore it was discovered, initially during the preparation of cytological material and subsequently confirmed in herbarium material, that *Hauya* has the sporogenous tissue in the anthers divided into numerous staggered groups surrounded by sterile tissue while *Xylonagra* has an undivided packet of sporogenous tissue in each locule of the anther. This illustrates the often overlooked fact that the cytologist can be uniquely placed to provide and assess morphological data on a scale which might otherwise be overlooked (*cf.* Tutin, 1963), if he is prepared to deviate periodically from looking solely at the chromosomes.

Although the karyotype is unlikely to help in such chromosomally uniform genera as *Eucalyptus* L'Hérit. and *Ribes*, L. there is a wealth of information on its use in delimiting intrageneric groupings, from which two interesting recent examples, both from the Southern hemisphere, are worth mentioning. Hair (1962) has shown that basic chromosome numbers differ in the three subgenera of *Cotula* L. and that it can consequently be suggested either that subgenus *Strongylosperma* ($x = 18$) was derived by amphidiploidy from subgenera *Cotula* ($x = 5$) and *Leptinella* ($x = 13$) or that *Leptinella* was similarly derived from the other two with a subsequent reduction in basic number from 14 to 13. In *Podocarpus* L'Hérit. ex Pers. variation in chromosome number and structure generally accords with the sectional subdivision, although suggesting that section Dacrycarpus be removed to *Acmopyle* Pilger, and the genus is of further note because its diversity in these chromosome characters makes it a conspicuous exception to the prevailing karyotypic stability in gymnosperm families and genera (Hair, 1963).

The bulk of karyotype data relates to taxonomic studies at the species level. Leaving aside their value in phylogenetic speculation, for which we can all cite our favourite examples in, for example, *Clarkia*, *Crepis*, *Gossypium*, *Nicotiana*, *Viola* etc., their use in phenetic studies can support specific limits, can point to the necessity for modification, or can be of little value. Nothing need be said about the first of these categories, which is much the commonest situation. In the second, chromosomal differences may suggest a re-evaluation of other types of data and the need to search for hitherto overlooked characters. Thus, differences in chromosome number initially pointed to the relatively minor petal characters which separate *Clarkia lingulata*, H. & M. Lewis, from the polytypic *C. biloba* (Dur.) Nelson & Macbr. (Lewis and Roberts, 1956) while in a "genus [in which] quite small morphological differences separate taxa with large biological and cytological differences" (Ingram, 1967: p. 283) chromosome data have shown that the Irish *Sisyrinchium* L., which is of considerable phytogeographical interest (Heslop-Harrison, 1952; Löve and Löve, 1958), belongs to *S. bermudiana* L. and is not conspecific with the Greenland plants. Similarly, to take another recent example, Pettet (1964*a*, *b*) has shown how constant differences in chromosome number indicated similarly constant differences in pollen characters which permit a satisfactory separation of the variable and frequently confused *Viola tricolor* L. and *V. arvensis* Murr., and their intermediates. In the third category mentioned, karyotype variability may exceed, or at least not be correlated with, variability in morphological and other characters, as in the multitude of chromosome numbers within *Cardamine pratensis* L. (Lövkvist, 1956) or of structural arrangements within *Clarkia tenella* (Cav.) H. & M. Lewis subsp. *tenella* (Moore and Lewis, 1966). In these instances karyotype data do not at present help the systematic treatment although they may be a great aid to understanding the evolution and history of the component populations, as for example the patterns of differential chromosome-segments in *Trillium* L. (e.g. Kurabayashi, 1958). On morphological grounds Lourteig (1964) has put all the Southern hemisphere populations of *Limosella* L. into a single species, *L. australis* R.Br., which also extends north through the Americas to Canada and which occurs in Africa and western Europe. Among the populations which she has thus united there are recorded somatic chromosome counts of 20 in Europe, 60 in New Zealand and 48 in southern South America (Moore, 1967). These data, although scant, seem to agree more with the earlier less conservative treatment of Glück (1934) and there seems to be a case for re-examining the taxonomic situation to determine whether it belongs to the second or third categories referred to above.

3. *Present Status of Karyotype Studies*

Apart from those groups where karyotype uniformity must obviously limit the taxonomic value of chromosome studies, it is clear from the examples quoted, and the very many others available, that karyotype data can support conclusions based on other data at all levels of the taxonomic hierarchy, as they do in a large number of instances, or they can provide a firm guide for the reassessment of other lines of evidence and a search for new characters, when they do not accord with other data. In some instances, such as the differentiation-hybridization cycles (Ehrendorfer, 1959), particularly where they involve diploid-polyploid "pillar-complexes", they may explain the taxonomic difficulty of certain groups without offering direct formal systematic solutions. Karyotype variation without comparable or correlated

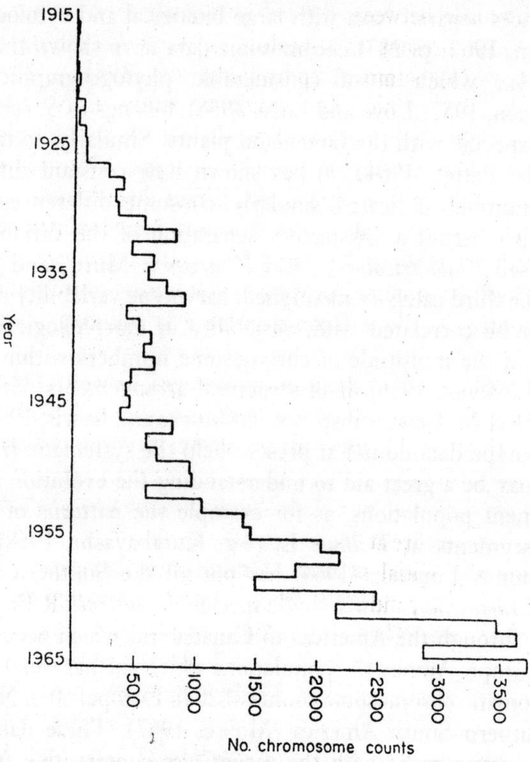

FIG. 2. Histogram showing the number of chromosome counts published per annum during the period 1915–65, according to the standard lists (for references see text). Despite an under-representation for the period prior to 1954, the figure shows the age-distribution of data currently available to most cytotaxonomists.

morphological variation, as occurs for example in many species containing diploid and polyploid populations, has tended to provide a focal point for those who hold extreme views about "biological" and "typological" species. However, most people would now accept that, whilst such situations are important for understanding the dynamics of the populations concerned, there is little chance of their formal recognition in mainstream taxonomy unless further data become available. Indeed, Lewis (1967) has cogently argued that the genetic continuity between diploid populations and their autopolyploid derivatives can be comparable to that between conspecific disjunct populations having the same chromosome number, so that "the ability to consistently identify polyploids from diploids on the basis of external morphology is not of itself a more valid basis for taxonomic recognition than the ability to distinguish between diploid populations or between tetraploid populations". The present relationship between karyotype studies and taxonomy is surely best summed up by Fernandes (1951: p. 187) who, after 30 years detailed taxonomic and cytological study of *Narcissus* L., says "Quelquefois, on constate qu'il y a un parallélisme étroit entre les caractères caryologiques et ceux de la morphologie externe. D'autres fois, un tel parallélisme n'existe pas. De cette façon, il faut être extrêmement prudent en ce qui concerne l'application des données caryologiques à la systématique. Cependant, lorsque les caractères caryologiques sont employés en connexion avec les données provenant des autres sources d'information, ils pourront contribuer d'une façon décisive à la solution de beaucoup des questions".

Assuming an awareness of the great advantages, as well as the limitations, of using karyotype data in taxonomic work, the principal problem facing us today concerns the available information, to which Webb (1965) has referred in a recent stimulating discussion. As has been emphasized previously, the bulk of karyotype data concerns chromosome number, to which the following remarks refer, but they are at least as equally applicable to other chromosome data. Following the initial studies of mitosis in flowering plants (*Lilium croceum* Chaix, $2n = 24$, and other species) by Strasburger (1882) information on chromosome numbers has accumulated rapidly and, as can be seen from Fig. 2, is increasing at the rate of about 3000 counts per annum. Current access to this information is given by the compilations of Darlington and Wylie (1955), Chiarugi (1960), Fabbri (1963), Cave (1958–1966) and Ornduff (1967), which also refer to earlier or more local lists, and there are now available chromosome numbers for about 20,000 angiosperm and 1300 pteridophyte species.

The immense labour of preparing such chromosome lists means that they must be uncritical and the figures given above undoubtedly overestimate our true knowledge. It is common experience that a significant proportion of the data given in the lists of chromosome numbers is, or may be suspected to be, inaccurate, either because of faulty cytology, faulty taxonomy, or both. Thus, in the Magnoliaceae, for example, Raven and Kyhos (1965: p. 245) observe "it is worth pointing out that not a single one of the chromosome numbers summarized by Darlington and Wylie for this family is accurate". Furthermore, it is frequently impossible to recheck records because of the absence of cytological and/or herbarium vouchers of the material upon which they were based. This is essentially the same problem as that facing the nomenclaturist —an unsupported number in a book has no more validity than an unsupported binomial, and the cytotaxonomist has no International Code to guide him. Furthermore, when we take the taxonomically and cytologically reliable chromosome counts we find that the sampling rarely exceeds a few individuals from a relatively restricted number of localities, or even of unknown provenance. Thus, even in a cytologically relatively well-known flora as that of Europe we find that moderately reliable chromosome counts are available for only 27% of the species in volume 2 of *Flora Europaea*, and for many species these are based on a single count unsupported by a voucher specimen, although all are on material from a known locality.

However, I do not wish to over-emphasize the problems. Although cytotaxonomists are still relatively thin on the ground, there are probably more now than there have ever been and their output still continues to increase (Fig. 2) with chromosome numbers annually being determined for *c*. 2900 species belonging to 915 genera (data for 1960–1965). It is becoming customary to deposit herbarium and cytological voucher specimens and to accompany published records by good photographs or camera lucida drawings, a trend pioneered by Manton (1950), so that the reported counts are less completely divorced from their sources. Furthermore, it is more usual now to find the combination of cytologist and taxonomist, whether in a single person or in a closely knit group, so that collaboration between the two approaches should be more and more fruitful. The aim now must be to attain at least representative sampling of groups over their range, both to confirm gametic numbers within species and to derive basic numbers for larger taxa. Most chromosome data have been obtained from temperate areas, particularly in the northern hemisphere, where cytologists have tended to congregate. It is surely unnecessary to stress the great need for information from the tropics, which is still extremely scanty despite the regional studies of Morton (1966) in West Africa and Larsen (1963) in Thailand, and the work on such largely tropical groups as the Dipterocarpaceae (Jong and Lethbridge, 1967) and Gesneriaceae (Ratter, 1963; Ratter and Prentice, 1964). That it is important to be able to base our knowledge of the role of the karyotype in taxonomy on other than largely Northern hemisphere groups is exemplified by Smith-White (1959), who has shown that in several hardwood families of Dicotyledons the endemic Australian genera show a much greater intergeneric diversity in chromosome numbers than would be suspected from work outside Australia. Now that cytologists and taxonomists have largely ceased arguing about the role of the karyotype in taxonomy we can surely look forward to the continued accumulation of the chromosome data which are of such mutual interest and value.

REFERENCES

AVDULOV, N. P. 1931. Karyo-systematische Untersuchungen der Familie Gramineen. *Trudy prikl. Bot. Genet. Selek.*, suppl., 43: 1–428.

BABCOCK, E. B. 1947. The genus *Crepis*. Pt. 1. The taxonomy, phylogeny, distribution and evolution of *Crepis*. *Univ. Calif. Publs. Bot.*, 21: 1–197.

BOSEMARK, N. O. 1956. On the accessory chromosomes in *Festuca pratensis* III. Frequency and geographical distribution of plants with accessory chromosomes. *Hereditas*, 42(1): 189–210.

CARLQUIST, S. and RAVEN, P. H. 1966. The systematics and anatomy of *Gongylocarpus* (Onagraceae). *Am. J. Bot.*, 53(4): 378–390.

CAVE, M. S. 1958–66 (ed.). *Index to Plant Chromosome Numbers 1956–64*. Univ. North Carolina Press, Chapel Hill.

CHIARUGI, A. 1960. Tavole chromosomiche delle Pteridophyta. *Caryologia*, 13: 27–150.

DARLINGTON, C. D. 1963. *Chromosome Botany and the Origin of Cultivated Plants*. Allen and Unwin, London.

DARLINGTON, C. D. and LA COUR, C. F. 1940. Nucleic acid starvation of chromosomes in *Trillium*. *J. Genet.*, 40: 185–213.

DARLINGTON, C. D. and WYLIE, A. P. 1955. *Chromosome Atlas of Flowering Plants*. Allen and Unwin, London.

DAVIS, P. H. and HEYWOOD, V. H. 1963. *Principles of Angiosperm Taxonomy*. Oliver and Boyd, Edinburgh and London.

DYER, A. F. 1963. Allocyclic segments and the structural heterozygosity that they reveal. *Chromosoma*, 13: 545–576.

EHRENDORFER, F. 1959. Differentiation-hybridization cycles and polyploidy in *Achillea*. *Cold Spring Harb. Symp. quant. Biol.*, 24: 141–152.

EHRENDORFER, F. 1964. Cytologie, Taxonomie und Evolution bei Samenpflanzen. *Vistas in Bot.*, 4: 99–186.

FABBRI, F. 1963. Primo supplemento alle Tavole chromosomiche delle Pteridophyta di Alberto Chiarugi. *Caryologia*, 16: 237–335.

FERNANDES, A. 1949. La problème de l'hétérochromatinisation chez *Narcissus bulbocodium* L. *Bolm Soc. broteriana*, Sér. 2, 23: 5–69.

FERNANDES, A. 1951. Sur la phylogénie des espèces du genre *Narcissus* L. *Bolm Soc. broteriana*, Sér. 2, 25: 113–190.

FERNANDES, A. 1966. Nouvelles études sur la section *Jonquilla* DC. du genre *Narcissus* L. *Bolm Soc. broteriana*, Sér. 2, 40: 207–248.

FRÖST, S. 1958. The geographical distribution of accessory chromosomes in *Centaurea scabiosa*. *Hereditas*, 44(1): 75–111.

GLÜCK, H. 1934. *Limosella*-studien. *Bot. Jb.*, 66: 488–566.

GRANT, V. 1963. *The Origin of Adaptations*. Columbia University Press, London and New York.

HAIR, J. B. 1962. Basic chromosome numbers in *Cotula*. *Chromosome Information Service*, No. 3: 41–42.

HAIR, J. B. 1963. Cytogeographical relationships of the southern podocarps. *In:* J. L. Gressitt (ed.) *Pacific Basin Biogeography*. Bishop Museum Press, Honolulu.

HESLOP-HARRISON, J. 1952. The North American and Lusitanian elements in the flora of the British Isles. *In:* J. E. Lousley (ed.) *The Changing Flora of Britain*, B.S.B.I., London., pp. 105–123.

HUBBARD, C. E. 1948. *In:* J. Hutchinson *British Flowering Plants*. Gawthorn, London.

INGRAM, R. 1967. On the identity of the Irish populations of *Sisyrinchium*. *Watsonia*, 6(5): 283–289.

JONG, K. and LETHBRIDGE, A. 1967. Cytological studies in the Dipterocarpaceae, I. Chromosome numbers of certain Malaysian genera. *Notes R. bot. Gdn Edinb.*, 27(2): 175–184.

KURABAYASHI, M. 1958. Evolution and variation in Japanese species of *Trillium. Evolution*, 12(3): 286–310.

KURABAYASHI, M., LEWIS, H. and RAVEN, P. H. 1962. A comparative study of mitosis in the Onagraceae. *Am. J. Bot.*, 49(9): 1003–1026.

LARSEN, K. 1963. Studies in the flora of Thailand, 14. Cytological studies in vascular plants of Thailand. *Dansk bot. Ark.*, 20(3): 211–275.

LESLEY, M. M. and FROST, H. B. 1927. Mendelian inheritance of chromosome shape in *Matthiola. Genetics, Princeton*, 12: 449–460.

LEVITSKY, G. A. 1924. *The Material Basis of Heredity*. State Printing Office, Kiev.

LEWIS, H. 1951. The origin of supernumerary chromosomes in natural populations of *Clarkia elegans. Evolution*, 5(2): 142–157.

LEWIS, H. 1967. The taxonomic significance of polyploidy. *Taxon.*, 16(4): 267–271.

LEWIS, H. and LEWIS, M. E. 1955. The genus *Clarkia. Univ. Calif. Publs. Bot.*, 20(4), 241–392.

LEWIS, H. and RAVEN, P. H. 1961. Phylogeny of the Onagraceae. *In: Recent Advances in Botany*. University of Toronto Press.

LEWIS, H. and ROBERTS, M. R. 1956. The origin of *Clarkia lingulata. Evolution*, 10(2): 126–138.

LEWIS, K. R. and JOHN B. 1963. *Chromosome Marker*. Churchill, London.

LOURTEIG, A. 1964. Étude sur *Limosella* L. *Bull. Comité National Français des Recherches Antarctiques – Biologie*, 1, No. 10: 165–172.

LÖVE, A. and LÖVE, D. 1958. The American element in the flora of the British Isles. *Bot. Notiser*, 111(1): 376–388.

LÖVKVIST, B. 1956. The *Cardamine pratensis* complex. Outlines of its cytogenetics and taxonomy. *Symb. bot. Upsal.*, 14(2): 1–131.

McKELVEY, S. D. and SAX, K. 1933. Taxonomic and cytological relationships of *Yucca* and *Agave. J. Arnold Arbor.*, 14: 76–81.

MANTON, I. 1950. *Problems of Cytology and Evolution in the Pteridophyta*. The University Press, Cambridge.

MARCHANT, C. J. 1963. Corrected chromosome numbers for *Spartina* × *townsendii* and its parent species. *Nature*, 199, No. 4896:929.

MELCHIOR, H. 1964. A. Engler's *Syllabus der Pflanzenfamilien*. Bd. II. Borntraeger, Berlin.

MOORE, D. M. 1967. Chromosome numbers of Falkland Islands angiosperms. *Bull. Br. Antarct. Survey*, No. 14: 69–82.

MOORE, D. M. and LEWIS, H. 1966. Variation and evolution in South American *Clarkia. Heredity, London*, 21(1): 37–56.

MOORING, J. S. 1960. A cytogenetic study of *Clarkia unguiculata*, II. Supernumerary chromosomes. *Am. J. Bot.*, 47(10): 847–854.

MORTON, J. K. 1966. The role of polyploidy in the evolution of a tropical flora. *Heredity, London*, 19, Supplement: 73–76.

MOSQUIN, T. 1961. Phylogenetic studies in *Clarkia*, section *Myxocarpa*. Ph.D. Thesis, Univ. of Calif. Library, Los Angeles.

ORNDUFF, R. 1967. Index to Plant Chromosome Numbers for 1965. *Regnum Veg.*, 50: 1–128.

ORNDUFF, R., MOSQUIN, T., KYHOS, D. W. and RAVEN, P. H. 1967. Chromosome numbers in Compositae. VI. Senecioneae II. *Am. J. Bot.*, 54(2): 205–213.

ORNDUFF, R., RAVEN, P. H., KYHOS, D. W. and KRUCKEBERG, A. R. 1963. Chromosome numbers in Compositae. III. Senecioneae. *Am. J. Bot.*, 50(2): 131–139.

PALIWAL, R. L. and HYDE, B. B. 1959. The association of a single B-chromosome with male-sterility in *Plantago coronopus*. *Am. J. Bot.*, 46(6): 460–466.

PAYNE, W. W., RAVEN, P. H. and KYHOS, D. W. 1964. Chromosome numbers in Compositae. IV. Ambrosieae. *Am. J. Bot.*, 51(4): 419–424.

PETO, F. H. 1933. The cytology of certain intergeneric hybrids between *Festuca* and *Lolium*. *Jl. Genetics*, 28: 113–156.

PETTET, A. 1964*a*. Studies on British pansies. I. Chromosome numbers and pollen assemblages. *Watsonia*, 6(1): 39–50.

PETTET, A. 1964*b*. Studies on British pansies. II. The status of some intermediates between *Viola tricolor* L. and *V. arvensis* Murr. *Watsonia*, 6(1): 51–69.

RATTER, J. A. 1963. Some chromosome numbers in the Gesneriaceae. *Notes R. bot. Gdn Edinb.*, 24: 221–229.

RATTER, J. A. and PRENTICE, H. T. 1964. Chromosome numbers in the Gesneriaceae, II. *Notes R. bot. Gdn Edinb.*, 25: 303–307.

RAVEN, P. H. 1964. The generic subdivision of Onagraceae, tribe Onagreae. *Brittonia*, 16(3): 276–288.

RAVEN, P. H. and KYHOS, D. W. 1961. Chromosome numbers in Compositae. II. Helenieae. *Am. J. Bot.*, 48(9): 842–850.

RAVEN, P. H. and KYHOS, D. W. 1965. New evidence concerning the original basic chromosome number of angiosperms. *Evolution*, 19(2): 244–248.

RAVEN, P. H. and LEWIS, H. 1960. Observations on the chromosomes and relationships of *Hauya* and *Xylonagra*. *Aliso*, 4(3): 483–484.

RAVEN, P. H., SOLBRIG, O. T., KYHOS, D. W. and SNOW, R. 1960. Chromosome numbers in Compositae, I. Astereae. *Am. J. Bot.*, 47(2): 124–132.

REES, H. 1961. Genotypic control of chromosome form and behaviour. *Bot. Rev.*, 27(2): 288–318.

REES, H. 1963. Deoxyribonucleic acid and the ancestry of wheat. *Nature, Lond.*, 198: No. 4875: 108–109.

SMITH-WHITE, S. 1959. Cytological evolution in the Australian flora. *Cold Spring Harb. Symp. quant. Biol.*, 24: 273–289.

SOLBRIG, O. T., ANDERSON, L. C., KYHOS, D. W., RAVEN, P. H. and RÜDENBERG, L. 1964. Chromosome numbers in Compositae. V. Astereae II. *Am. J. Bot.*, 51(5): 513–519.

STRASBURGER, E. 1882. Über den Theilungsvorgang der Zellkerne und das Verhältniss der Kerntheilung zur Zelltheilung. *Arch. mikrosk. Anat., EntwMech.*, 21: 476–590.

SWANSON, C. P. 1957. *Cytology and Cytogenetics*. Prentice Hall, Englewood Cliffs, N.J.

TUTIN, T. G. 1963. Scale effects and other subjective influences in taxonomy. *Bot. Notiser*, 116(2): 122–126.

WARBURG, E. F. 1938. Taxonomy and relationships in the Geraniales in the light of their cytology. *New Phytol.*, 37(3): 189–209.

WARMKE, H. E. 1941. Chromosome continuity and individuality. *Cold Spring Harb. Symp. quant. Biol.*, 9: 1–6.

WEBB, D. A. 1965. Chromosomes and plants. *Watsonia*, 6(2): 134–136.

7 | Fertility, Sterility and the Species Problem

OTTO T. SOLBRIG

Botanical Gardens, University of Michigan, Ann Arbor, Michigan, U.S.A.

Cytotaxonomy is the utilization of cytological characters and phenomena for the elucidation of taxonomic problems. However, most cytotaxonomic studies have largely made use of only two nuclear aspects: (1) number, shape and size of the chromosomes; and (2) the behavior of the chromosomes during meiosis and, to a lesser extent, their behavior during mitosis. These studies are particularly useful in the investigation of natural and artificial hybrids. In the present work the extent to which meiotic chromosomal behaviour is of use in assessing hybrids for taxonomic purposes will be considered. Space dictates that the review be limited to species with sexual reproduction. A second restriction is that only Angiospermous plants will be considered.

PRELIMINARY CONSIDERATIONS

The primary concern of taxonomists has been the delineation of species. Since the modern development of taxonomy in the seventeenth and eighteenth centuries, two major species concepts have been applied. These have not always been explicitly defined. We know them today as the *morphological species concept* and the *biological species concept*. The so-called morphological species concept is an application of the Aristotelian logical definition of the species to organisms. Species are defined strictly on the possession by their members of certain phenotypical characteristics not possessed by members of other species. This was the concept of species held by Linnaeus and most eighteenth- and nineteenth-century taxonomists. According to the biological species concept on the other hand, individuals of populations that under natural conditions are potentially capable of interbreeding form the species. The morphological species concept stresses similarity and difference, while the biological species concept stresses the breeding relationship. In general there is a high degree of coincidence between the species defined by these two concepts. Individuals that look alike tend to interbreed; and individuals that interbreed tend to look alike, since they have large numbers of common genes. Consequently, although some biologists look at these two criteria as non-complementary, opposing definitions, most taxonomists, starting with Linnaeus himself, have tended to use both criteria. So for example, sexual polymorphisms such as those found in heterostylous species, have not been used for differentiating species even by the most morphologically inclined taxonomists. Likewise, the genetically inclined systematist has usually been reluctant to recognize in a formal sense genetically isolated populations, even in cases where they constitute clear cases of sibling species (Grant, 1964; Solbrig, 1964).

At the basis of much discussion of the species problem lies a confusion of the aims and the tools of the taxonomist and the evolutionist, as well as a clouded understanding of what a taxonomist does. Fortunately the recent literature on the subject seems to be shedding some needed light.

Systematic studies involve three main general objectives: (1) to describe, delimit and define taxa; (2) to classify; (3) to explain the variability of organisms and the mechanisms that gave rise to it. Description is the basic operation performed by the taxonomist. Every good taxonomic work has to be based on precise and exact description. The powers of observation and discrimination of the taxonomist are nowhere more in evidence than at this first step. Description does not involve only the superficial morphology of an organism but each and every aspect of it: physiology, anatomy, cytology, genetics, ecology and so on. Occasionally data are obtained by experimentation, but basically data are obtained by observation and the application of the comparative method.

Classification is the erection of classes. There are many ways of erecting a classification, and *a priori* it is impossible to state that one classification is better than any other. One of the interesting modern findings in the field, is the realization that all classifications are special purpose classifications. Once the purpose of a classification has been established, then it is possible to ascertain its usefulness.

The third function of the taxonomist is to give an explanation for the diversity of nature. By definition, science is an intellectual pursuit that makes meaningful explanations with high predictive content. Consequently the taxonomist cannot be satisfied with describing and classifying but has to try to find ultimate and general causes for the phenomena he studies. These explanations are found in the analysis of the elements that underlie a structure or a process. Ultimately biological phenomena will be explainable in physicochemical terms. Although a trend in this direction is already clearly discernible even in taxonomic studies, for the moment many phenomena can only be explained at levels of integration higher than the chemical, such as in terms of nuclear phenomena or Mendelian genetics.

The problem surrounding the species concept and its application can in part be understood in terms of this trinity of purpose in taxonomy. Basically the explanation for the existence of discrete units has been too often confused with the description of the units themselves. Since there is no one single universal mechanism to explain the origin of species, and since there is more than one "kind" of species, no attempt at defining species in terms of the process of speciation can be very successful. Using a modern terminology (Davis and Heywood, 1963), the morphological species concept is the one employed in a horizontal, natural classification, while the biological species concept is an attempt to define the species of a vertical, phylogenetic classification. As these authors point out, with our present knowledge to put vertical classification before horizontal natural classification is to put the cart before the horse. Viewed in this way it is clear that the conflict is not one of semantics, but one of objectives and methodology.

Chromosomes and cytological phenomena can be used as additional characters in the erection of a horizontal classification. However, they are also the carriers of the genes, and their behavior in part directs and determines the fate of the plant. Since "chromosomes derive their prominence as a tool in taxonomy from their direct relation to the genetic system of which they are an integral part" (Lewis, 1957), they figure prominently in any genetically oriented taxonomic work. For the remainder of this discussion I will be concerned primarily with the behavior of chromosomes in meiosis, not as taxonomic characters, put as underlying causes of evolutionary phenomena. Likewise, I will be concerned, not with the species as a classificatory unit, but with the species as an evolutionary unit.

HYBRIDIZATION AS AN AID IN THE DELIMITATION OF SPECIES

"The theory of evolution introduced the concept of hereditary continuity in time, and hence genetic relationship, between species and other taxa derived from a common origin" (Lewis, 1957). Two corollaries that were drawn quite early (Anderson, 1931, 1937; Clausen, 1931; Heilborn, 1924; Manton, 1932; Smith, 1933) were that all the members of a species had to be interfertile, and that members of different species were genetically isolated. These concepts in turn led to the formulation of the biological species concept (Dobzhansky, 1937; Mayr, 1942).

Among animals, and particularly among vertebrates, natural hybridization seems to be a rare event (Mayr, 1963; Dobzhansky, 1951). When hybrids are produced through artificial insemination and other techniques they usually are sterile. Behavioral factors usually keep members of different genetic pools from interbreeding, and the chromosomal XY sex-determining mechanism is largely responsible for the high sterility of the hybrids.

In plants the situation is very different, particularly among the vascular plants. Natural hybrids are quite frequent, and furthermore artificial interspecific hybrids can be produced almost at will among many related species. This situation was seized upon by the early cytologists as an ideal one to provide an explanation of the relationship and the stage of speciation of two taxa. As Edgar Anderson (1937) said "Taxonomy and Cytology study the same phenomenon but from two angles. Cytology, or more properly karyology, concerns itself with the architecture of the germplasm; taxonomy with the adult forms which result from germplasms." This statement assumed a simple correlation between the structure of the chromosomes and the resulting phenotype. In such a case, the degree of difference in the karyotype would be an indication of the degree of difference of the phenotypes, and consequently of the species involved.

It was this kind of reasoning that led to the use of artificial hybrids as a tool to ascertain the degree of genetical differentiation between taxa. Not only could the chromosomes be studied and their structural differences analysed, but the hereditability of the various characters could be investigated and their relative importance ascertained, under the assumption that characters that show simple Mendelian inheritance are less reliable as indicators of genetical differentiation than characters showing a multifactorial type of inheritance. Although on occasion the importance of this technique has been overemphasized, it has proven so far to be the most powerful *experimental* approach available in the study of the process of speciation.

When the stigma of a plant is dusted with pollen from another plant that on morphological or other grounds is suspected of belonging to another species, the pollen may or may not germinate and produce a zygote. In turn the zygote if formed may mature into an embryo and seed that in turn will germinate and grow into an adult plant, or its growth may become arrested at any of various intermediate stages, usually before the mature seedling stage. In those cases where a more or less normal looking, mature, hybrid plant is formed, the gametes may be totally non-functional (sterile) or they may be partially or totally viable.

The genetic, developmental, structural and biochemical mechanisms that control the formation of a hybrid, allowing or blocking its development, are very poorly understood even today. On the other hand, the gross behaviour of chromosomes at meiosis is better understood and abnormal meiotic behaviour in hybrids due to chromosomal aberrations can usually be interpreted rather easily. Consequently it is usually possible to interpret the causes of hybrid sterility when they are solely or largely due to chromosomal differences between the parents. At present, the study of meiosis in hybrids is the most commonly employed technique to interpret species differences in genetical terms.

It can be seen then that the greatest drawback in realizing a truly genetic interpretation of the species, is the lack of understanding of genetical phenomena at a lower level of integration than the chromosome and of relatively simple techniques to study the gene at a physicochemical level. The new insights provided by molecular biology and molecular genetics, and the development of techniques such as gel electrophoresis may provide the necessary tools for an understanding of the species in genetic terms.

From the point of view of natural classification however, the crucial event is the development of sterility between species. By sterility we mean not a decrease in gene flow, but total and absolute cessation of gene exchange. However, when the hybrid is sterile a study of meiosis (the usual procedure) becomes superfluous to establish this fact. If the hybrid on the other hand is fertile, from a taxonomic point of view a study of meiosis again becomes superfluous. It can be seen then that a study of the behavior of chromosomes at meiosis in species hybrids is not very helpful in the delimitation of species.

Once the sterility or fertility of the hybrid has been established the study of meiosis is useful to provide an explanation for the mechanisms that control fertility. But, since the sterility is not always due to chromosomal differences, the study of meiosis can provide an explanation only in certain special cases. Sterility can be due to other causes such as non-complementary genes, lethal or semilethal genes, non-functional zygotes, or other still poorly understood causes.

Finally there is the case where the artificial hybrids between two species are completely fertile and nevertheless there is no gene flow because no hybrids are formed in nature. This is the case of species that are ecologically or geographically separated. In the first of these cases the hybrids do not get a chance to become established due to lack of an ecological niche; in the second there is no opportunity for the formation of a hybrid at all.

In summary, the study of meiosis is of little help in delimiting species, even when the taxonomist wishes to apply the so-called biological species concept. However, whenever there is an incomplete barrier to gene flow between two taxonomic entities (populations, subspecies, species, etc.), a study of meiosis in the hybrid often becomes very helpful to establish an explanation for the partial sterility in the hybrid.

MECHANISMS THAT REDUCE FERTILITY IN HYBRIDS

Table I lists the different kinds of isolation mechanisms found in plants. Prefertilization mechanisms are the most widespread, and any two unrelated species and many related ones are so separated. Since no hybrid is formed there is no opportunity to study the genetical differences. The same applies to prezygotic barriers. It is only in cases where hybrids are formed, that is, when there is a post-zygotic barrier that the study of meiosis becomes useful. Stebbins (1958) has discussed extensively the different kinds of mechanisms that reduce fertility in hybrids. Consequently the various mechanisms will be only briefly mentioned giving some new examples.

TABLE I. Isolation mechanisms.

Prefertilization	Reduction of contact	1. Geographical separation 2. Ecological separation
	Reduction of mating frequency	3. Pollen incompatibility (gametic isolation) 4. Inbreeding and asexual reproduction 5. Specific pollinators (ethological isolation) 6. Different flowering times (allochronic isolation)
Postfertilization	Reduction of zygote formation (prezygotic)	7. Genetic isolation
	Reduction of gene flow through hybrids (postzygotic)	8. Hybrid sterility 9. Hybrid weakness or breakdown 10. Lack of hybrid establishment (environmental isolation)

1. Hybrid Inviability and Weakness

Under this heading Stebbins included all of the mechanisms that prevent or retard the development of hybrids from the first division of the zygote up to the final differentiation of the reproductive organs and the spores which they produce. These mechanisms act at various stages in the development of the plant, retarding or impeding growth. Since the hybrid never reaches maturity, or if it does, it is weak and sterile, mechanisms producing hybrid inviability and weakness provide an effective barrier to gene exchange.

Stebbins has grouped the mechanisms that lead to hybrid inviability into three categories: (1) disharmonious interaction between the genetic material of the two species; (2) disharmonious interaction between the genetical material of the sperm and the cytoplasm of the egg; and (3) disharmonious interaction between the zygote and the tissues of the mother plant, or more commonly in higher plants, the endosperm.

Examples of these three kinds are not too numerous. Disharmonious interaction between the genetic material of the two species was found in a hybrid between hexaploid ($n = 15$) *Glandularia elegans* (L.) Small and diploid ($n = 5$) *G. stellaroid*· (Cham.) Schnack & Covas, produced by us. One seed germinated of approximately 20 produced as a result of several hundred pollinations. The hybrid plant was intermediate in leaf morphology, but its growth was severely retarded, the internodes were much shorter than in either parent, and when it flowered after two years of growth (normal flowering time for the parents three to five months), it produced inflorescences of one to ten flowers (normal for both parents 30–50 flowers) that were partly deformed.

An example of disharmonious interaction between the sperm and cytoplasm of the egg that has been extensively studied is in the genus *Epilobium* (Michaelis, 1954). A more recent example is found in crosses between species of isophyllous Campanulae in Europe, such as *C. garganica* × *C. poschaskyana* studied by Damboldt (1965). The F_1 of this hybrid was chlorotic and slow growing, meiosis however was normal. No such deleterious effect was found in the reciprocal cross, indicating that the negative interaction was between the pollen of *poschaskyana* and the egg cytoplasm of *garganica*. However, in the F_1 of the cross *C. garganica* × *C. fenestrellata* both the cross and the reciprocal cross showed chlorosis and slow growth with high seedling mortality. The surviving plants were highly retarded in growth and did not bloom. This kind of sterility does not appear to be too common, however.

As to sterility due to interaction between zygotic genotypes and maternal tissues, one of the best examples from the modern literature is that of the genus *Primula* studied extensively by Valentine (1961).

In cases of hybrid inviability or weakness the sterility is due to premeiotic factors. Consequently the study of meiosis in the hybrids is not of much consequence in ascertaining the causes of sterility, although in certain cases as with the investigations of Valentine (1961) and collaborators on *Primula* and of Damboldt (1965) on *Campanula*, the study of meiosis can be used to learn the degree of chromosomal changes that have taken place independent of this other cause of sterility.

2. *Reduced Fertility of the Hybrids*

In most cases of hybridization, the hybrids are normal in their vegetative development, but show some kind of reduction in fertility. There are numerous examples given by Stebbins (1950, 1958) and others in the literature. Following Renner (1929) and Müntzing (1930) two major types of mechanisms that reduce fertility will be recognized: haplontic or gametic sterility, which acts on the gametes or gametophytes, and diplontic sterility, which affects diploid tissue. Stebbins (1958) cites the reasons why this classification is to be preferred over that of Dobzhansky (1951) of genic and chromosomal sterility.

(a) *Diplontic Sterility* (*Genic Sterility*) This kind is common in animals but not in plants, and when present in plants it is often associated with haplontic sterility. A plant exhibits diplontic sterility when the sterility affects diploid tissue, such as gametangia or floral parts, either in the F_1 before or at meiosis, or in certain zygotes, and embryos of the segregating F_2 generation.

(b) *Haplontic Sterility* (*Chromosomal Sterility*) This is the most frequent kind of sterility observed in plants. It is the type that becomes apparent at the time of spore formation, and is not associated with any disturbances of development. It is due to structural differences in the chromosomes of the parental species.

There are numerous examples of this kind of sterility. In some cases the chromosomes are sufficiently large and the differences between the genomes of sufficient magnitude to permit an exact analysis of the differences in the parental karyotypes, such as in the case of three species of the genus *Chaenactis* studied by Kyhos (1965). *Chaenactis glabriuscula* ($n = 6$), *C. fremontii* ($n = 5$), and *C. stevioides* ($n = 5$) are morphologically very similar. One species, *C. glabriuscula* is common throughout California in mesic areas, the two others being confined to the dry areas of S.W. California and adjacent areas in the southwestern United States. By a very careful and detailed analysis of meiosis in interspecific hybrids between these three species, Kyhos showed conclusively that the two species with five pairs of chromosomes have originated independently from six-paired *C. glabriuscula*. Two translocations with a loss of a centric region in each case are involved, resulting in a configuration of 1 IV + 1 III + 2 II in the cross *C. glabriuscula* × *C. fremontii*, and in a configuration of 1 V + 3 II in the cross *C. glabriuscula* × *C. stevioides*. That the translocations in each case involve different chromosomal segments is indicated by the fact that in the cross *C. stevioides* × *C. fremontii* the maximum configuration found was 1 IV + 1 III + 1 II.

In most cases, however, such a detailed analysis is not possible or not desirable, either because the chromosomal aberrations are too complex or the chromosomes are too small to permit a detailed analysis. A major problem is the possible presence of minor structural changes—cryptic changes—that may not affect meiotic pairing, but do affect fertility due to deficiencies in the gametes. They also can be suspected when in spite of bivalent formation, the chiasma frequency is decreased. A recent example is the analysis of the *Agropyron scabriglume* complex in Argentina, by Hunziker (1966). Two Argentine species were analyzed: the hexaploid *A. scabriglume* (21 II) and the tetraploid *A. tilcarense* (14 II), presumed to be the ancestor of the first species. Crosses between the two species yielded hybrids with varying degrees of fertility (Table II). The fertility varied according to the geographic origin of the race of *A. scabriglume* employed in the cross. Highest fertility was obtained between a race of *A. scabriglume* from La Quiaca, Jujuy, a locality less than 100 miles from Tilcara (the sole known locality of *A. tilcarense*), and in approximately the same environmental and ecological conditions. These results led to an investigation through interspecific crosses and morphological and biochemical analyses of the various populations of *A. scabriglume* to determine the nature of the intraspecific changes in this species. This study showed that the race of La Quiaca and that of Balcarce (populations that are geographically separated by over 1000 miles) differ by at least three

TABLE II. Fertility of *Agropyron tilcarense*, *A. scabriglume* and their hybrids.

Species or hybrid and origin	Chromosome No.	% Fertile florets
A. tilcarense (Tilcara)	28	93·5
A. scabriglume (La Quiaca)	42	86·9
A. scabriglume (Volcan)	42	83·8
A. scabriglume (Balcarce)	42	90·9
A. scabriglume (Tupungato × Balcarce)	42	95·1
A. scabriglume (La Quiaca × Balcarce)	42	10·4
A. tilcarense (Tilcara) × *A. scabriglume* (La Quiaca)	35	36·2
A. scabriglume (Volcan) × *A. tilcarense* (Tilcara)	35	11·2
A. tilcarense (Tilcara) × *A. scabriglume* (Tafi)	35	28·5
A. tilcarense (Tilcara) × *A. scabriglume* (Balcarce)	35	6·4
A. scabriglume (Balcarce) × *A. tilcarense* (Tilcara)	35	8·3
A. scabriglume (Tupungato) × *A. tilcarense* (Tilcara)	35	8·2

Data from Hunziker, 1966.

reciprocal translocations, the hybrid between them being only 10% pollen fertile. On the other hand, the Tupungato and Balcarce races (separated also by almost 1000 miles) are chromosomally almost identical. A fourth race, Carancho (Hunziker, 1967) also differs chromosomally from that of Balcarce by several translocations and inversions. Analysis of the seed proteins, however, showed that the Carancho and Balcarce races (which are geographically and ecologically related) are virtually identical in their protein structure (and presumably in the genic makeup determining them) while those from Balcarce and La Quiaca are quite different. Chromosomal rearrangements and genic differentiation for seed proteins are not parallel in their evolution in these instances. This is a very important piece of experimental information particularly if it can be confirmed in other groups. It tends to confirm the thesis that genic differentiation and chromosomal repatterning are independent phenomena.

MORPHOLOGY AND CHROMOSOMAL DIFFERENTIATION IN *GLANDULARIA*

This genus (often considered as section *Glandularia*, of the genus *Verbena*; however see Schnack and Covas, 1944, 1946 and Schnack, 1944, 1964) is widespread in South America where it grows from Peru to Argentina and in North America where it is found from Guatemala to the Central and Eastern United States. The North American species number 19 (Perry, 1933); the number of South American species is closer to 50, of which some 20 are of the general habit and morphology of the North American ones, the rest being specialized Andean and Patagonian cushion plants. For a number of years Schnack and collaborators have been investigating the cytogenetical and cytotaxonomical relationships among the South American species, while more recently a comprehensive morphological, cytogenetical, and biochemical analysis of the North American species has been initiated in our laboratory at the University of Michigan. Some of the results obtained are of interest for the purpose of the present discussion.

Most species can be readily crossed, even in cases where the chromosome numbers of the species differ greatly. As can be seen from Table III various cytological and genetical mechanisms have operated to produce hybrid sterility. There is no way of predicting these mechanisms on the basis of morphological differentiation or of secondary plant products. Although the analysis of leaf protein is in its early stages it appears that in *Glandularia* the leaf proteins also do not correlate with chromosomal differentiation.

For example, the three species *G. peruviana*, *G. pulchella*, and *G. santiaguensis*, occupy approximately the same habitat of open, dry, and semidry grassland in northern and central Argentina. *Glandularia pulchella* and *G. santiaguensis* replace each other geographically, with the former growing in the cooler southern and eastern area, and the latter in the more subtropical northern and western areas. *Glandularia peruviana* stretches over all the territory occupied by both species. Morphologically, *G. peruviana* differs from the other species because of its entire leaves, a more appressed habit, larger internodes, large red flowers (due to the presence of pelargonidin) and absence of glandular appendages. The other two species have divided leaves, are more erect, with shorter internodes, blue flowers (the anthocyanin being cyanidin) with glandular appendages. *Glandularia santiaguensis* in turn is larger than *G. pulchella*, with larger leaves and flowers and less divided leaves. Hybrids between *G. santiaguensis* and *G. peruviana* are 50% pollen sterile. The reduced fertility is due to at least one translocation (the maximum association in the hybrid is 1 IV + 3 II) and probably also to small cryptic changes since the chiasma frequency of the hybrid is lower than the parent,

TABLE III. Fertility and meiotic configurations in species and species hybrids of the genus *Glandularia*.

Species or hybrid	% Univalents	% Bivalents	% Multivalents	Chiasma frequency	% Pollen fertility
peruviana (2*n*)	0	100	0	1·59–1·98	98
pulchella (2*n*)	2	98	0	1·88	94
peruviana × *pulchella* (2*n*)	3	97	0	1·31	42
peruviana × *pulchella* (4*n*)	5	83	12	—	70
megapotamica (2*n*)	0	100	0	—	99
peruviana × *megapotamica* (2*n*)	?	90	?	1·67	65
santiaguensis (2*n*)	0	100	0	1·43	89
santiaguensis × *peruviana* (2*n*)	4	86	10	1·07	51
santiaguensis × *pulchella* (2*n*)	7	79	14	—	50
moricolor (2*n*)	0	100	0	1·63	97
moricolor × *peruviana* (2*n*)	1	87	12	1·25	81
peruviana × *moricolor* (2*n*)	–	—	—	—	56
canadensis (6*n*)	0	100	0	—	98
elegans (6*n*)	0	100	0	—	97
elegans × *peruviana* (4*n*)	2	38	60	—	9
elegans × *pulchella* (4*n*)	3	61	36	—	24
maritima (6*n*)	–	—	—	—	85
canadensis × *maritima* (6*n*)	–	—	—	—	61
canadensis × *elegans* (6*n*)	—	—	—	—	98

leading to the formation of approximately 4% of univalents (although the reduced chiasma frequency could of course also be due to just the effect of the translocation). A very similar behavior was found by Schnack and Covas (1945*a*) in the hybrid between *G. santiaguensis* and *G. pulchella*. The cross between *G. peruviana* and *G. pulchella* on the other hand was 42% pollen fertile, but in spite of this showed almost normal meiotic pairing, the differences being almost entirely due to small cryptic differences, as was shown when the artificial amphiploid was produced, and fertility restored to 70% (Schnack and Solbrig, 1953). On morphological grounds, *G. santiaguensis* and *G. pulchella* are more related; chromosomally however *G. pulchella* and *G. peruviana* appear to be closer, although the genetical isolating barrier between all three species is about the same.

A correlation between cytology and morphology is found in crosses between these three species and *G. megapotamica*, a species morphologically related to *G. peruviana* although ecologically separated since it grows in the subtropical gallery forest of N.E. Argentina, Brazil and Paraguay. The hybrid between *G. peruviana* and *G. megapotamica* is 65% pollen fertile, the decrease in pollen fertility being due apparently to one or two small translocations (Schnack and Covas, 1945*b*). The hybrids between *G. megapotamica* and *G. pulchella* and *G. megapotamica* and *G. santiaguensis* are both completely sterile, meiosis presenting chains of many chromosomes and many univalents, indicative of big rearrangements (Schnack and Gonzales, 1945).

Finally the hybrid between *G. peruviana* and *G. moricolor* should be mentioned. *Glandularia moricolor* is a species of northern Argentina where it grows in the margins of the subtropical forest. It differs morphologically from *G. peruviana* in having elliptical leaves, erect habit, smaller and deep purple flowers and a more compact inflorescence. Hybrids between the two species are nevertheless quite fertile (80% pollen fertility), the pairing at meiosis being regular in most cells. The small amount of sterility is due to a small translocation that manifests itself in occasional tetravalents. In addition there might also be some cryptic structural differences, but these are impossible to detect cytologically. Although the morphological differences are sizeable, the chromosomal differentiation appears to be slight.

At the hexaploid level all the crosses performed to date between *G. canadensis* and other hexaploid species (*G. elegans*, *G. tampensis*, *G. ambrosifolia*, *G. bipinnatifida*, and *G. racemosa*) (Dermen, 1936; Solbrig, ined.) have proven to be quite fertile, no meiotic disturbance whatsoever having been discovered. The cross between *G. canadensis* and *G. elegans* grown in the experimental field was more vigorous and faster growing than either parent, indicative of some kind of luxuriance or hybrid vigor.

Crosses between diploid and hexaploid species were highly sterile (pollen fertility 5–20%). In crosses between *G. canadensis* and *G. peruviana* (Schnack *et al.*, 1959), and *G. elegans* and *G. peruviana* (Solbrig *et al.*, in preparation), a large number of tetravalent associations were observed; while in the hybrid *G. elegans* × *G. pulchella* (Solbrig *et al.*, in preparation) 6 to 10 II were found at meiosis. The interpretation of these findings is that the hexaploid is a segmental allohexaploid, derived from South American diploid species. On morphological grounds the two North American species involved in the crosses are similar to the South American ones crossed, particularly *G. pulchella*.

In summary, the hybrids here reported briefly (Fig. 1), are an indication that morphological and cytological differentiation are not directly correlated

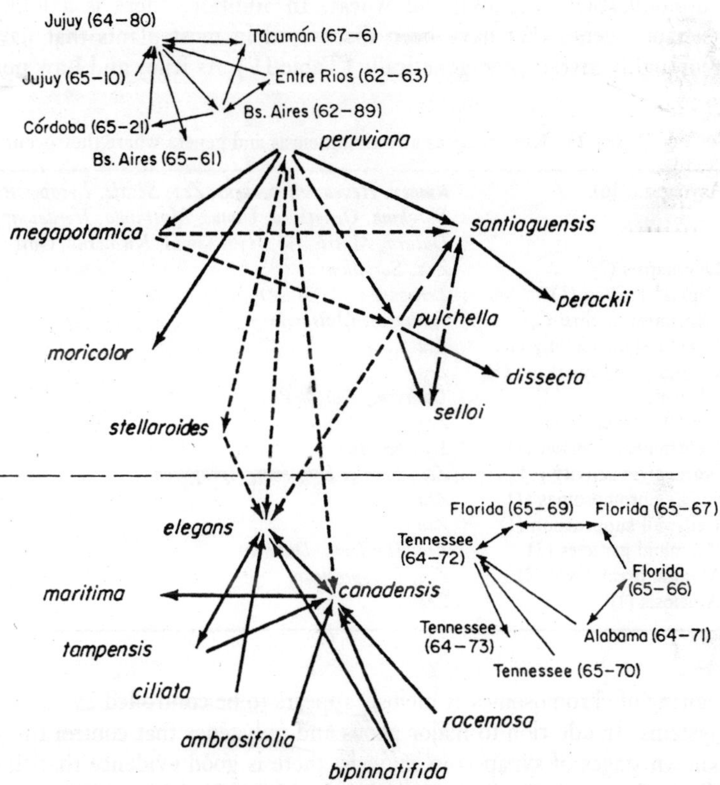

FIG. 1. Crossing relationships among species of *Glandularia*. ———, Fertility better than 60%; — —, fertility between 40 and 60%; - - -, fertility less than 20%, usually near 0. Species below the dividing horizontal line, North American hexaploids; species above the line, South American diploid species. Further discussion in the text.

in *Glandularia*. Meiotic analysis gives a good indication of the nature of the interspecific barriers and the degree to which they have developed. However, as the hybrid *G. peruviana* × *G. moricolor* or the hybrids among the hexaploid species show, isolation barriers other than genetic ones may be in operation, since the species involved in the crosses are morphologically distinct and little hybridization is observed in nature.

GENETIC CONTROL OF MEIOSIS

Evidence has accumulated from many sources (for a review see Riley and Law, 1965) that chromosome pairing at meiosis is under genic control. The most conclusive evidence is that produced by Riley and Chapman (1958) concerning a gene present in chromosome 5B of *Triticum*, that controls the "diploidization" of hexaploid wheat. In addition, there is a long list of asynaptic genes that have been discovered in most plants that have been thoroughly investigated genetically (Table IV). As Riley and Law point out,

TABLE IV. Known genes affecting meiosis and genera where they occur.

Asynapsis (16)	*Rumex*, *Hevea*, *Alopecurus*, *Zea*, *Secale*, *Lycopersicum* *Sorghum*, *Oenothera*, *Ulmus*, *Matthiola*, *Hordeum*, *Datura*, *Matricaria*, *Hyosciamus*, *Nicotiana*, *Gossypium*
Desynapsis (2)	*Zea*, *Sorghum*
Diploid Pairing (1)	*Triticum*
Chromosome Size (2)	*Lathyrus*, *Matthiola*
Chromosome Coiling (1)	*Zea*
Chromosome Breakage (1)	*Zea*
Chiasma position (2)	*Lathyrus*, *Matthiola*
Spindle structure (1)	*Zea*
Centromere division (1)	*Lycopersicum*
Neocentric activity (4)	*Zea*, *Secale*, *Hordeum*, *Agropyron*
Sticky chromosomes (1)	*Zea*
Cell-wall suppression (1)	*Zea*
Polyploid gametes (3)	*Zea*, *Hordeum*, *Datura*
Meiotic breakdown (2)	*Zea*, *Lycopersicum*
Ameiosis (1)	*Zea*

pairing of chromosomes at meiosis appears to be controlled by various genic systems. In addition to major genes and polygenes that control the yet unknown stages of synapsis in meiosis, there is good evidence that the specificity of synapsis can be narrowed and widened by gene action to permit the pairing of chromosomes distantly or closely related in genetic and evolutionary terms. This information reinforces what has been already pointed out, namely that degree of pairing is not necessarily a good indication of genetic differentiation. This is further borne out when the origin of sterility barriers is considered.

ORIGIN AND DEVELOPMENT OF GENETIC ISOLATION

In considering the origin and development of genetic isolation, three major alternatives suggest themselves: (1) that hybrid sterility evolves randomly and is independent of other factors; (2) that it is a byproduct of evolutionary differentiation; and (3) that it is the result of direct selection. Let us consider these three alternatives independently.

The idea held by Goldschmidt that one major mutation may lead to the formation of a new species has by now been demonstrated conclusively to be wrong. However, this does not rule out the sudden or accelerated origin of genetic isolation, through some simple genetical or cytological event.

Lewis (1962) has recently brought forward the idea of rapid speciation as a result of the establishment of a population with a new chromosomal rearrangement so that it is genetically isolated from the ancestral population. The novel idea of Lewis is that this rearrangement of the chromosomal segments is due not to the accumulation of small changes, but to one major "catastrophic" event. Lewis marshalls a great deal of evidence to back his theory (see also Lewis and Roberts, 1956; Lewis and Raven, 1958; Vasek, 1958; Mosquin, 1964). What causes these catastrophic rearrangements in the first place is not known, but mutator genotypes, and environmental stress have been suggested. In any case, catastrophic selection is an instance of the independent evolution of sterility. As Lewis points out, catastrophic selection is most likely to occur among annuals or short lived herbaceous perennials subjected to a great deal of environmental stress and to violent fluctuations in population size. It is also likely to occur at the periphery of the species range, where population extinction is at its greatest.

That genetic isolation is a byproduct of evolutionary and morphological divergence is the implicit assumption of the eco-geographic theory of speciation, that has been documented extensively in several books, most recently by Mayr (1963). It is of course true that most species are no longer able to interbreed, and that the greater the degree of differentiation the less the likelihood that the hybrid between two species will be fertile. However, there is a great deal of variation. In trees and long-lived perennials such as species of *Quercus*, *Acer*, *Populus*, *Betula*, *Vaccinium*, *Ceanothus*, *Arctostaphylos*, etc., related species are usually interfertile, and they are isolated by barriers other than genetical ones, while in annuals and short lived herbaceous perennials, genetic barriers often develop even before morphological differences have taken place. Stebbins (1950, 1958), Grant (1958), and others have pointed out repeatedly the importance of the life cycle in controlling the amount of hybridization that a species can withstand. It can be easily seen that if the developmental pathways of two organisms are different that a moment must come when their genetical messages are no longer compatible, leading to a non-functional hybrid. It is most likely that genetic isolation as a byproduct of phyletic differentiation has developed mostly in long lived perennials and in groups that have been completely isolated for their entire evolutionary history.

Direct selection for isolation has not been demonstrated for plants, although there seems to be enough evidence to indicate that it occurs among animals (Brown and Wilson, 1956; Wilson, 1965). However there is some indirect evidence (Grant, 1965, 1966) to indicate that some selection for prefertilization barriers might have occurred in the genus *Gilia*. Likewise in *Glandularia* populations of *G. stellaroides* will not cross with populations of

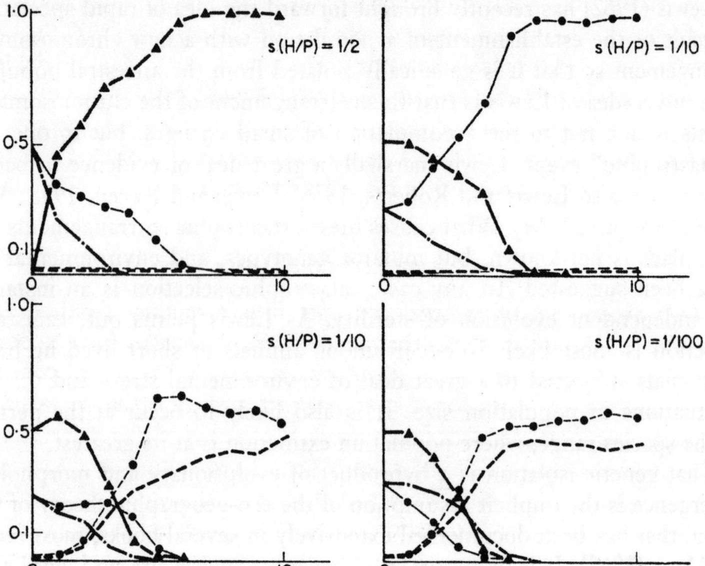

FIG. 2. Generalized results of a model for the development of prefertilization sterility. Population size: 10,000 plants; both species are supposed to be interfertile in 98% of its individuals; 2% are not. —▲—▲—, Hybrids; —●—●—, species A; ·—·—, species B; ●---●---● isolated components of A; ------ isolated components of B. When the postfertilization barrier is 50% under the conditions of the model, a hybrid population is established; when the postfertilization barrier is 90%, either one of the species gets established by outcompeting the other, or selection for prefertilization operates and both species coexist side by side; when the postfertilization barrier is 99%, selection for prefertilization will operate and both species coexist side by side. (Data obtained by Solbrig on an IBM 7090 computer at the University of Michigan Computation Center.) A grant from the NSF is gratefully acknowledged.

G. peruviana from the area where *G. stellaroides* grows but they will cross with populations from outside the area. Also, cases where related species flower at different times of the day, such as *Oenothera brevipes* and *O. clavaeformis aurantiaca* (Raven, 1962), might represent instances of direct selection for an isolation mechanism to control the degree of hybridization. It is very difficult, if not impossible, to design experiments to test the hypothesis of direct selection of isolation. Consequently we are in the process of developing computer simulation programs (Fig. 2) to develop a model that will give the parameters under which such selection might take place.

A final point should be mentioned. In addition to selection for isolation, it is very likely that selection to permit hybridization may also be taking place in certain cases. Anderson and Stebbins (1954) and more recently Raven (1960) have pointed out the advantages of hybridization as the source of new genetic recombination. Therefore, as Raven has expressed it "partly isolated populations may be at a selective advantage in the rapidity with which they are able to adjust to a changing environment". Since hybridization (as any other source of genetic variability) may constitute also a genetic load, it might be that in perennials which are able to carry a greater genetic load than annuals, the selection for the ability to hybridize is stronger than in annuals, in which case this may account for the lack of genetic isolation barriers between related species. This point also requires more investigation.

SUMMARY AND CONCLUSIONS

The introduction of genetic concepts into taxonomy resulted in the development of the hypothesis that all members of a species have to be interfertile and that members of different species are genetically isolated. When genetic experimentation showed that the species of the orthodox taxonomist were not always isolated from each other, nor necessarily interfertile within themselves, the concept of the species was modified so that a species was defined as a group of potentially interfertile individuals. Such a procedure does not alter the fundamental question, namely, why are there cases of discrete morphological groups of populations that are not genetically isolated from other such groups of populations; and also the complementary question, why in certain other cases the members of such groups are not all interfertile. What has to be investigated is the origin and development of genetic isolation. What are needed are crucial experiments that will show unmistakably what the relationship between isolation and evolution is, both in fostering speciation and in permitting genetical and morphological differentiation. There is already a large body of knowledge, and many observations and experiments that partly explain the process, and some of this evidence has been presented here. However, there are alternative explanations that have not yet been disproven conclusively. For example, it is altogether possible that the hypothesis of the genetic coherence of species has been overemphasized. Perhaps there is very little genetic exchange between populations, so that genetic coherence cannot counteract the action of strong selection even in a sympatric situation (see for example Thoday, 1958, 1959; Thoday and Boam, 1959). The breeding system undoubtedly also plays an important role, as well as population size and structure. It will take special thought to devise the crucial experiments and to choose the right materials and methods. However, it can be done and the development of the theory of catastrophic selection is an example of how it can be done. Strong inference should be as rewarding in biosystematics as it has proven in molecular biology.

REFERENCES

ANDERSON, E. 1931. Internal factors affecting discontinuity between species. *Am. Nat.*, 65: 144–148.

ANDERSON, E. 1937. Cytology in its relation to taxonomy. *Bot. Rev.*, 3: 335–350.

ANDERSON, E. and STEBBINS, G. L. 1954. Hybridization as an evolutionary stimulus. *Evolution*, 8: 378–388.

BROWN, W. L. and WILSON, E. O. 1956. Character displacement. *Syst. Zool.*, 5: 49–64.

CLAUSEN, J. 1931. Cytogenetic and taxonomic investigations on melanium violets. *Hereditas*, 15: 219–308.

DAMBOLDT, J. 1965. Zytotaxonomische Revision der isophyllen Campanulae in Europa. *Bot. Jahrb.*, 84: 302–358.

DAVIS, P. H. and HEYWOOD, V. H. 1963. *Principles of Angiosperm Taxonomy*. Oliver & Boyd, Edinburgh.

DERMEN, H. 1936. Cytological study and hybridization in two sections of Verbena. *Cytologia*, 5: 160–170.

DOBZHANSKY, T. 1937. *Genetics and the Origin of Species*, ed. 1. Columbia University Press, New York.

DOBZHANSKY, T. 1951. *Genetics and the Origin of Species*, ed. 5. Columbia University Press, New York.

GRANT, V. 1958. The regulation of recombination in plants. *Cold Spring Harb. Symp. quant. Biol.*, 23: 337–363.

GRANT, V. 1964. The Biological Composition of a Taxonomic Species in Gilia. *Adv. Genet.*, 12: 281–328.

GRANT, V. 1965. Evidence for the selective origin of incompatibility barriers in the leafy-stemmed Gilias. *Proc. natn. Acad. Sci., U.S.A.*, 54: 1567–1571.

GRANT, V. 1966. The selective origin of incompatibility barriers in the plant genus *Gilia*. *Am. Nat.*, 100: 99–118.

HEILBORN, O. 1924. Chromosome numbers and dimensions, species-formation and phylogeny in the genus *Carex*. *Hereditas*, 5: 129–211.

HUNZIKER, J. H. 1966. Diferenciación cromosómica en el complejo hexaploide *Agropyron scabriglume*. *Kurtziana*, 3: 127–149.

HUNZIKER, J. H. 1967. Chromosome and protein differentiation in the *Agropyron scabriglume* complex. *Taxon*, 16: 259–266.

KYHOS, D. W. 1965. The independent aneuploid origin of two species of *Chaenactis* (Compositae) from a common ancestor. *Evolution*, 19: 26–43.

LEWIS, H. 1957. Genetics and cytology in relation to taxonomy. *Taxon*, 6: 42–46.

LEWIS, H. 1962. Catastrophic selection as a factor in speciation. *Evolution*, 16: 257–271.

LEWIS, H. 1967. Comparative Cytology in Systematics. *Proc. Conf. System., Ann Arbor, Mich.* (in press).

LEWIS, H. and RAVEN, P. H. 1958. Rapid evolution in Clarkia. *Evolution*, 12: 319–336.

LEWIS, H. and ROBERTS, M. R. 1956. The origin of *Clarkia lingulata*. *Evolution*, 10: 126–138.

MANTON, I. 1932. Introduction to the general cytology of the Cruciferae. *Ann. Bot.*, 46: 509–554.

MAYR, E. 1942. *Systematics and the Origin of Species.* Columbia University Press, New York.

MAYR, E. 1963. *Animal Species and Evolution.* Harvard University Press, Cambridge, Mass.

MICHAELIS, P. 1954. Cytoplasmic inheritance in Epilobium and its theoretical importance. *Adv. Genet.*, **6**: 287–401.

MOSQUIN, T. 1964. Chromosomal repatterning in *Clarkia rhomboidea* as evidence for post-Pleistocene changes in distribution. *Evolution*, **18**: 12–25.

MÜNTZING, A. 1930. Über Chromosomes-Vermehrung in Galeopsis-Kreuzungen und ihre phylogenetische Bedeutung. *Hereditas*, **14**: 153–172.

PERRY, L. 1933. A revision of the North American species of *Verbena. Ann. Missouri Bot. Gdn.*, **20**: 239–262.

RAVEN, P. H. 1960. Interspecific hybridization as an evolutionary stimulus in Oenothera. *Proc. Linn. Soc.* (*Lond.*), 173: 92–98.

RAVEN, P. H. 1962. The systematics of *Oenothera* subgenus *Chylismia. Univ. Calif. Publ.* (*Botany*), **34**: 1–22.

RENNER, O. 1929. *Artbastarde bei Pflanzen.* Borntraeger, Berlin.

RILEY, R. and CHAPMAN, V. 1958. Genetic control of the cytologically diploid behavior of hexaploid wheat. *Nature, Lond.*, **182**: 713–715.

RILEY, R. and LAW, C. N. 1965. Genetic variation in chromosome pairing. *Adv. Genet.*, 13: 57–114.

SCHNACK, B. 1944. Nota preliminar sobre una modificación de la sistemática del género *Verbena. Anal. Inst. Fit. Santa Catalina* (*Argentina*), **4**: 17–22.

SCHNACK, B. 1964. Bases naturales de la separación genérica de *Verbena y Glandularia* (Verbenáceas). *Notas Com. de Inv. Científica Prov. Bs. Aires* (*Argentina*), **2**(2): 1–13.

SCHNACK, B. and COVAS, G. 1944. Nota sobre la validez del género *Glandularia* (Verbenaceas). *Darwiniana*, **6**: 469–476.

SCHNACK, B. and COVAS, G. 1945*a*. Hibridación interespecífica en *Glandularia* (Verbenáceas). *Darwiniana*, **7**: 71–79.

SCHNACK, B. and COVAS, G. 1945*b*. Un híbrido interespécifico del género *Glandularia* (*G. peruviana* × *G. megapotamica*). *Rev. Argentina de Agronomia*, **12**: 224–229.

SCHNACK, B. and COVAS, G. 1946. Nota taxonómica sobre el género *Glandularia* (Verbenáceas). *Bol. Soc. Argentina de Botanica*, **1**: 282–284.

SCHNACK, B., FEHLEISEN, S. and CUCUCCI, A. E. 1959. Estudio del híbrido interespecífico *Glandularia canadensis* (L.) Small × *G. peruviana* (L.) Small. *Rev. Fac. Agronomia* (*La Plata*), **35**: 113–121.

SCHNACK, B. and GONZALES, F. F. 1945. Estudio morfológico y citogenético del híbrido *Glandularia santiaguensis* × *G. megapotamica. Rev. Argentina de Agronomia*, **12**: 285–290.

SCHNACK, B. and SOLBRIG, O. T. 1953. El híbrido *Glandularia laciniata* × *G. peruviana* y su anfidiploide artificial. *Rev. Fac. Agronomia* (*La Plata*), **29**: 255–266.

SMITH, W. W. 1933. Some aspects of the bearing of cytology on taxonomy. *Proc. Linn. Soc. London*, **145**: 151–181.

SOLBRIG, O. T. 1964. Infraspecific variation in the *Gutierrezia sarothrae* complex. *Contr. Gray Herb., Harvard Univ.*, **193**: 67–114.

SOLBRIG, O. T., PASSANI, C. and GLASS, ROGER. 1968. Artificial hybridization between different polyploid levels in *Glandularia* (Verbenaceae). (In prep.)

STEBBINS, G. L. 1950. *Variation and Evolution in Plants.* Columbia University Press.

STEBBINS, G. L. 1958. The inviability, weakness, and sterility of interspecific hybrids. *Adv. in Genetics*, 8: 147–215.

THODAY, J. M. 1958. Effects of disruptive selection: the experimental production of a polymorphic population. *Nature*, 181: 1124–1125.

THODAY, J. M. 1959. Effects of disruptive selection. I. Genetic flexibility. *Heredity*, 13: 187–203.

THODAY, J. M. and BOAM, T. B. 1959. Effects of disruptive selection. II. Polymorphism and divergence without isolation. *Heredity*, 13: 205–218.

VALENTINE, D. H. 1961. Evolution in the genus *Primula*. *In:* P. J. Wanstall (ed.) *A Darwin Centenary*, pp. 71–78.

VASEK, F. C. 1958. The relationship of *Clarkia exilis* to *Clarkia unguiculata*. *Am. J. Bot.*, 45: 150–162.

WILSON, E. O. 1965. The challenge from related species. *In:* H. G. Baker and G. L. Stebbins (ed.) *The Genetics of Colonizing Species*, pp. 7–27.

8 | Phenotypic Plasticity with Particular Reference to Three Amphibious Plant Species

C. D. K. COOK

*The Hartley Botanical Laboratories, The University, Liverpool, England**

INTRODUCTION

To the taxonomist phenotypic plasticity is often difficult to distinguish from genetic variability and, in certain circumstances, has undoubtedly led to a large number of "paper species" which are without taxonomic value. It is not that the taxonomist is unaware of plasticity but is often unaware of the mechanism causing it.

Not all phenotypic variability should be called plasticity. If a group of genetically identical plants are reared under uniform environmental conditions they will not be phenotypically identical at any given time. The variation observable between them is due to epigenetic effects during the expression of genes and has been termed by Waddington (1957) "developmental noise". The effects of developmental noise are not entirely at random but are largely specific for particular plant characteristics and genotypes. In other words, some genes are expressed more easily than others.

The morphological changes that take place in a plant as it matures (heteroblastic development) are, from a practical point of view, difficult to separate from plasticity. The concept of maturity is difficult to apply in plants because there are virtually two superposed kinds of ontogenetic development: the overall ontogenetic development from the fertilized egg cell through to the production of gametes and the continuous ontogenetic development of organs by apical growth. One is thus presented with an apparent paradox: the flowering portion of a stem is regarded as mature but, in terms of apical development, it is much younger than the juvenile portions of the stem that support it. This means then that as far as heteroblastic development is concerned it is only possible to regard parts of the plant as being mature or having reached their ontogenetic climax.

The juvenile and adult phases of a plant have the same genotype but their

* Now at Institut für Systematische Botanik der Universität, Zürich, Switzerland

phenotypes are different. Such ontogenetic changes are not only epigenetic effects since environmental stimuli also have a role to play. In some heteroblastic species such as *Podocarpus dacrydioides* A. Rich, *Elaeocarpus hookerianus* Raoul. (Rumball, 1963) and *Hedera helix* L. (Brink, 1962), the external stimuli have little effect as both juvenile and adult stages develop and persist together on single plants at the same time. In other species such as *Synnema triflorum* (Roxb. ex Nees) O. Kuntze and *Proserpinaca palustris* L. (McCallum, 1902; Burns, 1904) the heteroblastic development is largely under environmental control.

It is virtually impossible to give a precise definition of plasticity because there is so much interaction between the genotype and the phenotype, but as a general rule, the amount by which expressions of individual characteristics are altered by environmental stimuli is a measure of the plasticity of these characteristics. Bradshaw (1965) in a review of the evolutionary significance of phenotypic plasticity points out that a plastic response is: (1) specific for a particular plant characteristic; (2) specific in relation to a particular environmental influence; (3) under genetic control and therefore capable of being altered by selection which means that plastic responses may vary from individual to individual within a species.

THE MECHANISMS OF PLASTICITY

It is convenient to recognize two different kinds of plastic response. The first kind is the "dependent response" or "dependent morphogenesis" as Schmalhausen (1949) called it, where the intensity and direction of the environmental stimulus is directly related to the intensity and direction of the plastic response. For example, Barnes (1936) tested the effects of temperature, soil moisture and photoperiod and the interrelationships of these factors on the growth and carotene concentration of the *Daucus carota* L. cultivar "Red Cored Chanteney". He found that the length of the root and the carotene concentration were directly related to temperature and to soil moisture. Root shape, however, was modified more by temperature than soil moisture. Photoperiod had almost no effect on root length, root shape and carotene concentration but produced marked changes in leaf characteristics. An almost linear relationship existed between the stimuli (temperature and soil mosture) and the plastic responses (root length and carotene concentration), showing that these responses were linked in both intensity and direction to the environmental stimuli. This kind of plastic response is usually obviously correlated with a particular environmental effect and has not, on the whole, led to undue taxonomic confusion.

Many plastic responses are not related in direction and intensity to the environmental stimuli. These responses are self-regulating once the required threshold amount of stimulus is applied. This kind of response has been called an "autoregulatory dependent morphogenesis" by Schmalhausen (1949).

The environmental stimulus bringing about a self-regulating response may be indirect and may need to be applied only once. For example, Chouard (1960) found that *Geum urbanum* L. needs a cold treatment before it will come into flower and that this cold treatment need only be given once; after the cold treatment *G. urbanum* will flower under a wide range of photoperiods and temperatures. Similarly, Hendricks *et al.* (1959) demonstrated that the seeds of *Lactuca sativa* L. cultivar "Great Lakes", after they have imbibed water, will not germinate unless they have been exposed, at least briefly, to light. If the light intensity is great enough, the soaked seeds require an exposure to light of only a few seconds. This kind of self-regulating response has, occasionally, led to taxonomic confusion and will be discussed later.

To demonstrate the complex interaction between environmental stimuli and plastic responses I shall give an account of some of the factors that influence the leaf-shape in three heterophyllous, amphibious plants that are essentially similar in morphology and ecology. I must stress that the changes described in this chapter are in leaf-shape while those changes in anatomy and absolute size which may or may not be correlated with leaf-shape are disregarded.

SYNNEMA TRIFLORUM (ROXB. EX NEES) O. KUNTZE

Synnema triflorum (Roxb. ex Nees) O. Kuntze is a member of the *Acanthaceae* and grows in India and S.E. Asia where it is a weed of rice fields. The leaf-shape of this species is very variable and it is convenient to describe leaf-shapes under three categories:

(1) The entire leaf which is ellipitical to ovate with an undulate to distinctly toothed margin in the submerged state (Fig. 1(f)) or with a serrate margin in the terrestrial state (Fig. 1(c)).

(2) The divided leaf which is somewhat irregularly bipinnatifid with almost linear segments (Fig. 1(a) and (d)).

(3) The intermediate leaf which is irregularly pinnatifid (Fig. 1(b) and (e)).

The leaf shapes found in this species show an almost continuous spectrum of variation from entire to divided and the three leaf-shape categories described were chosen merely as reference points.

Arber (1920) showed that the linear leaf or the divided leaf with linear segments, is an almost universal feature of submerged aquatic plants and she developed a convincing argument, based largely on circumstantial evidence, that this kind of leaf is advantageous in the submerged environment. In nature, it may be true that the divided leaf of *S. triflorum* is usually developed in the submerged environment but in cultivation it has been found that submergence in water does not necessarily initiate the development of divided leaves and that other environmental factors have a role to play.

When *S. triflorum* is cultivated submerged in clear water that is kept saturated with oxygen, not more than 25 cm deep, at 25°C and with a 12 hour photoperiod in normal daylight it is found that entire leaves develop. If virtually any one of these factors is altered, intermediate or divided leaves

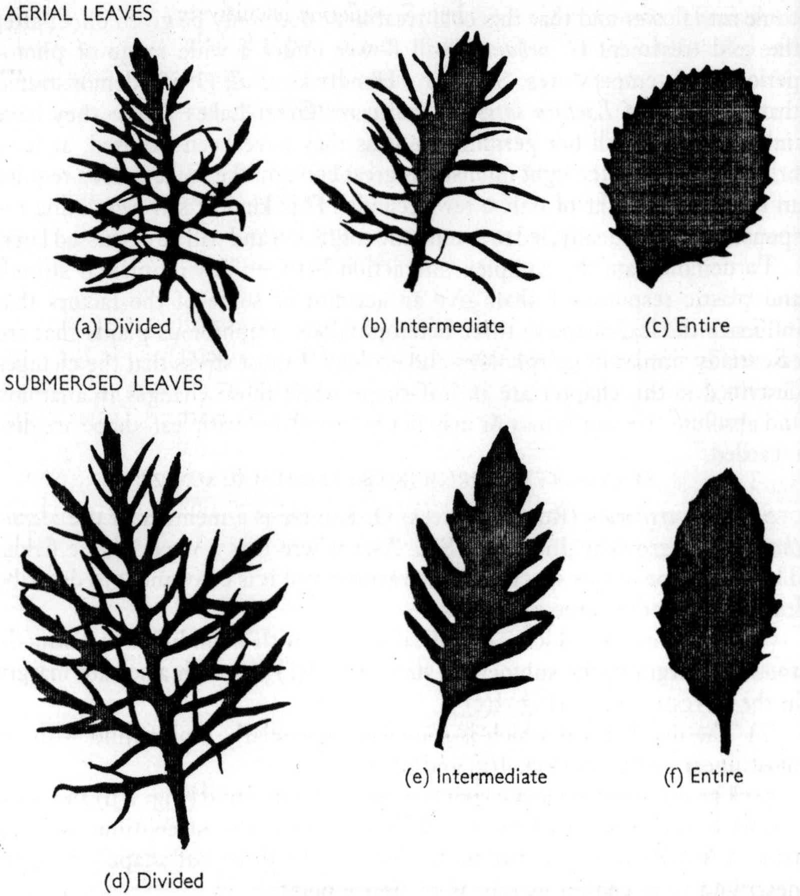

FIG. 1. Leaf silhouettes of *Synnema triflorum*. (a), (b) and (c) grown terrestrially, (d), (e) and (f) grown under water.

develop. For instance, if the photoperiod is reduced to eight hours or less or increased to 16 hours or more while the other factors are kept constant the new leaves that develop will initially be of the intermediate kind and later of the divided kind. Similar results are found if the temperature is lowered to 18°C or less or raised to 32°C or more. If the light intensity reaching the plant is reduced by shading, cloudiness of the water or by increased depth, the result is invariably the development of divided leaves. The first response when the concentration of oxygen is reduced or carbon dioxide increased is an immediate stem elongation which usually brings the apex above the surface of the water.

A similar pattern is shown when *S. triflorum* is cultivated terrestrially but the entire kind of leaf is much more stable, and it is only when the plant is placed in circumstances approaching conditions of stress such as temperatures of 15°C or less, photoperiods of six hours or less, or extremely low light intensities that divided leaves develop.

In both submerged and terrestrially cultivated plants interactions between these factors also modify the leaf-shape but, at the present stage of investigation, it is not possible to say whether any factor has more effect on one aspect of leaf development than any others, or whether the degree of division of the leaf is controlled by a single morphogenetic process or by several.

The divided leaves on both submerged and terrestrially grown stems of *S. triflorum* develop from small apices (± 60 μ diameter) while the entire leaves develop from larger apices (± 120 μ diameter). Troll (1937, 1939) showed that changes in morphology associated with maturity (heteroblastic development) were, in many plants, accompanied by changes in apical size and that, as a general rule, apical size increases with ontogenetic advancement. *S. triflorum* develops flowers only in the axils of emergent entire leaves and if the flowering portion of the stem is to be regarded as mature then this species must be considered as showing some degree of heteroblastic development. It is not yet possible to say whether the divided leaves are characteristic of young seedlings which must be regarded as juvenile, because *S. triflorum* appears to be self-incompatible and I have not been successful in obtaining seed.

Allsopp (1954, 1964) has suggested that apical size is a reflection of the nutritional status of the plant which, he argues, is linked with heteroblastic development. Working on *Marsilea drummondii* A. Braun he found that experiments with carbohydrate and nitrogen nutrition and with metabolic inhibitors caused big changes in the heteroblastic development of the sporlings. Carbohydrate or nitrate starvation induced reversion to juvenile foliage but this change was only readily induced in sporelings and that once past the sporeling stage the nutritional status of *Marsilea* was not necessarily correlated with sexual maturity. However, more recent work by Gaudet and Malenky (1967) suggests that, in *Marsilea vestita* Hook & Grev., there is no correlation between the size of the shoot apex and the production of adult leaves because sporelings supplied with a carbon source would grow and develop adult leaves in total darkness even when the shoot apices were as small as those of 12 day old juvenile plants. The better known heteroblastic aquatic species, such as *Marsilea drummondii* A. Braun, *Sparganium erectum* L., *Aponogeton distachyos* L. fil., *Heteranthera reniformis* Ruiz & Pavon and *Nymphaea alba* L., all complete their ontogenetic leaf-form changes in sporeling or seedling stages so that the mature kind of leaf is developed long before the sexually mature stage is reached. In the mature state these species require environmental stimuli of high intensity that bring the plants into conditions of stress or metabolic upset, before stems bearing mature foliage can be induced to revert and develop juvenile foliage.

In *S. triflorum* it is seen that changes in leaf-shape are readily brought about by environmental stimuli of low intensity that do not necessarily bring the plant into conditions of stress. This indicates that nutritional status alone does not control the leaf-shape in spite of the fact that the apical size is correlated with leaf-shape. However, stems of *S. triflorum* that are actually flowering (at their ontogenetic climax) do require environmental stimuli of a higher intensity to bring about changes in leaf-shape than stems that are not flowering. It could be argued that *S. triflorum* has not quite escaped the rigorous morphogenetic changes that are the usual result of heteroblastic development.

The flowering stage of *S. triflorum* is remarkably constant in the morphology of its leaves. Herbarium taxonomists tend to work amost entirely on flowering material so that this species, because of its lack of morphological variation in the flowering state, appears to have attracted only one specific epithet. It has, however, been placed in two families (Scrophulariaceae and Acanthaceae) and in four different genera (*Adenosma*, *Cardanthera*, *Ruellia* and *Synnema*). There is no doubt, however, that it belongs to the Acanthaceae, and the possession of numerous ovules with hook-shaped retinacula places it in the subfamily Acanthiodeae. Generic delimitation in the Acanthaceae is still in a somewhat confused state but changes in generic limits do not hide the identity of the species.

S. triflorum, however, is a very popular plant amongst aquarists. Under the special conditions of aquaria, which are far from ideal for the cultivation of most aquatic species, *S. triflorum* usually develops only divided leaves and thus remains in a sterile state. In this state, with its divided leaves it is a very decorative plant but because it is sterile, it is not easily identified and is often offered for sale under the names *Glycine* and *Wisteria*, two genera of the Fabaceae (Leguminosae). This then is a real, if somewhat bizarre, case of taxonomic confusion which is the direct result of the plasticity.

RANUNCULUS FLABELLARIS RAF.

Ranunculus flabellaris is a widespread species of temperate North America which is found growing in shallow water, marshes or on mud flats.

It resembles *S. triflorum* in having the ability to develop divided, intermediate and entire leaves. The basic pattern of the leaves is, however, palmate rather than pinnate (Fig. 2).

The effects of various environmental stimuli on the leaf-shape of this species have been studied by Glück (1923, 1924), Bostrack and Millington (1962) and Johnson (1967). It has been shown that a decrease in temperature or a decrease in photoperiod causes an increase in leaf dissection, whether

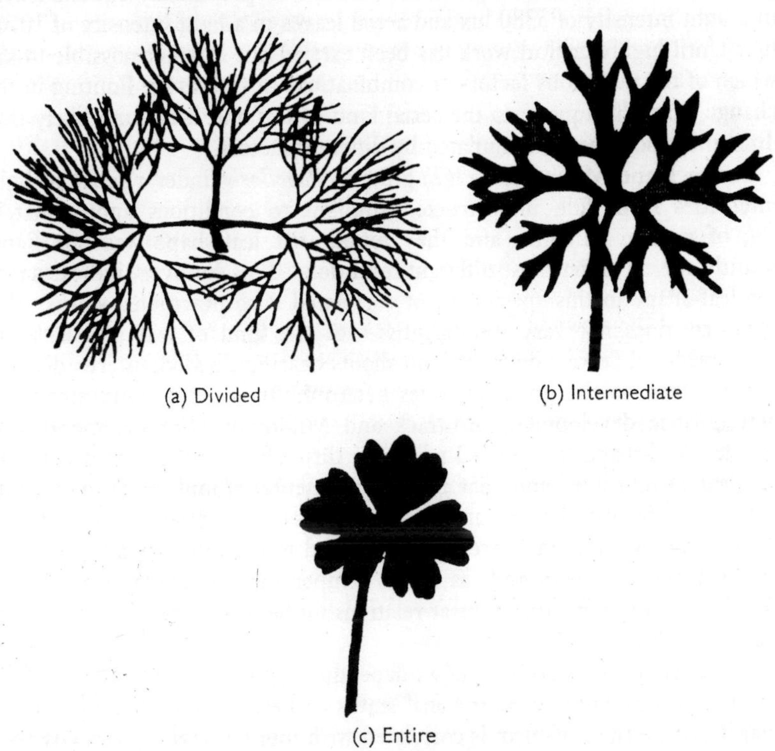

FIG. 2. Leaf silhouettes of *Ranunculus flabellaris*. (a) Divided, (b) Intermediate, (c) Entire.

the plants are cultivated terrestrially or submerged, and that of these two stimuli temperature has the more marked effect. However, at a given temperature, the leaves that develop on stems grown under water are more dissected than those that develop in a terrestrial environment. Johnson (1967) suggests that this is because the temperature of the submerged leaves approximates ambient temperature more closely than does that of the terrestrial leaves but Bostrack and Millington (1962) argue that water *per se* affects leaf-shape and its effect is additive to the low temperature effect because leaves developed terrestrially in a sealed container achieve the same degree of dissection as leaves developed under water. Neither of these arguments are entirely convincing and it is possible that differences in oxygen and carbon dioxide tension may have a role to play as suggested by Gessner (1940) who worked with *Ranunculus baudotii* Godron, and McCully and Dale (1961) who worked with *Hippuris vulgaris* L. The latter workers also found that when grown in a saturated atmosphere *Hippuris* developed mainly aquatic leaves in a light intensity of 5380 lux and aerial leaves in a light intensity of 10760 lux. Until highly critical work has been carried out it is not possible to say which of the numerous factors or combinations of factors are limiting in the change from the aquatic to the aerial kind of leaf and it is quite likely that different genotypes are stimulated by different factors.

Bostrack and Millington (1962) grew *R. flabellaris* under a range of light intensities from full, unobstructed sunlight to conditions approximately 2% of normal sunlight, and they found that leaf-shape was not significantly altered under these different conditions. This discovery indicates that the leaf-shape in this species is not correlated with nutritional status. The apex-size does not vary significantly with the kind of leaf that is being developed and flowers develop from shoots bearing divided, intermediate or entire leaves, so that in this species heterophylly is not a manifestation of heteroblastic development. Bostrack and Millington (1962) demonstrated that the leaf-shape is regulated primarily through control of cell division in the leaf primordium and that the environmental stimuli or stimulus acts directly on the leaf primordium. In this species the intensity and direction of the environmental stimuli are directly related to the intensity and direction of the plant responses and, as far as temperature is concerned, Johnson (1967) has shown an almost linear relationship between temperature and the number of leaf-lobes.

In leaf-shape *R. flabellaris* shows dependent plastic responses that are not altered by differences in nutritional status or heteroblastic development so that in nature the leaf-form is correlated with more or less obvious environmental conditions. It is, therefore, not surprising to find that no undue taxonomic confusion surrounds this species. The following list gives all the known synonyms of *R. flabellaris* at the rank of species (taken from Benson (1948))

Ranunculus flabellaris Raf., Am. Monthly Mag. 2: 344, March (1818).
Syn:

R. multifidus Pursh, Fl. Am. Sept. 2: 736 (1814), non Forskål, Fl. Aegypt., 102 (1775).

R. fluviatilis Bigelow, Fl. Bost. 139 (1814), non Willd., Sp. Pl., 2: pt. 2, 1333 (1800).

R. delphinifolius Torrey in A. Eaton, Man. Bot. N. Am., ed. 2, 395, May or later (1818) date of publication given by Fernald, Rhodora 38: 171–173 (1936).

R. lacustris Beck & Tracy in A. Eaton, Man. Bot. N. Am., ed. 3, 423 (1822).

R. missouriensis Greene, Erythea, 3: 20 (1895).

The first of the above names, *R. multifidus* and *R. fluviatilis*, although illegitimate (because they are later homonyms of species based on different types) were both described in 1814 and based on the same plant. *R. flabellaris* was based by Rafinesque on the type specimen of *R. fluviatilis* Bigelow. *R. delphinifolius* Torrey was published only a few weeks later than *R. flabellaris* but no type was cited in the original description so that, perhaps, *R. lacustris* Beck & Tracy and *R. missouriensis* Green represent the only taxonomic synonyms in the sense that they were not deliberately based on the same living entity.

It would appear therefore, that *R. flabellaris* has only two, or perhaps three, synonyms that could be attributed to its plasticity. In north temperate plants it is common to find about five synonyms per species so that the extreme plasticity in leaf-shape shown by this species does not seem to have led to undue taxonomic difficulty.

RANUNCULUS AQUATILIS L.

Ranunculus aquatilis L. is a plant of temperate regions that is found in temporary or disturbed aquatic habitats such as ponds, drainage ditches, newly dug pits, slowly flowing canals small streams and artificially maintained pools. It is a very widespread species occurring almost throughout Europe except the extreme north, North Africa, the Altai region of Mongolia, North and West China, Japan and the New World, where it is confined in the west between the latitudes 50°–30° north and 10°–75° south. In tropical South America it is found between 2700 m and 4000 m altitude. In leaf-form and general habit it superficially resembles *R. flabellaris* but there is no close patristic relationship; *R. flabellaris* belongs to section *Hecatonia* in subgenus *Ranunculus* while *R. aquatilis* is included in subgenus *Batrachium*.

R. aquatilis develops divided and entire leaves but, unlike *R. flabellaris* and *Synnema triflorum*, it does not develop an intermediate kind of leaf. The behaviour of *R. aquatilis* and other species of *Ranunculus* subgenus *Batrachium* is discussed in detail by Cook (1966); the following is an outline of the heterophylly in *R. aquatilis*.

1. The Divided Leaf

The first seedling leaf of *R. aquatilis* is always divided, at any time of year regardless of whether the seed germinates terrestrially or submerged. When seedlings or mature plants are cultivated terrestrially, only divided leaves are developed and these aerially produced leaves differ from the submerged ones in having long primary divisions with short ultimate divisions (Fig. 3(a)) and in being light green with short, thick segments that are oval in transverse

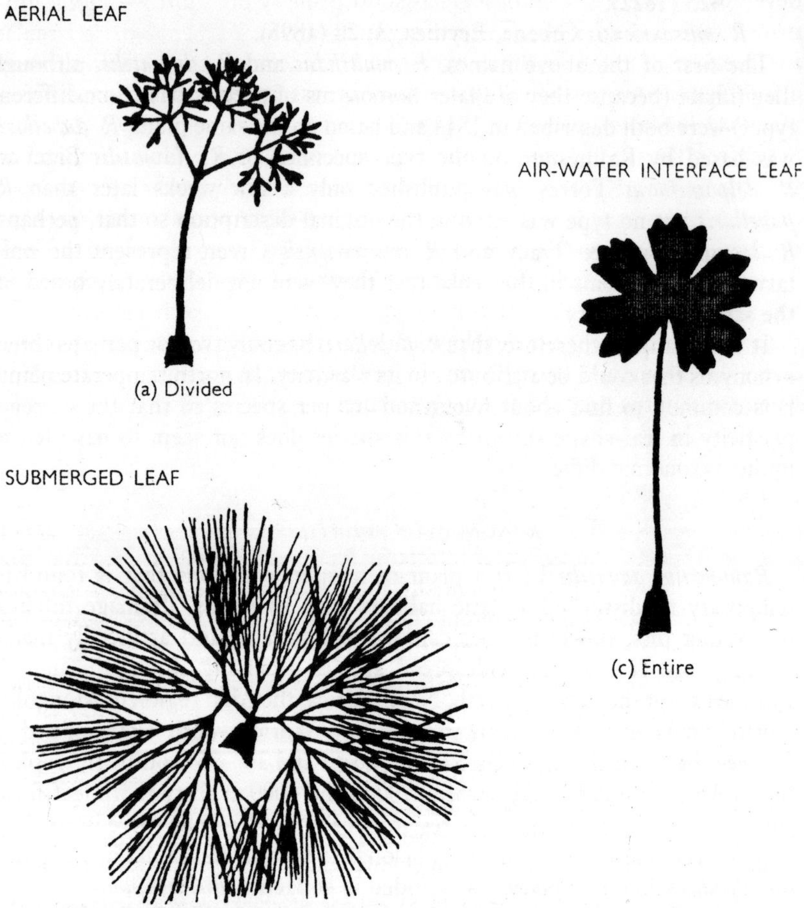

FIG. 3. Leaf silhouettes of *Ranunculus aquatilis*. (a) Terrestrial divided leaf, (b) Submerged divided leaf, (c) Entire leaf.

section. The epidermis is made up of small cells that contain very few chloroplasts; the mesophyll contains many chloroplasts and shows a certain amount of differentiation with a marked palisade on the adaxial side.

The remarkable thing is that these terrestrial divided leaves develop only under a regime of long photoperiods. If plants are grown in a saturated atmosphere (in a mist propagation unit not in a sealed container) under long photoperiods the terrestrial kind of divided leaf is developed but under the same conditions in short photoperiods the submerged kind of divided leaf develops. This latter kind of leaf has short primary divisions with long ultimate divisions (Fig. 3(b)). The leaf segments are dark green and are circular in transverse-section with a large-celled epidermis that contains many chloroplasts and a mesophyll that is more or less undifferentiated containing relatively few chloroplasts.

Under a regime of long photoperiods the development of the terrestrial or submerged kind of leaf is a direct plastic response dependent on the presence or absence of water. In short photoperiods, however, there is some "switch" mechanism that prevents the development of the terrestrial kind of leaf. There is, therefore, an autoregulatory mechanism switched by photoperiod which controls the competence of the plant to develop terrestrial leaves while the actual initiation of the development of terrestrial leaves is regulated by a direct environmental stimulus on the leaf primordium which operates only when the plant is exposed to long photoperiods.

2. *The Entire Leaf*

The entire leaves (Fig. 3(c)) develop only from stems growing under water and occupy the air-water interface. The change from divided to entire leaves is abrupt and the sequentially intermediate leaves that are characteristic of almost all other heterophyllous aquatic plants are not developed at all in *R. aquatilis.* The apex that produces entire leaves must be under the water surface: if it is raised above the surface divided leaves of the terrestrial kind develop. The depth of the apex below the surface of the water is not critical and entire leaves may be produced at a depth of up to 1m in clear water. As shown by Cook (1966) the action of water as merely a selective filter of light does not in itself alter leaf-shape.

The switch from divided to entire leaves in *R. aquatilis* is regulated by photoperiod. Entire leaves are formed only in a regime of long photoperiods. The evidence for this photoperiodic control is derived not only from experiments done in normal daylight, but also from work in controlled environment cabinets (Cook, 1966). The change to entire leaves is rarely synchronous for all the branches of a single plant and there are usually some branches that never develop entire leaves in any single growing season. Very low light intensities do, however, inhibit the production of entire leaves but only at intensities that adversely affect growth of the plant as a whole. Temperature, as long as it is within the range 6–20°C, does not effect the change from divided to entire leaves.

3. The Control of Heterophylly

As in the case of the production of the terrestrial leaves, the competence of the plant to develop entire leaves is controlled by an autoregulatory mechanism, switched by photoperiod, but it is effective only when the apex is below the water surface. The nature of the stimulus that regulates, or perhaps inhibits, the development of entire leaves on some stems but not on others is not known. However, different races of *R. aquatilis* switch their competence to develop entire leaves at different photoperiods and it is interesting to note that this is not necessarily correlated with latitude as might be expected. Also there are some races of *R. aquatilis* that seem to have lost altogether the ability to develop entire leaves.

The size of the apex does not vary significantly with the kind of leaf that is being produced. Flowers may be borne on stems bearing terrestrial or submerged divided leaves or entire leaves. Considering the above facts and the nature of the photoperiodic regulation of the competence to change leaf-form it would appear that heterophylly in *R. aquatilis* is not the direct result of changes in heteroblastic development or nutritional status. This is also borne out by the quite abrupt change from divided to entire leaves and the complete absence of sequentially intermediate leaves.

Although the precise nature of the mechanisms regulating the leaf-form in *R. aquatilis* is not known these mechanisms are complicated enough to give a flexibility that enables the plant to complete its generative cycle with an array of different phenotypes. This means that this species is not only capable of reproducing in changing environments but is also capable of exploiting different habitats at the same time. The adoption of autoregulatory mechanisms tied to stimuli such as photoperiod mean that *R. aquatilis* is capable, in some circumstances, of "anticipating" changes in the environment.

4. The Taxonomic Confusion

The taxonomic confusion surrounding *R. aquatilis* is, as one might expect, considerable and is cited in full in Cook (1966). There are 26 synonyms that can be definitely attributed to this species and to no other and a further 28 synonyms that, for one reason or another, are not typifiable but may in part be attributed to *R. aquatilis* sensu lato. Many synonyms represent nomenclaturally invalid names while others are what one might call "geographical" synonyms; for example, when it occurs in Japan it has the specific epithet *nipponicus* or in Mongolia, *mongolicus*. Most synonyms are, however, due to genuine confusion between the different phenotypic states of this species and this is often reflected in the choice of specific epithet; the following are good examples: *submersus*, *diversifolius*, *hypotrichus*, *circinnatoides*, *allophyllus*, *longifolius*, *pantothrix*, *caespitosus*, *triphyllos*, *pumilus*, *succulentus*, *intermedius*, *rigidus*, *eleophilus*, *capillaceus*, *trisectus* and *heterophyllus*.

DISCUSSION AND CONCLUSION

The degree of phenotypic plasticity shown by different plant characteristics is largely ignored, or perhaps it is better to say rejected by plant taxonomists because of their obvious need to search for "good" characters. By "good" characters is often meant those characters that are least phenotypically variable and that show discontinuities in variation. Characters that are easily modifiable by environmental factors are usually considered "bad" but this is largely because they are difficult to handle in keys and descriptions and not because they necessarily fail to show discontinuities in variation. Some biologists feel that the different phenotypic expressions of a single gene are of less consequence in evolution than more rigidly canalized phenotypes but this is not necessarily so. Amphibious plants, for example, are capable of exploiting the submerged, air-water interface and terrestrial environments simultaneously so that the plastic responses that permit these plants to exploit more than one environment are, at the same time, the very features that have enabled these plants to adapt to this disruptive environment. Bradshaw (1965) has cited many situations where phenotypic plasticity is of evolutionary significance, but although plasticity *per se* is important it is the mechanism of plasticity that ultimately confers what evolutionary success the plants concerned possess.

Synnema triflorum, *Ranunculus flabellaris* and *R. aquatilis* were chosen as examples in this review because they are superficially similar in morphology and ecology but show different mechanisms of plasticity. They are all dicotyledonous plants from predominantly terrestrial families and each has evolved the aquatic habit independently. The two species of Ranunculus are not closely related patristically and each shows more affinities to different terrestrial species of *Ranunculus* than they do to each other. Some of the terrestrial species of *Ranunculus* and *Synnema* are heterophyllous so that the ancestral stocks of the aquatic species were probably already heterophyllous, Arber (1919, 1920) argued that heterophylly is a prerequisite for the ability of a terrestrial plant to inhabit an aquatic environment. However, today the control of the mechanism of heterophylly is different between each of these aquatic species and between them and their terrestrial progenitors.

Synnema triflorum can be considered the least flexible of these three species in that flowering occurs only in the axils of entire leaves and during the flowering phase external stimuli of higher intensity are required to bring about modifications of the leaf-shape than are needed on non-flowering stems. This means that *S. triflorum* is able to complete its generative cycle only in a limited environment, while it is capable of growing vegetatively in a much wider range of environments. These observations considered together with the changes in apical size that are associated with leaf-shape, indicate that heteroblastic development exerts some influence on heterophylly in *S. triflorum* so that, it could be said that it has not yet shed its "evolutionary apron strings".

Excepting the flowering phase of *S. triflorum*, the leaf-shape in both *S. triflorum* and *R. flabellaris* is regulated by direct dependent stimuli and thus changes in leaf-shape are always one step behind changes in the environment; the intensity of the stimulus is related to the intensity of the plastic response. *R. aquatilis*, however, is controlled by autoregulatory mechanisms that are tied to indirect and reliable stimuli such as photoperiod where the intensity of the stimulus is not related to the intensity of the plastic response. The photoperiod stimulus merely activates a "switch" that changes the morphogenetic pathway. This means that *R. aquatilis* is capable, in some circumstances, of anticipating changes in the environment.

Over many generations an autoregulatory morphogenetic mechanism that is tied to a dependable environmental stimulus, such as photoperiod, is probably more reliable in maintaining a high level of adaptation in plants that occupy a constantly changing environment. It is very dangerous to argue evolutionary advantage but I feel it is worth pointing out that *R. flabellaris* is confined to temperate North America, while *R. aquatilis* is widespread in Europe, Asia, South America and North America; both species are sympatric in western North America. At the species level *R. flabellaris* is distinct and more or less isolated while *R. aquatilis* is part of a complex of biotypes that are difficult to delimit taxonomically not only because of their plasticity but also because they are genotypically variable.

In conclusion, I would not like to suggest that taxonomists should start classifying plants on the basis of their phenotypic variability because of its importance in evolution, but rather that they should be a little more critical in their description of plasticity and that where correlations in variation exist they should be pointed out. Plastic responses are specific in relation to particular environmental influences and it is rather shocking that so little information on phenotypic modification is presented in formal taxonomic works.

REFERENCES

ALLSOPP, A. 1954. Juvenile stages of plants and the nutritional status of the shoot apex. *Nature, Lond.*, **173**: 1032–1040.

ALLSOPP, A. 1964. Shoot morphogenesis. *A. Rev. Pl. Physiol.*, **15**: 225–254.

ARBER, A. 1919. Heterophylly in water plants. *Am. Nat.*, **53**: 272–278.

ARBER, A. 1920. *Water plants, a study of aquatic angiosperms*. Cambridge University Press.

BARNES, W. C. 1936. Effects of some environmental factors on growth and color of carrots. *Cornell Univ. Agric. exp. Stn, Memoir*, **186**: 1–36.

BENSON, L. D. 1948. A treatise on the North American Ranunculi. *Am. Midl. Nat.*, **40**: 1–261.

BOSTRACK, J. M. and MILLINGTON, W. F. 1962. On the determination of leaf form in an aquatic heterophyllous species of *Ranunculus*. *Bull. Torrey Bot. Club*, **89**(**1**): 1–20.

BRADSHAW, A. D. 1965. Evolutionary significance of phenotypic plasticity in plants. *Adv. Genet.*, **13**: 115–155.

BRINK, R. A. 1962. Phase change in higher plants and somatic cell heredity. *Q. Rev. Biol.*, **37**(1): 1–22.

BURNS, G. P. 1904. Heterophylly in *Proserpinaca palustris* L. *Ann. Bot. Lond.*, **18**: 579–589.

CHOUARD, P. 1960. Vernalization and its relation to dormancy. *A. Rev. Pl. Physiol.*, **11**: 191–238.

COOK, C. D. K. 1966. A monographic study of *Ranunculus* subgenus *Batrachium*. *Mitt. Bot. Staatss., München*, **6**: 47–237.

GAUDET, J. J. and MALENKY, R. K. 1967. Changes in the shoot apex during the early development of the fern *Marsilea vestita*. *Nature, Lond.*, **213**: no. 5079, 945–947.

GESSNER, F. 1940. Beiträge zur Biologie amphibischer Pflanzen. *Ber. dts. Bot. Ges.*, **68**: 2–22.

GLÜCK, H. 1923. Systematische Zusammenstellung der Standortsformen von Wasser- und Sumpfgewächsen. *Beih. Bot. Centralbl.*, 39(2): 289–398.

GLÜCK, H. 1924. *Biologische und morphologische Untersuchungen über Wasser- und Sumpfgewächse*, **4**: Jena.

HENDRICKS, S. B., TOOLE, E. H., TOOLE, V. K. and BORTHWICK, H. A. 1959. Photocontrol of plant development by the simultaneous excitations of two interconvertable pigments. III Control of seed germination and axis elongation. *Bot. Gaz.*, **121**(1): 1–8.

JOHNSON, M. P. 1967. Temperature dependent leaf morphogenesis in *Ranunculus flabellaris*. *Nature, Lond.*, 214, No. 5095: 1354–1355.

MCCALLUM, W. B. 1902. On the nature of the stimulus causing change of form and structure in *Proserpinaca palustris*. *Bot. Gaz.*, **34**: 93–108.

MCCULLY, M. E. and DALE, H. M. 1961. Heterophylly in *Hippuris*: a problem in identification. *Can. J. Bot.*, **39**: 1099–1116.

RUMBALL, W. 1963. Wood structure in relation to heteroblastism. *Phytomorph.*, **13**(2): 206–214.

SCHMALHAUSEN, I. I. 1949. *Factors of Evolution*. McGraw-Hill (Blakiston), New York.

TROLL, W. 1937. *Vergleichende Morphologie der höheren Pflanzen* 1: pt. 1, Borntraeger, Berlin.

TROLL, W. 1939. *Vergleichende Morphologie der höheren Pflanzen* 1: pt. 2, Borntraeger, Berlin.

WADDINGTON, C. H. 1957. *The Strategy of the Genes*. Allen and Unwin, London.

9 | Hybridization, Taxonomy and Evolution

W. H. WAGNER, JR.*

Botanical Gardens, University of Michigan, Ann Arbor, Michigan, U.S.A.

INTRODUCTION

There have been many different attitudes towards hybrids and their importance in evolution and taxonomy. The literature is enormous and I cannot hope to give a full review. Any writer on this subject would tend to see the problems from his own viewpoint. Since the evidence is drawn largely from experience with the pteridophytes, to what extent my conclusions and suggestions are applicable to all plants remains to be seen. Two main questions will be dealt with: What is the role of certain types of interspecific hybrids in producing divergent lines of evolution? How should hybrids be handled by taxonomists? An attempt is made to compromise between widely different viewpoints: the cytogeneticist's, the taxonomist's, and the phylogeneticist's.

Some preliminary definitions are called for, since our words are used in so many senses and communication has become increasingly difficult. The word hybridization will be limited to the taxonomist's usage, not the geneticist's, namely the crossing between species and genera and not the combination and recombination within species or between different cytogenetic and compatibility forms of the same taxon. To Stebbins' definition (1959) of hybridization as "crossing between individuals belonging to separate populations which have different adaptive norms" I would add "and which would be separated in ordinary taxonomic practice as readily defined phenetic species". Also, I shall confine myself to "genomic hybrids" of the A^1B^2 sterile type or $A^1A^1B^2B^2$ amphidiploid type, in which the pairing between genomes from different species is interfered with or stopped altogether. Some natural hybrids of this type have been reproduced experimentally by laboratory crossings which match with remarkable fidelity the wild taxon. In my

* Grateful acknowledgment is given to B. V. Barnes, Leslie D. Gottlieb, Raymond C. Jackson, E. L. McWilliams, Melvern Tessene, F. S. Wagner and others who have made helpful suggestions for the preparation of this paper and to N.S.F. Project GB–2025.

notation, different letters represent genomes from phenetically and taxonomically distinct species, and the numerical superscripts refer to different pairing factors. For example, A^1B^1 refers to genomes from different species which have no factors to interfere with pairing; A^1B^2 genomes from different species with factors 1 and 2 that keep normal pairing from occurring; and A^1A^2 genomes from the same species that do not pair normally with each other. Either A^1B^2 or A^1A^2 will tend to pair normally if doubled, i.e., $A^1A^1B^2B^2$ or $A^1A^1A^2A^2$.

What role gene exchange and introduction of possible new variability by introgression actually plays in the broad picture of evolution and taxonomy is not clear, and the reader is referred to the various reviews of Anderson (1948, 1953), Heiser (1949, 1963) and Davis and Heywood (1963) *cf.* bibliography. It can be imagined that interspecific hybridization contributes by extensive reshuffling of genetic materials to the evolutionary process by remodeling one or both of the parental species or by producing distinct taxa. As to what fraction of the overall picture of diversification of plants introgression actually causes, we can only guess at this time. As often as not what might be interpreted empirically as introgression might with equal justification be regarded as divergence, depending upon whether we are predisposed to imagine the extremes of divergence as coming together or pulling apart. On the contrary, the genomic hybrids which form the framework of the present discussion are usually unquestionable hybrids, and many of them have become sexual by doubling their chromosomes.

By the word evolution I shall not mean what might make evolution possible, such as geographical or reproductive barriers. The word is used here to mean the actual biological changes of populations which lead to new extremes of adaptation, specializations that result in new attributes. As I visualize it, the central phenomenon in all evolution is divergence, as measured by the number of specializations which have occurred, the characters which have undergone change, and the sequence and branchings of these changes (Wagner, 1962*b*, 1964). Characters and character-complexes develop new capacities and properties not present in the primitive state—leafiness to leaflessness, autotrophy to heterotrophy, insect pollination to wind pollinaation, and so on. Evolution, as used here, corresponds approximately to the word phylogeny but is concerned only with the amount, direction and sequence of biological change or divergence, without reference to when the events occurred. What I shall call "normal species" are those produced by divergence from common ancestors. The term "hybrids" will be limited to crosses between normal species, i.e., reticulation.

The word taxonomy as used here represents the art of classification, a matter of subjective taste; it is not the same as systematics, the scientific study and explanation of diversity. Taxonomy involves such paraphernalia s setting up categories, identification, nomenclature, and keys, all aimed toward finding the most useful way of expressing in names the products of evolution. Taxonomists are not likely to treat as distinct species populations differing only in chromosome number or compatibility factors any more than populations differing only in whether they grow in Florida or southern California. On the contrary, no taxonomist would be likely to treat *Platanus orientalis* as the same species as *P. occidentalis*, because these taxa have diverged in large numbers of characters.

INTEREST IN HYBRIDS

Interspecific hybridization has long been important in the study of cultivated plants, but it has been only in the past half century that very intensive attention has been given to natural hybrids in wild situations. Taxonomists varied widely in their reactions. I remember one of my mentors, after I had shown him a fern cross from the Pacific islands, protesting: "Please don't have hybrids in tropical ferns!" At the opposite pole, during the thirties and forties one of our prominent "biosystematists" was contemptuously alleged to find "hybrids under every bush". Herbarium botanists tended to regard hybrids as oddities not worthy of attention, but cytogenetic systematists found hybrids to have great interest and importance. Through the years amateur botanists and hobbyists enjoyed searching for wildflower and fern hybrids; after acquiring a fairly intimate familiarity with the local flora, there seemed to be little else to do but discover unusual specimens, interspecific hybrids, colour forms of flowers, and so on. Butterfly enthusiasts sought for odd wing patterns or aberrations. It was all part of the "game", and added to the thrill of the chase, like adding a new stamp to the collection. Only as the field of plant "biosystematics" unfolded did natural species crosses begin to arouse really serious botanical interest.

Investigation of hybrids became a point of emphasis for those concerned with cytogenetics of species, chromosome numbers and pairing behavior, devices for apomixis, hybrid sterility, and related problems. Flushed with the knowledge that such taxa as *Drosera anglica*, *Nicotiana tabacum*, and *Triticum aestivum* might be hybrids, the search was on. The hybrids were especially intriguing because they tended to pull out all the stops: one could find polyploidy, apomixis, and interesting ecological and geographical patterns. In the greenhouse and field plot researchers made extensive breeding tests, and extremists even argued that the definition of species should not be based on phenetic differences (or characters) but on the ability to interbreed and form viable offspring. Often the majority of other systematic data, morphology, anatomy, physiology, and chemistry, were given only fleeting notice. Strikingly differentiated species were said to be undifferentiated genetically if they interbred. And plants otherwise nearly or quite indistinguishable, if th happened not to be able to interbreed, were called "genetic species" or worse, "biological species". Only with the advent of numerical taxonom during the 1950's was stress once again placed upon all biological attribute in the conceptualization of taxa. By the 1960's the numerical taxonomists ha a greater claim of being truly biosystematists than did the biosystematists who were concerned almost exclusively with cytogenetic and breedin properties.

THE NATURE OF NATURAL HYBRIDS

Usually a field botanist familiar with a flora can spot a species hybrid at a glance. It simply does not fit either of the parental species. It lies between them in its characteristics. I can illustrate this by our experience in Hawai when we found the whiskfern cross for the first time in 1961. A few of the prominent characteristics of *Psilotum complanatum* and *P. nudum* are given in Table I and shown in Fig. 1. The terrestrial forms of these well-known

TABLE I. A comparison of the gross features of the terrestrial forms of whiskferns occurring in the Hawaiian Islands (cf. Fig. 1).

	Psilotum complanatum	P. nudum
Habit	Branches strongly arched	Branches straight, erect
Branchlet shape	Entirely flat	Three-angled, triangular in cross-section
Branchlet width	Uniform to apex 2·3–3·0 mm wide	Diminishing in distal 5–10 cm 0·7–1·2 mm wide
Sporangial position	Parallel on opposite sides of the branchlets	Alternately borne around the branchlet
Length of primary stalk	40 (25–70)% of length of branch cluster	170 (100–300)% of length of branch cluster

pantropical species grow together over approximately a half mile area on the Kanehoa Trail in the Waianae Mountains of Oahu. Four clumps were discovered, their branches only partially arched, in some places flat like *P. complanatum*, in others triangular like *P. nudum*. The stalk length of the clumps in question was intermediate, and the sporangia were borne either in two parallel rows or in scattered positions around the branches. It was impossible to place the intermediates in one or the other of the familiar species, because it was obviously neither one nor the other. Later the spores were examined. Three collections of *P. complanatum* showed 94% good spores; two of *P. nudum* 96%. But the intermediates yielded only 8% good. On the basis of this evidence we concluded that the intermediates were hybrids. Subsequent chemical examination supported this conclusion. Any other hypothesis, e.g., that the intermediates were ancestors which had become sterile, seemed unlikely.

The establishment of a hybrid diagnosis for a plant or population in the

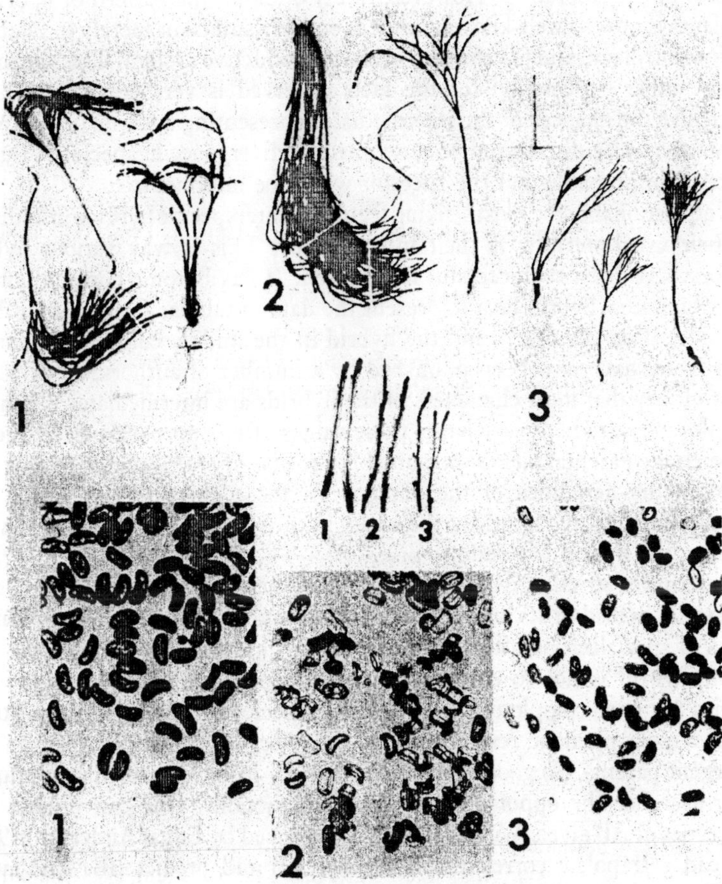

FIG. 1. Habits, branches and spores of Hawaiian whiskferns, *Psilotum*. 1. *Psilotum complanatum*. 2. Hybrid. 3. *P. nudum*.

natural environment usually follows from gross observations to microscopical observations, followed, where necessary, by experimental resynthesis. The steps are: (1) recognition that the evident characters are mainly intermediate between two well-known taxonomic species; (2) observation that some change has occurred in the reproductive apparatus which differs from one or both of the parental species—abortive pollen grains, or large (polyploid) pollen grains, failure of fruit set or sporangial maturation, or some form of apomixis; cytological peculiarities such as failure of normal meiosis especially faulty pairing, or doubled chromosome number; and (3) the re-creation of the taxon by artificially combining the gametes of the parental species in the laboratory. However, in my opinion, the pattern of hybridity in plants is now so well-known that it is entirely sufficient merely to establish (1) intermediacy, and (2) changes in the reproductive system. The likelihood of any other explanation for the facts observed is very small indeed. In particular the chances of the intermediate representing a sterile or polyploid ancestor may be discounted because unequal divergence in species evolution makes it improbable that the ancestor would be intermediate.

Unquestionably the most striking feature of interspecific hybrids in addition to their usual sterility, is their intermediacy. This is why various hybrid indices have been so successful in research. It has become conventional in describing new hybrid taxa to present the data in tabular form, parent A on one side, B on the other, and the hybrid in the middle column. Dozens of such tables have now been published by a number of authors, all of which show clearly that most characters of the hybrids are intermediate. When the samples measured are very large, these values often approach perfect means between the parental means, presumably because of multigenetic interactions between the heredities of the parents. In those characters in which the parents are alike, the hybrid is the same; in those which have diverged in the parents, the hybrid is a compromise.

In terms of strictly divergent evolution no new extremes are achieved in the sterile hybrid or its amphidiploid modification. Indeed the evolutionary peaks of specialization attained by the parents are now leveled off—the leaf of the hybrid is not so extremely needle-shaped, for example, as it is in species A, nor so completely ovate as it is in species B. The hybrids thus tend to have more phenetic resemblance to the ancestral conditions than do the divergent parents. So good is this intermediacy in most cases that it affords the researcher the opportunity to make predictions. One can predict the properties of AB given AA and BB. As emphasized by Edgar Anderson (1953) one can extrapolate correlates of AA and AB and predict BB. The latter technique makes a fine classroom exercise in systematic botany courses, and I have used from the local Michigan flora such intermediates as *Quercus* × *runcinata* (*Q. imbricaria* × *rubra*), *Apocynum* × *medium* (*A. androsaemifolium* × *cannabinum*), *Betula* × *purpusii* (*B. alleghaniensis* × *pumila*), *Lycopodium* × *habereri* (*L. flabelliforme* × *tristachyum*), *Equisetum* × *litorale* (*E. arvense* × *fluviatile*), and many others in many vascular plant families. I have even used my classes, without their knowing it, to help me out in my research. A small group of six students was given specimens and 15 minutes

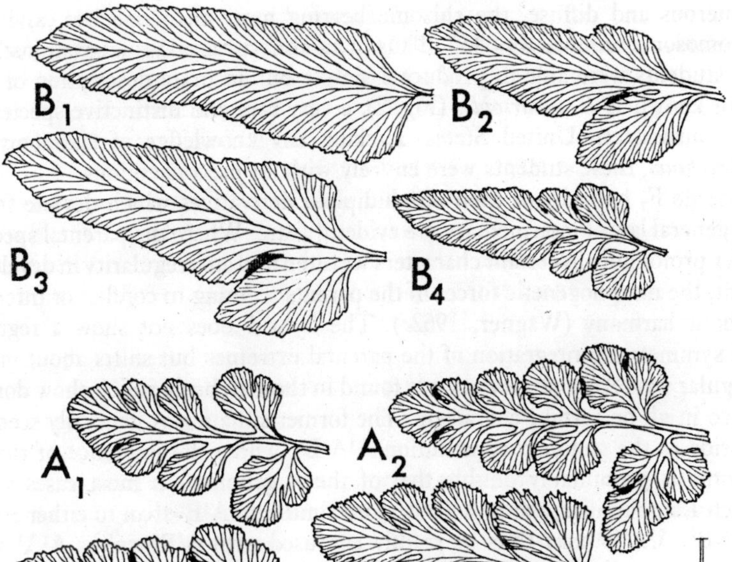

FIG. 2. Segments of Botrychiums from southeastern United States, showing subtle intermediacy between extremes. B. *Botrychium biternatum*, $n = 45$. A. *B. alabamense*, $n = 90$. L. *B. lunarioides*, $n = 45$.

to answer the following questions: Assuming that A^1A^1 = *Botrychium biternatum* (Fig. 2) with *n* known to be 45, and $A^1A^1B^2B^2$ = *B.* × *alabamense* (Fig. 2) with $n = 90$, tell me what B^2B^2 was like. They quickly came up with the following brief description: "Plant smaller than either, with more strongly dissected pinnae, the pinna margins wavy, not finely toothed, the segments shorter and rounder, and lacking central midribs, the roots more numerous and diffuse, the rhizome bearing more than one leaf, and the chromosome number $n = 45$". Thus in an almost casual visual examination, the students were able to produce a simple but accurate description of the plant known as *B. lunarioides* (Fig. 2), a very rare and distinctive species of the southeastern United States. Lacking any knowledge of taxonomy of *Botrychium*, these students were entirely without preconceptions.

Sterile F_1 hybrids, or their amphidiploid derivatives which deviate from the general law of intermediacy are evidently rare. Where the parental species differ profoundly in certain characters we often find an irregularity in development, the morphogenetic forces of the parents seeming to conflict or interact without harmony (Wagner, 1962*c*). The hybrid does not show a regular and symmetrical integration of the parental extremes but shifts about in an irregular way. Another deviation is found in those hybrids which show dominance in all or certain characters. The former situation is generally seen in hybrids of the genomic constitution $A^1A^1B^2$, where the influence of the A genome is presumably double that of the B genome. In most cases with which I am familiar, the $A^1A^1B^2$ is more similar to A^1B^2 than to either A^1A^1 or B^2B^2. $A^1A^1B^2$ may thus be readily confused with A^1B^2 unless A^1A^1 and B^2B^2 are very divergent. In those cases where certain characters of one of the parents dominate over those of the other, it is not usually the conspicuous and gross ones that are involved. For example, practically all of the eastern American hybrids involving the glandular diploid *Dryopteris intermedia* inherit the glandular character, with only one or two possible exceptions. The glands are usually not noticeable without magnification and that is why certain hybrids have been confused in the field. Rarely are gross and readily observable features dominant in the hybrid. Maternal dominance occasionally occurs at least in flowering plants. A remarkable example in ferns discovered by Panigrahi and Manton (1958) showed that F_1 artificial hybrids between species of *Cyclosorus* yielded dominance in such characters as creeping rhizome vs. erect, and uniform length of the lower pinnae vs. gradual reduction. They also found, of course, characters in which such dominance was not displayed, in which the hybrid expression was the expected intermediate. In general, however, we can expect the totality of characters, taken together, to show intermediacy. The possibility that occasional expressions of dominance may arise calls attention to the need of using a broad spectrum of comparisons in setting up hybrid indices.

IMPORTANCE OF HYBRIDS IN PLANT EVOLUTION

On this subject we have numerous writings, of which those of Heiser (1949) and Stebbins (1959) are especially valuable. Grant (1957) wrote that the consensus of zoologists is that hybridization in animals is a "rare and abnormal event which usually ends with the production of a few interesting but inconsequential freaks" while in plants "hybridization has profoundly affected the course of plant evolution". Camp and Gilly (1943: 346) stated that "As a dynamic population, the alloploidic species is a potent element in evolution; with time and selective habitats available, it may be a veritable fountain of new and markedly different species", thus assigning them an important role in evolution. Whether or not this view is actually true is subject to debate. In my own attempt to gain some idea of the relative importance of species hybridization in the general picture of divergent evolution of plants, I have come to the not new conclusion that hybrids are probably of little significance. This is at variance with much of the emphasis of my own work that has focused on hybrids, especially in ferns, found in the North American flora. I do not wish to say that hybrids may not have great importance ecologically in filling niches, and floristically in providing additional "kinds" of plants in any given area. A number of hybrids in many plant families are indeed widespread, fertile populations, and even sterile hybrids are found often enough to be picked up regularly by collectors. The ferns, for example, are notorious for the extent of their hybridization, and I am not sure that certain other plant groups, e.g., grasses, are not comparable. The question at issue is to what extent hybrids actually influence the broad pattern of evolutionary divergence, the creation of new types and new adaptive peaks. The modern concern with amounts of evolution (patristic distance) and the patterns (cladistic relationships) emphasizes concepts of primitiveness and specialization.

The amount of attention and publication which has been devoted to hybrids in genetic and evolutionary literature tends to make a misleading impression. Agriculturists, horticulturists and foresters have engaged in extensive studies of hybrids because of the possibility of producing economically important cultivars. In some cases they have succeeded in producing stocks of major value. Also, as indicated previously, cytogeneticists investigating natural populations have held hybrids to be of special interest because of their reproductive characteristics and for the possible bearing that they might have on the understanding and interpretation of the parents. Some of the natural hybrids which have been discovered have been so striking that they have become classroom subjects the world over.

Much of our knowledge of natural populations has come from civilized areas where disturbances and man's influence have stimulated more hybrids to be formed than would be expected in the pristine state of precivilization. The effects of Pleistocene glaciation have likewise probably produced hybrid taxa which have remained in the northern hemisphere and have attracted attention in the major botanical centers.

Our best source of judging the importance of hybrids in divergent evolution should come from taxonomic monographs. One cannot judge by looking at lists of taxa and their chromosome numbers, because, as will be discussed below, we have no way of knowing whether many of the polyploids are derived from normal diploid species or are, in fact, derived from interspecific hybrids. Unfortunately, in many taxonomic monographs, authors make little or no attempt to show the patterns of divergence, except in grouping species into subgenera, etc. I have for a number of years endeavoured to develop objective methods for interpreting and expressing evolutionary relationships by groundplan/divergence analyses (Wagner, 1962*b*, 1964) which have been adopted in the same or modified form by a number of workers (e.g., Brown, 1964; DeLisle, 1963; Evans, 1964; Graham, 1963; Hardin, 1957; Hauke, 1959; Keener, 1967; Iltis, 1959; Lellinger, 1965; Mickel, 1962; Miller, 1965; Scora, 1964; Stern, 1961). From these and what judgements I can make in perusing numerous monographs, I conclude that the majority of species in most groups are allopatric and divergent, i.e., "normal species" produced by ordinary processes of mutation, recombination, selection and isolation. Strong limitations must have worked against genomic hybrids through the ages; after four or five hundred million years of vascular plant evolution, normal diploid sexual species, all of which must trace their ancestors to the earliest land plants, are still prevalent on the earth. If *Psilotum* can be used as a model we can speculate that even the simplest and most primitive vascular plants were capable of forming interspecific hybrids.

The majority of plant species which are potentially able to hybridize are separated geographically and have no opportunity, unless man interferes, to form hybrids. Given a plant genus of numerous species, the chances of any one species coming into contact with others is limited to the small percentage of species which are sympatric. The number of possible hybrid combinations is thus limited. In the eastern United States one thinks of the numerous white oak species which grow sympatrically, but in terms of the total distribution of members of the white oak group, the number of species with which a given species can potentially hybridize is only a small minority. Those species which are in contact sympatrically do hybridize, of course, abundantly today (although in pre-settlement times such hybridization may have been rare).

Where species closely enough related to hybridize do coexist sympatrically, various factors enter into the picture which tend to inhibit hybridization or, if it does occur, to prevent the establishment of successful populations. This generalization rests upon the fact that there are factors which operate to (a) prevent cross-fertilization or to limit it, (b) hinder normal development of the embryo and/or seedling, (c) make for weakness of the mature plant in competition and survival, and (d) cause the hybrid to be sexually sterile. Many hybrid combinations which would be expected on taxonomic grounds simply do not occur or are exceedingly rare and sporadic. We must conclude, therefore, that the parent species have become reproductively isolated. Even where hybrids are common, as in certain tree genera, e.g., *Quercus*, *Acer*, the basic species manage to maintain themselves as distinct entities, suggesting that the hybrids, no matter how great their incidence, play no role in developing new aggressive populations or in breeding their parents out of existence. Stebbins has written (1959, p. 232), that "most interspecific hybrids are sterile". There are, of course, cases known in which the usual situation is reversed and the hybrids are not only more vigorous and competitive, but are fertile.

High incidence of polyploidy alone is not necessarily indicative of interspecific hybridization. Polyploidy can just as well arise within sectors of normal divergent species. The great extent of polyploidy in the angiosperms has been suggested by some workers to be the result of interspecific and intergeneric hybridization. One often cited example involves the Pomoideae tribe of the Rosaceae, supposed to have arisen by hybridization between a member or ancestor of Spiraeoideae and a member or ancestor of Prunoideae. One may agree with Stebbins (1950) that "since the basic number most commonly found in the Spiraeoideae is $x = 9 \ldots$, while that in the Prunoideae is $x = 8 \ldots$, the basic haploid number of the Pomoideae, $x = 17$, is directly explained by this hypothesis". However, there are also other ways to explain the same chromosome data (*cf.* Stebbins, 1950). Furthermore, the appearance of new characters which are clearly specialized in the Pomoideae such as the massive and adnate floral tube is at variance with our other knowledge of hybrids.

We are finding more and more examples of polyploids, aneuploids, and apomicts within the bounds of the same taxonomic species (*cf.* Lewis, 1967). Thus in reviewing the question of why there should be such high incidence of polyploidy we must ask how many of the polyploids included in the sample are accompanied at the present time by diploids of the same species and genus; and how many of the polyploids which have no known diploid sectors in the same species today are not, in fact, merely surviving polyploid forms which arose directly from diploid members, now extinct, of the same taxon at some time in the past.

The pteridophytes, in spite of their present-day high basic numbers (e.g., 22, 30, 36, 41, 45) are not necessarily the result of reticulate evolution. The high numbers of pteridophytes may have arisen first within normal species as polyploid sectors (just as today we find a number of fern species—the number of known cases increasing each year—with two or more polyploid levels). As Klekowski and Baker (1966) have pointed out, features of the homosporous pteridophyte reproductive system seem to call for storage of variation in the form of some higher level of polyploidy than is average for heterosporous plants. In their words, "Ultrafrequent establishment of polyploidy in the homosporous Pteridophyta appears to be necessary to create and maintain genetic variation in the face of the homozygotizing effects of habitual self-fertilization in the monoecious gametophytes of these plants". The higher numbers of pteridophytes could very well have been achieved by simple genetic changes in pairing factors resulting in $A^1A^1A^2A^2$ and more commonly $A^1A^1A^2A^2A^3A^3$ genomic constitutions, assuming base numbers in ancient times of $x = 8$–13.

Most interspecific hybrids have a reduced rate of potential evolution. Except for the A^1B^1 type of hybrid, the chances are strong that the evolutionary rate of interspecific hybrids will be reduced or nil. Even in the A^1B^1 type, the forces of adaptation are likely to be such in the respective parents that the hybrid will tend to occupy a hybrid habitat resulting from the breakdown of the normally differentiated community. How often this type of hybrid contributes significantly to further divergence either of itself or of A^1A^1 and B^1B^1 is not known. With respect to the A^1B^2 type of hybrid, there are several possible pathways. The most successful would be for it to double its chromosomes, becoming an amphidiploid with $A^1A^1B^2B^2$, and proceeding to follow a life cycle remarkably like a normal species. However, we cannot predict in a given case whether this will happen. Even if it does, the evolutionary rate of the amphidiploid is probably considerably less than that of its diploid parents because of the reduced opportunities for new mutations to express themselves.

Why more A^1B^2 hybrids do not become fertile tetraploids remains a mystery. In the case of the American *Asplenium* × *ebenoides* one experiment (Wagner and Whitmire, 1957) showed that doubling of the A^1B^2 constitution took place readily and produced a new amphidiploid, but this apparently occurs very rarely in nature. Indeed, to date we still have only one out of many localities (*cf.* Fig. 3A) where the $A^1A^1B^2B^2$ plants have formed in nature and succeeded in producing viable populations. In American *Dryopteris* there are numerous hybrids which would seem to have the potentiality of becoming amphidiploids, for example, of the A^1B^2 constitution *D. dilatata* × *marginalis*, *D. intermedia* × *marginalis*, *D. goldiana* × *intermedia* and *D. goldiana* × *marginalis*, the last a relatively common plant wherever the parents coexist; and of the $A^1B^2C^3$ constitution, *D. celsa* × *marginalis*, *D. celsa* × *intermedia*, *D. carthusiana* × *marginalis*, *D. cristata* × *marginalis*, and *D. cristata* × *intermedia*, the last two among the most common woodfern hybrids in the eastern United States.

The point is that many hybrids retain the A^1B^2 or $A^1B^2C^3$ constitution and fail to generate any sexual forms at all. The only pathway for them, if they are to persist and spread at all, is by vegetative propagation. The same probably applies with equal if not more force to the tetraploid × diploid interspecific hybrids of the constitution $A^1A^2B^3$. Thus all of these hybrids,

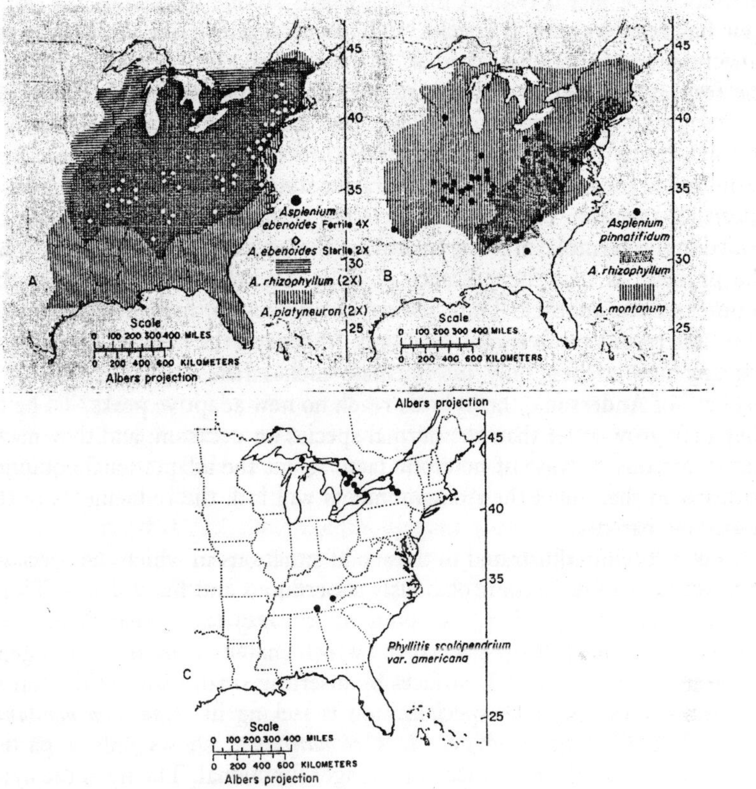

FIG. 3. Distribution of selected A^1B^2, $A^1A^1B^2B^2$, and $A^1A^1A^2A^2$ genome types in eastern United States. A. *Asplenium ebenoides*, stars A^1B^2; dots $A^1A^1B^2B^2$. B. *A.* × *pinnatifidum*, $A^1A^1B^2B^2$. C. *Phyllitis scolopendrium*, $A^1A^1A^2A^2$.

because they are incapable of sexual reproduction, may be regarded as terminal as far as evolution is concerned—even if they become common in the ecological community and flora, either by frequency of formation *de novo* or by vegetative reproduction. All of the forms of apomixis, whether fragmentation, rhizome spread, unreduced parthenogenesis, meiotic apogamy, nucellar embryony, and so on, are devices which may contribute to a hybrid's success ecologically. But in evolutionary terms the potential of apomicts is very low.

Interspecific genomic hybrids contribute little or nothing to the basic pattern and process of divergence which underlies all of evolution. Interspecific hybrids are intermediate between the evolutionary extremes achieved by their parental normal species. To evaluate the role of hybrids in evolution, therefore, this fact is critical, for the hybrids are merely blends and nothing new has been created. If species "A" has diverged in one set of characters, structural and functional, and "B" in another, so that both have changed in the respective amounts and directions from the ancestral stock, their hybrid is neither so specialized as "A" or "B", since the hybrid characters are diluted. If anything, the hybrid will tend to show a phenotype more nearly primitive in the expression of its character states than its most divergent and specialized parent. The phenetic facts are perhaps misleading because contained in the germplasm of the hybrid are the extremes shown by the parents; the problem is that the presence of both conditions tends to make them counteract one another. Thus, in fact, the process of hybridization tends to blunt the results of divergent evolution. In general the hybrids are not entering any new niches, only intermediate niches (the "hybridization of the habitat" of Anderson); the hybrids reach no new adaptive peaks. To be true they may grow taller than the normal species on occasion, and they may be more vigorous because of heterotic factors, but the adaptational equipment with which they meet their environments will lack the refinements of their respective parents.

This is forcibly illustrated in those odd situations in which the specializations of the parents become obviously superfluous and functionless. Thus in the American walkingfern, *Camptosorus rhizophyllus*, a remarkably exaggerated threadlike leaf tip has evolved which enables a single plant to generate large colonies over rock surfaces by inserting small plants at the leaf tips into moist crevices. This specialization is lacking in *Asplenium montanum*. Their $A^1A^1B^2B^2$ intermediate, *A.* × *pinnatifidum*, shows only a partially developed threadlike tip which is no longer functional. The tip of the hybrid produces proliferations of a distorted form and variable location, in spite of their apparent inability to reproduce the plant (Wagner, 1960*a*, *b*). A similar situation occurs in the hybrid bulblet fern, *Cystopteris* × *tennesseensis*. One of its parents is surely the eastern *C. protrusa*, a fairly unspecialized member of the *C. fragilis* complex. The other parental species is the bulblet fern, *C. bulbifera*, which shows a highly specialized mechanism of vegetative reproduction involving unique rounded bulbils which are produced by the rachises and fall off and roll into crevices to spread the plant. In the intermediate, *C.* × *tennesseensis*, functionless and deformed bodies arise along the rachises. They have little likelihood of reproducing the plant, but their presence is so characteristic that they may be used for identification purposes (Wagner and Hagenah, 1956). Both of the hybrids mentioned here are conspicuously successful, having doubled their chromosome numbers and spread over wide ranges, in spite of having to produce these seemingly pathological proliferations.

When Heiser discussed the question of the importance of hybrids in evolution some years ago (1949) he asked "If a 'new' species is created through hybridization, has anything really new been created in the process?" Does hybridization simply recombine previous genetic material with no new characters being created? According to Heiser, part of the controversy centers on how one would define a new character. If by "new character" one means a basic divergence or specialization from the ancestral condition, then there is little evidence that any new characters are produced.

EVOLUTIONARY DIVERGENCE AND CROSSABILITY

One cannot predict whether two species of plants will interbreed, or, if they do, whether the progeny will survive or reproduce. Who would have expected that *Platanus occidentalis* and *P. orientalis* would interbreed so successfully, in spite of their host of taxonomic differences. Why do such very distinctive trees as *Acer nigrum* and *A. saccharum* cross freely in the eastern United States, while such similar species as *Picea glauca* and *P. mariana* in the northern forests form only casual hybrids? In his monograph on pines, Mirov (1967: 326) comments as follows regarding hybridization: "Many species, often belonging to distant generally accepted groups, intercross freely and produce fertile hybrids; but there are also many species, often of the same taxonomic group, that possess strong barriers to intercrossing . . ." Barriers to intercrossing may in general be less associated with evolutionary differentiation than with sympatric co-existence. Writing about selection against hybrids. Ehrlich and Holm (1963) stated that "It is entirely possible that, when more is known about the processes of differentiation, it will be discovered that hybridization between individuals of rejoining segregates is almost universal, in other words, that mechanisms preventing exchange of genetic material between differentiated forms usually arise only through relatively unsuccessful hybridization after sympatry has been reestablished".

In normal species of plants (which are obviously not hybrids) it is particularly disturbing that more and more have revealed distinct pairing races or sectors, of which the formulae are evidently A^1A^1, $A^1A^1A^2A^2$, and $A^1A^1A^2A^2A^3A^3$. Many species of flowering plants and ferns have been shown to have $2x$, $3x$, $4x$ and so on forms within the same taxonomic entity. Some authors wish to designate these cytological forms of the same species with separate names, but W. H. Lewis (1967) comments that he can "find little purpose in tagging morphologically similar individuals with formal Latin designations. A simple form such as '*Hedyotis purpurea* (L.) T. and G. ($2x$) and *Hedyotis purpurea* (L.) T. and G. ($4x$)' ought to suffice for both taxonomist and cytogeneticist". If we do not follow this suggestion, as research on chromosome numbers continues we may well be led into taxonomic chaos. For example, a single cliff in Giles Co., Virginia, will have maidenhair spleenwort, *Asplenium trichomanes*, in $2x$, $3x$, and $4x$ forms, i.e., A^1A^1, $A^1A^1A^2$ and $A^1A^1A^2A^2$. The triploid shows 36 pairs and 36 singles. The same is true for *Polypodium virginianum* over a wide range in the eastern United States. When a triploid form of *Athyrium asplenioides* was discovered with the formula $A^1A^1A^2$ it was predicted that we would find a tetraploid $A^1A^1A^2A^2$ which, pairing with the normal A^1A^1 diploid, would give the triploid condition. Instead, the form A^1A^2 was discovered with practically no pairing of its 80 chromosomes ($x = 40$). (Wagner and Wagner, 1966.) The most striking example comes from the state of Missouri, where the same subspecies, *Pellaea glabella* var. *glabella* grows sympatrically on limestone cliffs in two forms—a sexual diploid and an apogamous tetraploid (Wagner *et al.*, 1965). Similar conditions are being found by workers in many parts of the world. Thus plants with identical or nearly identical morphology and ecology and belonging to the same phenetic species (or even variety!) are unable to breed with each other.

The conflict between the cytogeneticist who wants to base the use of binomials on sterility factors and chromosome numbers and the general biological systematist who wants binomials used only for taxa which differ in numerous characters is at its worst when undifferentiated forms of the A^1A^1 and $A^1A^1A^2A^2$ constitution are awarded binomials, and wholly different phenotypes of the A^1A^1 and B^1B^1 constitution are referred to as "genetically undifferentiated". The most striking case of A^1B^1 hybrids known in ferns was discovered by Trevor Walker in Ceylon. Two very distinct species, *Pteris quadriaurita* and *P. multiaurita* cross with each other where the natural habitat has been disturbed by man, and form hybrids which are apparently nearly 100% fertile diploids, the meiotic process being normal and the spores viable. These hybrids are capable of interbreeding among themselves and presumably with their parents to produce many degrees of intermediacy connecting the parental species. Taxonomists had given a formal name to plants combining the parental characters in a certain way, but Walker concluded that this "*P. otaria*" is not a valid species but is really only a particularly common member of a hybrid swarm. Walker did not merge the two parental species, even though they were interfertile and lacked the usual interference in pairing expected in interspecific hybrids. The A^1B^1 hybrids described by Walker represent the single example thus far brought to light among ferns, although such hybrids are well-known among angiosperms.

The so-called "biological species concept" if based only upon breeding barriers is a serious misnomer. In overall biology, the plants involved often belong to exactly the same taxonomic species. Genetically, therefore, they must be essentially homologous. The only differences pertain to factors influencing steps in the reproductive system. Workers may go out of their way to discover fine differences, but in most cases these are far less than standard for species discrimination of the other members of the same genus. Breeding tests of species relationships will test only breeding relationships. I am inclined to think that such phenomena as polyploidy, hybrid sterility, and the various expressions of apomixis, represent readily fluctuating factors which tend to interfere with the normal flow of successive generations of normal species, except when some especially fit genotype appears and is perpetuated by them. When a hybrid fitted for an intermediate habitat arises, these deviations from the normal life cycle come in handy to perpetuate the hybrid and to enable it to hold the ground. Thus I see the role of interspecific hybrids of the $A^1A^1B^2B^2$ constitution and other variations, A^1B^2, $A^1B^2C^3$, and so on, as more relevant to problems of ecology and floristics than to evolution. Chromosome pairing factors by themselves are insufficient either for classification purposes or for assaying divergent patterns of evolution. The taxonomist and evolutionist should be concerned with extent of total differentiation, involving all characters, structural and functional, and should not exaggerate the importance of cytogenetic peculiarities.

HYBRIDS AND TAXONOMY

The complicated problems of hybrids in numerical taxonomy and in working out divergences in monographic studies will not be dealt with here. Suffice it to say, estimates of phenetic relationships and clustering techniques of Sokal and Sneath (1963) are likely to give unexpected positions to hybrids. If a hybrid is intermediate between two different subgenera, it might group with one or the other or with neither, because the clustering technique employed was not designed to accommodate taxa of this nature. Likewise, in studies of divergence, hybrids tend to blur and confuse the conclusions unless they are removed before the analysis because the character states of hybrids are usually averages of the values attained by divergences of the parents—if one parent has state a, the other b, the hybrid will generally show $\frac{a+b}{2}$. Members of the same line may not hybridize, while members of widely different lines will, thus obscuring the basic evolutionary patterns (*cf.* Wagner, 1966*c*; Camin and Sokal, 1965; Hauke, 1959; Mickel, 1962).

I wish to ask four practical taxonomic questions here: To what extent should we recognize, describe and key out hybrids in Floras and manuals? How should we designate them? What role does hybrid abundance play? What role the ability to reproduce? There is much variation in practice and viewpoint regarding these questions. Valentine (1963) has discussed the treatment of hybrids in connection with the *Flora Europaea*, and now we are confronted with the planning for the *Flora North America*.

RECOGNITION OF HYBRIDS

Valentine (1963) writes that "From the practical point of view hybrids have to be identified, described and named". If hybrids are not recognized in Floras and manuals it will lead to confusion for both students and professionals. I recall a professor of Forestry who taught for years the trees and shrubs to students without even mentioning their hybrids. How he managed to do this was extraordinary, for hybrids are so common in some groups, e.g., oaks, maples, that a student simply cannot know eastern American woody plants unless he is able to distinguish the hybrids. Nevertheless, generations of students attempted to fit everything they saw into *Quercus alba*, *Q. macrocarpa*, *Q. bicolor*, and so on. If a student had trouble with a given specimen it was blamed on his incompetence at learning the characters or using the key. The professor floated along for years on a dreamy and erroneous idea that the species of oaks given in his manuals were somehow inviolate. The story simply points out the need for facing the facts. We must recognize hybrids, and in my opinion the more important ones should appear in keys. Manual writers should not rely on the idea that if you know your basic species the hybrids will be evident. If they are not described or keyed, they may be confused or missed simply because the collector or student does not know the parents or knows only one of them. Hybrids, including sterile ones, may exist separately from one or both parents. We have 36 different cytomorphotypes of *Dryopteris* in the eastern United States as shown in the studies of S. Walker (1962) and others. The sterile *Dryopteris* hybrids often occur separately; indeed, in the area of Rochester, New York, we encountered one locality—"Cedar Swamp"—where there were only sterile hybrids, *D.* × *celsa* × *goldiana* ($A^1A^1B^2$) and *D.* × *celsa* × *cristata* ($A^1B^2B^2C^3$). (Wagner and Wagner, 1965).

In the genus *Dryopteris* and, to a lesser extent, in *Asplenium*, the result of not recording and describing hybrids in Floras and manuals has led to chaos in a number of American herbaria. The local taxonomist, using only the literature available to him which was usually floristic manuals, was simply incapable of classifying his specimens.

We must ask "Who uses our names?" "Who uses Floras and manuals?" I would answer that they are used mainly by professional biologists and fairly advanced students. There are numerous popular handbooks and guides available to the amateur; in these not only hybrids but rare or subtle normal species are generally eliminated. However, any hobbyists who take their avocations seriously will soon graduate beyond the popular handbooks, and will invade the Floras of the professional. Thus there is no reason to ignore or throw out hybrids from Floras, even if they are rare. Even the most sporadic hybrid should be included, therefore, at least by formula.

NAMES OF HYBRIDS

The trouble with hybrid nomenclature, at least in the United States, is that there is so much variation in usage. One author favors binomials, another formulae. Some authors are inclined to leave off the × in the binomial if the plant is common, even if it can reproduce only vegetatively. Perhaps the problem is that we have too many alternatives to choose from. Obviously we cannot legislate in matters of pure taxonomy, but it is conceivable that traditions may develop which would promote greater uniformity. I am convinced, for example, that the use of the taxonomic category "variety" has become more uniform and meaningful over the past three decades. In regard to hybrids, Grassl (1963) attempted to bring some uniformity into the picture by proposing that all hybrids be designated by formulae and that the code of nomenclature be changed accordingly. His proposals were rejected at the Xth International Botanical Congress. My conclusion is just the opposite of Grassl's; I believe that we should designate all hybrids, except for the very rarest, with binomials and the ×.

From a purely pragmatic point of view in nomenclature the use of the binomial calls attention to hybrids. It tends to make both collector and student aware that such "kinds" of plants exist and can be keyed. Binomials are also, because of our nomenclatural customs, more likely to be indexed. Whenever a new binomial is created for a hybrid, the taxon will have a type, just as if it were a normal species. Being based precisely upon a type, the taxonomy of the hybrid name can be fixed. For many years, for example, it was thought that the eastern American *Dryopteris* × *leedsii* Wherry was *D. goldiana* × *marginalis*. But this interpretation did not accord with later discovered facts—geography, habitat associations, morphology and cytology —all of which indicated that *D.* × *leedsii* is actually *D. celsa* × *marginalis* (Wagner and Wagner, 1966).

Most commonly an author uses two criteria to decide whether to use a particular designation. If the hybrid reproduces sexually, whether or not it is a frequent plant, the author usually gives a simple binomial without the ×. If the plant is "sterile", the question of treatment depends on abundance: if reasonably common or likely to be encountered, the binomial with the × is used; if rare or not likely to be encountered, the taxonomic formula is used. Either of these latter methods is preferable to designating hybrids as varieties or subspecies of one of the parents, e.g., *Dryopteris cristata* × *goldiana* = "*D. cristata* var. *clintoniana*" or *Polystichum mohrioides* × *munitum* = "*P. mohrioides* var. *scopulinum*". Advocates of this form of taxonomic designation maintain that the "hybrids look more like one parent than the other". And indeed in some cases they do (e.g., where the origin is $A^1A^2 \times B^3$ or $A^1B^2 \times C^3$); but in others the one-sided resemblance is an illusion (*cf.* Evans and Wagner, 1964).

THE FERTILITY OF HYBRIDS

I see no reason why the use of the × in binomials should not be extended to all hybrids, regardless of whether they are fertile or sterile, diploids or polyploids. Practically all of the hybrids under discussion in this paper are of the nature of F_1's; the diploid and tetraploid forms are essentially indistinguishable phenetically except for chromosome numbers, cell sizes, and extent of spore, pollen or seed abortion. Applying Lewis's suggestion for different cytological forms of the same species (given above) to the same hybrid, there would seem to be little purpose in tagging morphologically similar individuals with different designations. "*Planta* × *hybrida* L. (sterile 2*x*)" and "*P.* × *hybrida* L. (sexual 4*x*)" ought to suffice for both cytogeneticist and taxonomist.

In the Great Lakes area of North America we find *Drosera* × *anglica* Huds. mainly as a sterile diploid with undeveloped fruits (Wood, 1955). However, fertile forms are found as well. Since we cannot predict whether the sterile diploid state may not give rise again and again to the fertile tetraploid, and since the two forms are distinguishable only in respect to a single cytogenetic modification, it seems to me that the convenient name for filing and keying would be *D.* × *anglica*. If we accept the idea of having sterile members within the populations of normal species (e.g., A^1A^2 or $A^1A^1A^2$) and using the same taxonomic designation for them as the fertile (e.g., A^1A^1 or $A^1A^1A^2A^2$), then there is no obvious reason why we should not follow the same procedure with respect to hybrids.

The nomenclature problem is beautifully illustrated in the holly-ferns, *Polystichum*, of the western United States (Wagner, 1966*a*, and unpublished). Here we have four basic species, *P. dudleyi*, *P. munitum*, *P. mohrioides*, and *P. lonchitis*, all very distinctive and obviously divergent, diploid, and sexual normal species. Taxonomists have been confused by the array of intermediates; in fact, one pair of authors (in a manuscript fortunately withdrawn from publication) went so far as to suggest that the entire complex was but one polymorphic species. According to my data, on the other hand, the relationships of the forms may be explained simply by the following hybrids: *P. dudleyi* × *munitum* = *P.* × *californicum*; *P. mohrioides* × *munitum* = *P.* × *scopulinum*; and *P. lonchitis* × *mohrioides* = *P.* × *kruckebergii*. In all three of the hybrid combinations the $A^1A^1B^2B^2$ constitution is achieved by at least some individuals, and indeed the rarest of the hybrids, *P.* × *kruckebergii*, has been discovered thus far only in the fertile state. Where the fertile forms grow with their parents, back-crossing produces plants with the $A^1A^1B^2$ pairing behavior. In some localities where *P.* × *californicum* and *P.* × *scopulinum* are found, all of the possible cytomorphotypes occur: A^1A^1, B^2B^2, A^1B^2, $A^1A^1B^2B^2$, $A^1A^1B^2$ and $A^1B^2B^2$. Plants of the A^1B^2 and

$A^1A^1B^2B^2$ constitution are inseparable except on the familiar differences of spore abortion and cell sizes. Consequently I would argue here, as with *D.* × *anglica*, that both sterile and fertile forms of the same phenotype should receive the same nomenclatural designation.

ABUNDANCE OF HYBRIDS

The reason for the consideration of abundance in deciding whether we designate a hybrid by binomial or formula is the dictum that a hybrid should be named "whenever it seems useful or necessary". The question is this: Will the user of my Flora find the hybrid in question commonly enough that it warrants being designated by a binomial? My answer is that any taxa, whether normal species or hybrids, should be included if there is a good likelihood that they will be encountered by collectors.

In Table II, I took the twenty most common *Dryopteris* species of the "*D. spinulosa* complex" and Aspleniums of the eastern United States north of Alabama and estimated the relative likelihood of their being encountered by a biologist traveller from Quebec to Minnesota down to Missouri and east to the Carolinas. (The order of abundance is probably correct, but the scale has been greatly telescoped—for example, *Asplenium platyneuron* is probably more like a thousand times as common as *Phyllitis scolopendrium*.) What the table shows is that sterile hybrids can be more numerous than fertile hybrids and that hybrids may be more frequent than normal species (*cf.* also Fig. 3). The fertile hybrid, *Asplenium* × *pinnatifidum* is much more common and likely to be collected than the normal species, *Phyllitis scolopendrium*. As shown in the map in Fig. 3, the sterile form of *Asplenium* × *ebenoides* is much more often collected than the fertile form of the same taxon. If we designate the fertile form with a binomial then it would be illogical not to recognize the sterile form the same way. In Michigan the fertile form of *D.* × *anglica* is so far known from only a single collection, but the sterile form has been collected over a dozen times (Wood, 1955). The rarest of all the Aspleniaceae in the eastern United States is a fertile normal species, *Phyllitis scolopendrium*, but a number of sterile spleenworts are more common. Obviously we will include the hart's-tongue in the flora of the eastern United States, so we should include the more abundant hybrids as well.

Some floristicians object to plotting hybrids on range maps, assuming that wherever the parents coincide the hybrid will develop. This is not necessarily true. Indeed the incidence of hybrids in different parts of the range where they are expected may be strikingly different, as shown in the work of Barnes (1961) on hybrid aspens in the Lower Peninsula of Michigan. The hybrids *Populus grandidentata* × *tremuloides* are much more common in southern Michigan than in the north, although the parents coexist equally at both

TABLE II. Estimated relative abundance of normal species and hybrids of selected eastern North American ferns. Scale: 0–10, from least common to most common.

	Taxa		
Abundance Index	Dryopteris	Appalachian aspleniums	Other aspleniums
10	*D. intermedia* A^1A^1	*A. platyneuron* A^1A^1	
9	*D. marginalis* A^1A^1		
8			*A. trichomanes* A^1A^1, $A^1A^1A^2$ and $A^1A^1A^2A^2$
7		*Camptosorus rhizophyllus* A^1A^1 = *Asplenium rhiz.*	
6	*D. spinulosa* = *D. carthusiana* $A^1A^1B^2B^2$*		*A. ruta-muraria* $A^1A^1A^2A^2$
5	*D. cristata* $A^1A^1B^2B^2$*	*A. montanum* A^1A^1 *A.* × *pinnatifidum* $A^1A^1B^2B^2$	
4	*D.* × *triploidea* $A^1A^1B^2$*	*A.* × *bradleyi* $A^1A^1B^2B^2$	*A. resiliens* $A^1A^2A^3$**
3	*D.* × *boottii* $A^1B^2C^3$		
2	*D.* × *clintoniana* $A^1A^1B^2B^2C^3C^3$		
1	*D.* × *campyloptera* $A^1A^1B^2B^2$ *D.* × *celsa* $A^1A^1B^2B^2$	*A.* × *ebenoides* A^1B^2 and $A^1A^1B^2B^2$ *A.* × *trudellii* $A^1A^1B^2$	
0			*Phyllitis scolopendrium* $A^1A^1A^2A^2$

* Data modified from S. Walker in part.
** Plant obligately apogamous, and could be as well $A^1A^2B^3$ or $A^1B^2C^3$.

latitudes. The difference in floristic histories of the respective areas probably bears upon why the hybrid is more common in southern Michigan (Barnes, 1961: 314). It is interesting to note that a collector is far more likely to encounter the hybrid aspen in routine collecting in southern Michigan than the very rare normal species, swamp cottonwood, *P. heterophylla*.

SUMMARY AND CONCLUSIONS

(1) Interspecific hybrids exclusive of introgressants are discussed in terms of their influence on the patterns of divergent evolution, and in terms of how they should be treated taxonomically.

(2) Hybrids of the A^1B^2 or $A^1A^1B^2B^2$ and related types are characterized by an intermediacy which is normally so dependable that they may be readily recognized in the field and accurate predictions of parents can be made. Significant deviations from the general law of intermediacy are uncommon, at least in ferns and probably in other plant groups as well.

(3) As nearly as can be judged from existing monographs, the majority of taxa on the earth represent normal species which evolved by ordinary processes dependent upon allopatric isolation and mutation. Hybrids are in the minority.

(4) Wholesale hybridization between normal species is usually prevented by some barrier to crossing, either geographical or reproductive.

(5) Interspecific hybridization should not necessarily be interpreted as the cause of a high incidence of polyploidy in a given group of plants. Polyploids may result just as well from mutations within sectors of normal species.

(6) Most interspecific hybrids have a greatly reduced rate of potential evolution. They are usually sterile. If they reproduce at all, they are mostly apomictic or polyploid.

(7) Phenetically hybrids do not form new specializations; they tend to counteract the extreme character-states evolved by their respective parents. Hybridization thus tends to blunt and dilute the results of divergent evolution by producing compromise characters.

(8) The correlation between the overall phenetic differentiation of taxa and the cytogenetic behavior of their chromosomes is far from good. Members of the same species may have strong pairing factors (e.g., A^1A^1, $A^1A^1A^2$, and $A^1A^1A^2A^2$), members of very different species may show no differentiation of pairing factors (e.g., A^1B^1). Total divergence in all characters is more important in assaying evolution and taxonomy than pairing factors alone.

(9) Taxonomically it is important that hybrids be recognized in Floras and manuals. If hybrids are excluded, identification becomes confusing and the value of taxonomic treatments to students and professional botanists is reduced.

(10) It is suggested that the most convenient way to solve the present problem of wide variation in nomenclatural usage would be to adopt the hybrid binomial for all except the rarest hybrids, thus calling attention to them and providing proper typification.

(11) If a hybrid of the A^1B^2 constitution becomes fertile and sexual by

chromosome doubling, this is not considered a sufficient reason for giving it another name. Since the two forms are practically indistinguishable phenetically and since the fertile form may arise from the sterile repeatedly, they should be placed within the same taxon. The designations may be distinguished, if desirable, by parenthetical postscripts, e.g., "*Planta* × *hybrida* L. (sterile 2*x*)" and "(sexual 4*x*)".

(12) The matter of abundance or rarity of hybrids must be weighed carefully in assigning binomials. If there is a good likelihood that a hybrid will be encountered by students and professional biologists, then it is useful and necessary to name it. A very rare normal species does not warrant a binomial more than a frequent or common hybrid.

REFERENCES

ANDERSON, E. 1948. Hybridization of the habitat. *Evolution*, 2: 1–9.

ANDERSON, E. 1953. Introgressive hybridization. *Biol. Rev.*, 28: 280–307.

BARNES, B. V. 1961. Hybrid aspens in the Lower Peninsula of Michigan. *Rhodora*, 63: 311–324.

BROWN, D. F. M. 1964. A monographic study of the fern genus *Woodsia*. *Beih. Nova Hedwigia*, 16: i–x; 1–154; pl. 1–40.

CAMIN, J. H. and SOKAL, R. R. 1965. A method for deducing branching sequences in phylogeny. *Evolution*, 19: 311–326.

CAMP, W. H. and GILLY, C. L. 1943. The structure and origin of species. *Brittonia*, 4: 323–385.

DAVIS, P. H. and HEYWOOD, V. H. 1963. *Principles of angiosperm taxonomy*. Van Nostrand Co., Princeton, N.J., pp. 462–481.

DELISLE, D. G. 1963. Taxonomy and distribution of the genus *Cenchrus*. *Iowa St. J. Sci.*, 37: 259–351.

EHRLICH, P. R. and HOLM, R. W. 1963. *The process of evolution*. McGraw-Hill Co., New York.

EVANS, A. M. 1964. Interspecific relationships in the *Polypodium pectinatum-plumula* complex. Ph.D. Thesis, Univ. of Michigan.

EVANS, A. M. and WAGNER, W. H., Jr. 1964. *Dryopteris goldiana* × *intermedia*—a natural woodfern cross of noteworthy morphology. *Rhodora*, 66: 255–266.

GRAHAM, S. A. 1963. Systematic studies in the genus *Cuphea* (Lythraceae). Ph.D. Thesis, Univ. of Michigan.

GRANT, V. 1957. The plant species in theory and practice. In *The Species Problem*, *Am. Ass. Adv. Sci*. Washington, D.C.

GRASSL, C. O. 1963. Proposals for modernizing the international rules of nomenclature for hybrids. *Taxon*, 12: 337–347.

HARDIN, J. W. 1957. A revision of the American Hippocastanaceae. *Brittonia*, 9: 145–171.

HAUKE, R. 1959. A taxonomic monograph of the genus *Equisetum* subgenus *Hippochaete*. *Beih. Nova Hedwigia*, 8: 1–123.

HEISER, C. B. 1949. Natural hybridization with particular reference to introgression. *Bot. Rev.*, 15: 645–687.

HEISER, C. B. 1963. Modern species concepts: Vascular Plants. *Bryologist*, 66: 120–124.

ILTIS, H. H. 1959. Studies in the Capparidaceae—VI. *Cleome* sect. *Physostemon*: Taxonomy, Geography and Evolution. *Brittonia*, 11: 123–162. (Cf. p. 129 ff.)

KEENER, C. S. 1967. A biosystematic study of *Clematis* subsection *Integrifoliae*. *J. Elisha Mitchell Soc.*, 83: 1–42.

KLEKOWSKI, E. J. and BAKER, H. G. 1966. Evolutionary significance of polyploidy in the Pteridophyta. *Science*, N.Y. 153: 305–307

LELLINGER, D. B. 1965. A quantitative study of generic delimitation in the adiantoid ferns. Ph.D. Thesis, Univ. of Michigan.

LEWIS, W. H. 1967. Cytocatalytic evolution in plants. *Bot. Rev.*, 33: 105–115.

LÖVE, Á. 1964. The evolutionary framework of the biological species concept. In *Genetics Today, Proc. XI Int. Congr. Genetics*, pp. 409–415.

MICKEL, J. T. 1962. The fern genus *Anemia*, subgenus *Coptophyllum*. *Iowa St. J. Sci.*, 36: 349–382.

MILLER, C. N., Jr. 1965. The evolution of the fern family Osmundaceae. Ph.D. Thesis, Univ. of Michigan.

MIROV, N. T. 1967. *The genus* Pinus. Ronald Press Co., New York.

PANIGRAHI, G. and MANTON, I. 1958. Cytological and taxonomic observations on some members of the *Cyclosorus parasiticus* complex. *J. Linn. Soc. Lond.*, 60: 729–743.

SCORA, R. W. 1964. Interspecific relationships in the genus *Monarda* (Labiatae). Ph.D. Thesis, Univ. of Michigan (cf. pp. 85–98).

SOKAL, R. R. and SNEATH, P. H. A. 1963. *Principles of Numerical Taxonomy*. W. H. Freeman Co., San Francisco and London.

STEBBINS, G. L. 1950. *Variation and Evolution in Plants*. Columbia University Press, New York and London.

STEBBINS, G. L. 1959. The role of hybridization in evolution. *Proc. Am. phil. Soc.*, 103: 231–251.

STERN, K. R. 1961. Revision of *Dicentra* (Fumariaceae). *Brittonia*, 13: 1–57. (Cf. p. 13 ff.)

VALENTINE, D. H. 1963. The treatment of hybrids in *Flora Europaea*. *Webbia*, 18: 47–55.

WAGNER, W. H., Jr. 1954. Reticulate evolution in the Appalachian Aspleniums. *Evolution*, 8: 103–108.

WAGNER, W. H., Jr. 1958. The hybrid ragweed, *Ambrosia artemisiifolia* × *trifida*. *Rhodora*, 60: 309–315.

WAGNER, W. H., Jr. 1960*a*. Evergreen grapeferns and the meanings of infraspecific categories as used in North American pteridophytes. *Am. Fern J.*, 50: 32–45.

WAGNER, W. H., Jr. 1960*b*. The proliferations of *Asplenium pinnatifidum*. *Castanea*, 25: 74–79.

WAGNER, W. H., Jr. 1962*a*. The endemic Botrychiums of the Southeastern United States. (*Abstract.*) *ASB Bull.*, 9: 40.

WAGNER, W. H., Jr. 1962*b*. The synthesis and expression of phylogenetic data. *In:* L. Benson (ed.). *Plant Taxonomy. Methods and Principles*, pp. 273, 276, 277, 415–417. Ronald Press, New York.

WAGNER, W. H., Jr. 1962*c*. Irregular morphological development in hybrid ferns. *Phytomorphology*, 12: 87–100.

WAGNER, W. H., Jr. 1963. Biosystematics and taxonomic categories in Lower Vascular Plants. *Regn. veget.*, 27: 63–71.

WAGNER, W. H., Jr. 1964. The evolutionary patterns of living ferns. *Torrey Bot. Club Bull.*, 21: 86–95.

WAGNER, W. H., Jr. 1966*a*. Two new species of ferns from the United States. *Am. Fern J.*, 56: 3–17.

WAGNER, W. H., Jr. 1966*b*. New data on North American oak ferns, *Gymnocarpium*. *Rhodora*, 68: 121–138.

WAGNER, W. H., Jr. 1966*c*. Modern Research on evolution in the ferns. Chapt. 10, *in* W. A. Jensen and L. G. Kavaljian (eds): *Plant Biology Today. Advances and Challenges*, pp. 164–185. Wadsworth Publishing Co., Belmont, Calif.

WAGNER, W. H., Jr. and CHEN, K. L. 1965. Abortion of spores and sporangia as a tool in the detection of *Dryopteris* hybrids. *Am. Fern J.*, 55: 9–29.

WAGNER, W. H., Jr. and HAGENAH, D. J. 1956. Observations on some bulblet-producing populations of the *Cystopteris fragilis* complex. *Am. Fern J.*, 46: 137–146.

WAGNER, W. H., Jr. FARRAR, D. R. and CHEN, K. L. 1965. A new sexual form of *Pellaea glabella* var. *glabella* from Missouri. *Am. Fern J.*, 55: 171–178.

WAGNER, W. H., Jr. and WAGNER, F. S. 1965. Rochester area log ferns (*Dryopteris celsa*) and their hybrids. *Proc. Rochester Acad. Sci.*, 11: 57–71.

WAGNER, W. H., Jr. and WAGNER, F. S. 1966. Pteridophytes of the Mountain Lake area, Giles Co., Virginia: Biosystematic Studies, 1964–65. *Castanea*, 31: 121–140.

WAGNER, W. H., Jr. and WHITMIRE, R. S. 1957. Spontaneous production of a morphologically distinct fertile allopolyploid by a sterile diploid of *Asplenium ebenoides*. *Torrey Bot. Club Bull.*, 84: 79–89.

WALKER, S. 1962. Further studies in the genus *Dryopteris*: the origin of *D. clintoniana*, *D. celsa* and related taxa. *Am. J. Bot.* 49: 497–503. (*cf.* bibliography).

WALKER, T. 1958. Hybridization in some species of *Pteris* L. *Evolution*, 12: 82–92.

WHITE, R. A. 1963. Tracheary elements of the ferns. I. Features which influence tracheid length; correlation with evolutionary divergence. *Am. J. Bot.*, 50: 447–455.

WOOD, C. E., Jr. 1955. Evidence for the hybrid origin of *Drosera anglica*. *Rhodora*, 57: 106–130.

Biochemistry, Computers and Taxonomy

10 | Chemosystematics with Emphasis on Systematic Serology*

DAVID E. FAIRBROTHERS

Department of Botany, Rutgers-The State University, New Brunswick, New Jersey, U.S.A.

INTRODUCTION

With the development of natural products chemistry, botanists and chemists have expressed the opinion that it should be possible to employ chemical constituents in helping to characterize, describe, and classify taxa. The making of correlations between morphological and chemical groupings of taxa is a very old one. As early as 1699, J. Petiver published on such correlations between medicinal (chemical) properties and certain morphological groupings (Umbelliferae, Labiatae, etc.). Although the concept of employing chemical data in systematic investigations is an old one, a genuine interest in an understanding of the possible correlations between plant constituents and classification has been relatively recent. The interest in this type of investigation has increased with the expanded data coming from biochemical, immunochemical and organic chemical research, and the development of relatively quick and simple analytical techniques. The coming of age of phytochemistry was aptly presented by Harborne (1967*a*) when he indicated that chemotaxonomy is one aspect of the subject of phytochemistry. He also indicated the rapidly developing interest in phytochemistry, and that the recognition of it as a discipline somewhat distinct from pure organic chemistry or pure biochemistry was now an actuality.

Tétényi (1967) discussed the various names that have been applied to phytochemical research concerned with the taxonomic importance of plant products and the systematic application of the accumulated data. He also indicated that the international taxonomic and chemical associations have established a joint committee on chemotaxonomy (Alston, 1965). Without reviewing the definitions and history of all the names that have been applied

* Financial aid from National Science Foundation Grant GB-3847 and Rutgers Research Council is gratefully acknowledged.

to such taxonomic research, I would like to endorse the use of the terms chemotaxonomy and/or chemosystematics as appropriate nomenclature.

I do not desire to present a prolonged discussion with respect to the place of phylogeny in taxonomy, nor the non-phylogenetic position regarding the objectives of taxonomy. Neither do I want to discuss the role of orthodox *v.* experimental taxonomy. The pro and con statements have been frequently expressed by numerous taxonomists within the last fifteen years. However, I will endeavour to bring two publications to the attention of botanists that bear on the above mentioned subject matter (Boyden, 1965, 1966*b*). I offer these to the reader for consideration because I have found them to be thought-provoking and because they were printed in publications not readily available to botanists.

The 1965 publication deals with the basic principles of systematics as a discipline of biology. Professor Boyden put forth his ideas about various ways organisms have been classified. He suggested that the Present Nature Method is a more scientific approach to taxonomy than most others, and is worthy of serious study and development. This method uses all kinds of biological characteristics and weighs them regarding their consistency or conservatism and their systematic distribution. Then the correlations of attributes are placed into patterns. It does not claim that such a classification reveals phylogeny adequately nor does it confuse genealogic with genetic relationship.

The title of the article by Professor Boyden (1966*b*) was "*Un-Natural*" *History*. He indicated that un-natural history differs from natural history in significant ways: (1) terms are confusing; (2) fact and theory are confused; (3) logic and reason are set aside. He contends that needless confusion resulting from carelessness in the use of terms relating to fundamental concepts has made much of our "natural history" un-natural.

Taxonomists constantly group organisms on the basis of general resemblances based upon a range of characteristics. These types of comparative relationships have proven valuable in classification in the past as well as the present. Recognized taxa do not express the same degree of distinctiveness, therefore data obtained from diverse disciplines should help reflect this state of flux. By revising the taxonomic systems, taxonomists strive to develop a more accurate and more readily understandable explanation of the nature and status of taxa. Thus tentative conclusions sometimes must be expressed because of incomplete data; and the search for more facts must be continued, using different methods and characteristics, if the systematic classification of taxa is to be improved. The various kinds of chemical tests used in chemotaxonomic research often yield data that can be included as additional characteristics which contribute to the construction of taxonomic profiles.

The current interest in chemotaxonomy is indicated by the reports frequently published in scientific journals, symposia and books (Hegnauer,

1962; Alston and Turner, 1963; Swain, 1963, 1966; Leone, 1964; McNair, 1965). Upon reading a large portion of the chemosystematic publications the writer discovered that many of the statements and claims presented have been conservative and warn that these data offer no panacea. However, in contrast to these conservative statements, various radical claims have been presented in written and/or oral communications about chemotaxonomy by a few taxonomists engaged in this kind of research as well as some not engaged in it directly. Most of such claims refer to primary constituents and/or about methods and information not yet available. I do not mean to indicate that chemotaxonomic researchers have command of the overall perspective which is still needed. Alston (1967) in writing about this same problem aptly referred to it as, "aura of immaturity" and "an assumption of naïveté" on the part of some biologists.

The same criteria should be used when evaluating chemical (*sensu lato*) data in systematics as must be used in evaluating data from any other method. This means the relative value of employing a certain chemical technique should be judged only after analyses of the appropriate chemical concepts and principles have been made and related to those concepts, principles and guides employed in classification. Therefore a knowledge of some of the basic principles and concepts of the specific chemical approach, as well as of systematics contributes to a better appreciation of the role of a chemotaxonomic endeavor.

Some researchers have indicated that data obtained by the various chemical techniques should only be used in the same fashion data obtained from other techniques have been used in taxonomy. This is the way most of the data obtained by these methods have been interpreted to date. However, this does not mean comparable data will or should be treated similarly in the future. I believe it is important not to limit the future role of such information, because chemosystematics is a very young field of investigation in taxonomy, and the development should not be burdened or curtailed by *a priori* reasoning. The proper application of new techniques requires a willingness to review present tenets, methods and guides, and to assess and modify them where necessary (Rogers *et al.*, 1967).

M. Greshoff (1893) was a very energetic worker in the discipline of chemotaxonomy. He pointed out several important basic tenets. One indicated that biochemists and phytochemists had much to investigate about evolutionary tendencies of metabolic pathways and groups of chemically related plant constituents before they would achieve an understanding of evolution comparable to that of morphologists.

The employment of chemical data in an attempt to reach phylogenetic conclusions is extremely complex with the present state of knowledge. However, the employment of chemical data in an attempt to help characterize the nature of the taxa has proven successful. The information obtained to date has shown that chemical data are not always of value in systematic investigations. The diversity of techniques poses problems for those interested in the use of chemical data. However, the more sophisticated methods do not necessarily provide the most useful data for taxonomic purposes. Some investigations require team research while others can be performed satisfactorily by individuals.

McNair's (1965) book is a reprinting of his papers published between the years 1916 and 1945. Within the text, taxonomy was considered in relation to: oils, fats and waxes; oil and starch in seeds; and alkaloids. The 1935 reprinted paper included in this book entitled "Angiosperm Phylogeny on a Chemical Basis" has been discussed by various botanists (Mirov, 1963; Gibbs, 1958, 1965; Alston, 1967). One of the conclusions that McNair presented in this 1935 publication is, that in general, the evolution of chemical substances in plants has followed evolution of the plants themselves. This would mean that advanced taxa possess the most complicated chemical substances. Gibbs (1958) has warned that judging biogenetic complexity of a molecule by its size may be faulty, and the drawing of phylogenetic conclusions from limited chemical data may be folly.

Which chemical tests and products should be investigated is a question that is often asked. Hegnauer (1965) has indicated that inorganic and organic compounds should both be considered in chemotaxonomic research. He also stated, "in the future macromolecules will play a prominent role".

The problems related to the taxonomic levels to which such data can be applied has been discussed several times. Gibbs (1965) presented convincing evidence that chemotaxonomy has much to offer at various taxonomic levels (races to orders). He also expressed the opinion that the contribution to phylogenetic diagrams was long in coming. The merit of such information seems to be in helping taxonomists to decide between alternate views as to relationships.

The great need for organizing information for ready usage has been advocated in numerous biological reports. Mentzer (1966) in writing about this problem in relation to classification of plant constituents has offered a workable suggestion. His classification should also help taxonomists answer the frequently asked question as to what the various authors mean by macromolecules, secondary constituents etc. Mentzer indicated that researchers have needs for various kinds of classification. However, he suggested a "biogenetic classification" had merit for chemotaxonomical research. This method is based on the natural relationships between the various plant constituents. The three large classes of substances are: (1) the primary or basic constituents (proteins, nucleic acid derivates, chlorophylls, and polysaccharides); (2) the secondary constituents (ones that lack nitrogen and are not involved in the basic metabolism of the cell), and (3) "miscellaneous" substances. These classes can then be divided into groups, families, subfamilies, sections and etc., to meet the various requirements. It is important to realize that this classification like any *one* classification has non-conformers, as for example the alkaloids.

SECONDARY CONSTITUENTS

1. Amino Acids

Early workers in the area of chemical taxonomy largely disregarded the amino acids. However, with the development of chromatography and ionophoresis and the detection of "non protein" amino acids (Fowden, 1964), the significance of amino acids has been realized (Bell, 1966). Investigations of the distribution of the less commonly occurring "non-protein" amino acids would potentially be of more value for taxonomic purposes because, irrespective of concentrations, they provide evidence of inherited differences.

2. Betacyanin and Betaxanthin

Notable advances in the chemistry of the betacyanin and betaxanthin pigments ("nitrogenous anthocyanins") have yielded significant insights into the use of these data for phylogenetic implications (Mabry *et al.*, 1963; Mabry, 1964, 1966). These investigations have led to a recognition of ten betacyanin plant families (Centrospermae) and the proposed separation of the two families Caryophyllaceae and Illecebraceae from the Centrospermae. Several taxonomists have studied this group of plants and are not prepared to follow the suggestion of Mabry. Cronquist (1965) has indicated defining the Centrospermae by the presence or absence of betacyanin or betaxanthin pigments in the flowers, thus excluding the Caryophyllaceae and Illecebraceae and including the Cactaceae and Didiereaceae, would require the investigator to ignore the evidence available from other disciplines. He also indicated that, "One-character taxonomy, like a one-mouse experiment, is always suspect". Mabry (1966) indicated that not just a single chemical characteristic was evaluated in arriving at such a suggested taxonomic change. He indicated two factors must be considered, as they are of particular taxonomic significance: (1) the totally different structures of the two classes of pigments (betacyanin and anthocyanin) indicate different biosynthetic pathways, and (2) the betacyanin distribution is limited to only a few taxa.

I believe this clearly expresses the pit-fall of evaluating all "one-character" bits of information the same in taxonomic investigations.

3. Terpenes

Terpenes have been successfully employed as taxonomic characteristics in the genus *Salvia*. Emboden and Lewis (1967) indicated that a survey of monoterpenes in nineteen species of *Salvia* aided greatly in their identification. They also discovered that the study of terpene composition was as valid as morphological characteristics in the analysis of introgression within this group. These data from *Salvia* are of interest since attempts to use terpenes as chemical characteristics in other taxa of flowering plants have had little significance for taxonomists. The advent of gas chromatography has increased the value of studying terpenes in taxonomic investigations; and since terpenes are frequent in many groups, such chemical compounds are worthy of additional taxonomic consideration. Those readers wanting additional information about terpenoids should find the twelve papers published, by members of the phytochemical group, in the book, *Terpenoids in Plants* (Pridham, 1967) valuable reading.

4. *Flavonoids*

The use of flavonoids in taxonomic research has been reported for several diverse taxa: (1) *Pinus* (Erdtman, 1963); (2) Centrospermae (Mabry, 1966); (3) *Malus* and *Pyrus* (Williams, 1966); (4) Gesneriaceae (Harborne, 1966, 1967*b*); (5) Leguminosae (Alston and Turner, 1963; Alston, 1967); (6) Eucryphiaceae (Bate-Smith and Swain, 1966); etc. Taxonomists wanting information about the comparative biochemistry of the flavonoids have available a very informative book, *Comparative Biochemistry of the Flavonoids* (Harborne, 1967*b*). Those that do not desire to consider the biochemical aspect in detail will still find chapter 10 (Chemical Taxonomy of Flavonoids) of value.

In this book Harborne indicated that flavonoids can be used as taxonomic markers, because they possess the necessary requirements for chemical characteristics to be useful in plant taxonomy. These requirements are: (1) structural variability; (2) chemical stability; (3) widespread distribution in the plant kingdom, and (4) easy and rapid identification.

Harborne (1967*b*) also dealt with the evolutionary aspect of flavonoids and indicated some conclusions were possible although knowledge of flavonoid biosynthesis is still limited. The two general trends he mentioned were: (1) the change in flavonoid pattern in leaves with the replacement of woody by herbaceous habit, and (2) the correlation of anthocyanin type with natural selection for flower color.

5. *Reliability of data*

Written and oral comments are often presented questioning the value, reliability and usefulness of employing non-identified chemical compounds in taxonomic research. The publications by Turner and Alston (1959), Alston and Turner (1964) and several others are excellent examples in which chemical data played a critical role, although practically no knowledge of the identities of the various species-specific compounds was known.

Biochemical genetics has been one of the most investigated areas of chemotaxonomy. Various kinds of biochemical data have made important contributions to the investigation of the differentiation of putative hybrids and parental taxa. Such data have been used to document hybridization, and when combined with morphological data have been of great value in the analysis of complex hybrid populations.

The role of chemical data in the study of hybridizing populations, and in the analysis of past hybridization and introgression, has been aptly demonstrated by Alston and Turner (1962), Smith and Levin (1963), Torres and Levin (1964), Garber and Strommaes (1965), plus several other publications. These research projects have demonstrated that in selected cases, chemical characteristics yield very valuable and provocative data for the taxonomist's evaluation.

I want to present a few comments about the recent monograph of the genus *Pinus* by Mirov (1967). This book is a model, demonstrating how data from diverse disciplines (paleobotany, geography, genetics, morphology, chemistry, ecology, etc.) can be combined to construct a profile for a taxon. The chapter concerned with the chemical aspects indicates that even though the extraneous substances (polyphenols, seed fats, waxes and terpenes) were found in relatively small amounts, they were more useful than cell wall components for taxonomic investigations within the genus. Most extraneous substances (terpenes, polyphenols) were found in many of the species; others, such as alkaloids, were only detected in a few species. Mirov also showed that what had been learned about the extraneous substances has been useful in taxonomic studies of the genus. He indicated future investigations should be concerned with species, population and individual variation of chemical characteristics.

The chapter entitled "Chemical Geography" shows that several chemical compounds found in pines have a definite distribution pattern. These chemical patterns in context with other evidence indicate that the genus *Pinus* originated in the north (Bering Sea), and also suggest possible migration routes of pines from the northern region.

There are various additional secondary constituents and miscellaneous substances that have also contributed to our understanding of chemotaxonomy. The reliability of chemical data and intraspecific variation are also matters of interest in modern systematics. These and other topics could be reviewed in more detail. However, because of various limitations of time, space, interest and knowledge, I will commence my review of the use of primary or basic constituents in chemosystematics.

PRIMARY CONSTITUENTS

1. Introduction

The many important contributions of comparative macromolecular chemistry can be found scattered among various publications. However, the following two recently published symposia yield valuable, provocative and concise data for the taxonomist to evaluate.

The symposium on *Evolving Genes and Proteins* (Bryson and Vogel, 1965) relates to the role of the accomplishments in biochemistry and genetics in drawing evolution into the context of a molecular biology. Those interested in evolutionary and phylogenetic implications of selected chemical data would find these twenty-seven topics on the evolution of pathways, proteins and genes extremely profitable.

Topic IV entitled "Evolution of Protein Characters" (Landa, 1966) was a portion of a symposium held in Czechoslovakia in 1965. The emphasis in this book is on the use of serological methods as means for revealing the distribution of protein characteristics, especially in the plant kingdom. The general theme expressed in the eight papers was that serological investigations can reveal a system (pattern) of similarities and dissimilarities within a class of chemical characters (proteins) which have not been sufficiently utilized in systematic investigations.

Even if one only looks briefly at macromolecular chemistry as related to studies of evolutionary biology, one soon realizes that major developments in the methods of analysis of proteins and nucleic acids have led to achievements that offer promise in helping to better appreciate and to modify certain basic concepts.

Comparative studies of similar enzymes potentially could yield information about differences in primary structure that may be traced to evolutionary adaptations. This statement indicates a great need for a better understanding of the variations found within a protein.

2. DNA and RNA

The relatively simple, effective, procedures for measuring the relative homology of DNAs and/or RNAs of different organisms have been described by Bolton and McCarthy (1962), McCarthy and Bolton (1963), Hoyer *et al.* (1964).

The methods described for this "DNA hybridization" or measure of homologies have been shown to be useful in taxonomic investigation. The early results were obtained from studies employing bacteria and related phages, then comparing bacteria species and still later comparing higher organisms, including mammals. Until recently this method had not been applied with much success to higher plants because of a lack of a satisfactory method for the preparation of high molecular weight plant DNA. This absence of data from higher plants has now changed and a publication discussing relatedness among plants as measured by the DNA agar techniques has appeared (Bendich and Bolton, 1967). In this publication quantitative species comparisons based on DNA–DNA hybridization were reported for taxa belonging to the Leguminosae and for *Secale*, *Hordeum* and *Triticum*. The interactions of the DNAs of the cereal fruits showed wheat DNA (75%) to be more similar to rye DNA (100%) than was barley DNA (57%). The researchers in this area have stated that, "the presence of genes in common may be taken as a guide, not only to taxonomic relationships among organisms, but also to probable evolutionary relationship" (McCarthy and Bolton, 1963). Researchers from this same laboratory also stated, "DNA–DNA hybridization may be useful in systematic and evolutionary study of plants, and also as a possible screening procedure for interfertility of species" (Bendich and Bolton, 1967). I believe that such data can provide one type of index of "genetic relativeness" among living organisms, and look forward with anticipation to additional publications.

I should like to bring to the attention of readers two publications concerned with the taxonomic implications of DNA pairing. Alston (1967) not only informed us of the value of such data but also warned that no one can say the extent of binding is actually a measure of similarities (or homology) in the total linear organization of the DNA. Boyden (1965) described in detail his reservations, indicating that there can be no simple and direct conversion of indices of genetic relationship based on DNA homologies into general classification. Boyden (1965) also stated, "Gene mutations, even those resulting in drastic alterations of the phenotype, may not alter the process of pairing among the alleles so produced: Yet such genes and the chromosomes which include them would be considered "homologous" by the DNA–RNA hybridization procedure". He also indicated there is not a simple proportionality between homology in the genetic codes and amounts of similarity in the phenotypes of the formed organisms known at the present time.

In mid-September of 1967 (*Symposium on Chemotaxonomy and Serotaxonomy* held at the University of Birmingham, England) D. E. Kohne (Carnegie Institution of Washington) presented a stimulating paper entitled, "Taxonomic Applications of DNA Hybridization Techniques". The participants attending this symposium interpreted his statements as a need for a cautious assessment of the value of the DNA hybridization techniques. One of the points raised during the discussion of Dr. Kohne's paper was the need for a better understanding of the problem of repetition of sequences and how this understanding would effect the use of this type of information in taxonomic interpretations.

Pollard (1964) reported on the specificity of ribosomal RNA in corn, cauliflower, cabbage, parsnip, and celery. He presented evidence that both 28S and 18S ribosomal RNA from these taxa had distinct base compositions which were characteristic of the species. The data also indicated that in higher plants the precision of formation of the nucleic acids is preserved during certain mutations.

SYSTEMATIC SEROLOGY

1. *Introduction*

Serology is that portion of biology which is concerned with the nature and interactions of antigenic material and antibodies (Boyden, 1964.) It is also proper to include in serology the reactions of the lectins named and studied extensively by Boyd (1962). Researchers in comparative serological systematics endeavor to reveal dissimilarities or similarities which are detectable only by serological procedures. Taxonomists employing appropriate serological and various electrophoretical methods believe these types of data will help detect and compare non-morphological characteristics (proteins) which will contribute knowledge to the field of systematics (Moritz, 1960, 1964, 1966; Fairbrothers, 1966*a*, *b*).

Some botanists have indicated that they believe the evaluation and interpretation of serological methods and data are too vague (poorly understood) to be useful in botanical research at the present time. Therefore, I would indicate a few recent books which are available in addition to textbooks, for those searching for information to help them evaluate and/or select serological methods for research (Ackroyd, 1964; Kwapinski, 1965; Nerenberg, 1966).

Kloz *et al.* (1960) and Klozova and Kloz (1966) have demonstrated, that owing to their specificity, the antigenic components of plants represent characteristics of value in taxonomic studies. They also indicated that serological methods are suitable when the specificity of protein characteristics, whether individual or a whole complex, is being investigated. Thus relationships discovered by such methods are comparative and are based on degree of cross reactivity caused by structural similarity of antigenic material, such as the correspondence of a greater or lesser number of determinant groups on the protein molecule (Klozova, 1966; Kloz, 1962, 1966; Moritz, 1966).

It has been adequately demonstrated that both reserve and structural proteins can be employed in systematic studies, as long as the same group of proteins and same organs are always compared (Kloz *et al.*, 1960; Ghetie, 1966*a*, *b*). It has also been shown that proteins from pollen, seeds, tubers, algal cells, and fern spores can also be used as satisfactory antigenic material for systematic investigations (Hagman, 1964; Sakaguchi and Arai, 1965; Sakaguchi, 1965; Lester, 1965; Hawkes and Lester, 1966; Brown and Lester,

1965; Lee and Fairbrothers, 1967; Petersen and Fairbrothers, 1967). However, the literature will reveal that a large portion of the data has been based upon proteins extracted from mature seeds. Kloz and his co-workers have indicated that, with some organisms, reserve proteins gave weaker cross reactions between genera than did structural proteins. However, regardless of the type of proteins used, the serological placement (ranking) of the taxa remained unchanged.

There are several serological tests available; however the various precipitin methods have been used most frequently in systematic studies. Of the other types of serological tests, the agglutination reaction has made the most contributions to systematics. The agglutination test may be used directly to test the agglutinogens naturally associated with the cells of organisms, as is done regularly in the blood-typing of men and cattle. It may also be used as a foundation system upon which to absorb soluble antigenic materials and compare them by a very sensitive and economical (amounts of material) system (Boyden, 1964). In our plant serology laboratory we have used the second type of testing to help resolve some difficult problems.

Since the vast majority of plant systematic serological data has been obtained from various precipitin tests further discussion will be limited to data obtained from such precipitin reactions. The precipitin reaction, almost from the time of its discovery (Kraus, 1897), has been used as a tool in taxonomic research. Nuttall (1901), the English zoologist, was the first to use it for indicating possible relationship between species. The pioneering plant studies by Mez (1922), Gilig and Schurhoff (1926, 1927) and their students in phytoserology are well documented and reviewed by Chester (1937). Presently a vast majority of the workers in this discipline have come to view the early contributions of Mez, Gilig and Schurhoff as of historical interest only (Fairbrothers, 1966*a*). Present day plant serological research should not be compared with this early work, because of the numerous advances in immunochemistry, serology, and biochemistry since the 1920s and 1930s.

2. *Annotated Terminology*

Before proceeding with discussions of methods, data, and interpretations, it is essential to define some terms, and present reasons and explanations for such usage. Most of the terms have been defined, but in an array of publications (much of it serological). Also, some definitions are in need of modification as additional data have been reported.

The use of the two terms "homologous" and "heterologous" antigens and reactions should be replaced by the more descriptive terms *reference antigenic material* (*antigen*) and *cross-reacting antigenic material* (*antigen*) respectively (Williams, 1964; Boyden, 1966*a*; Fairbrothers, 1966*a*). I have also suggested extending this terminology to include *reference reaction* and *cross* *reaction* replacing the terms "homologous" and "heterologous" reactions respectively (Fairbrothers, 1966*a*; Pickering and Fairbrothers, 1966*a*). The term homologous, as first and most often used in comparative morphology, applies to structurally corresponding parts (structure, development, relative position, and connections) of organisms (Owen, 1847, 1848; Boyden, 1947). The term as used in serological publications does not fit the definition and is in need of correction. The terms reference and cross are truly descriptive of the antigenic material and reactions in both plant and animal comparative systematic serological studies. The term reference reaction indicates that it is the standard of reference in comparative studies. *Reference antigenic material* (*antigen*) (*R-Ag*): the antigenic material used to immunize the antibody producer. *Reference reaction* (*R-Re*): the reaction between an antiserum and the antigenic material used to stimulate its formation. *Cross reacting antigenic material* (*antigen*) (*X-Ag*): the antigenic material, other than the reference antigenic material which will react serologically with the antibodies. *Cross reaction* (*X-Re*): the reaction between an antiserum and any antigenic material (or hapten) other than the antigenic material used in its formation.

Since the word antigen presently conveys various connotations, it may be difficult to comprehend an author's usage. Antigen as now being used can mean antigenic material, antigenic system, or a single antigenic molecule (complete or incomplete (hapten)). In most comparative serological systematic research antigenic material would be the appropriate term to replace antigen. *Antigenic material*: substance capable of inducing the formation of antibodies (serum globulins) under appropriate conditions and which has the ability to react with the antibodies.

Precipitins: antibodies capable of combining with and reacting upon antigenic material (or haptens) because of certain determinant groups. *Precipitin reactions*: serological reactions between antibodies and soluble antigenic material resulting in the formation of a precipitate. *Immunoprecipitating systems*: precipitin reactions observable as bands lines, or arcs in diffusion gel techniques. Such systems are also present in liquid medium, but detection of individual systems is difficult.

Specificity and reactivity are terms which are often confused and interchanged. *Specificity of antiserum* is the ability to react (antiserum reactivity) only with determinants contained in the reference antigenic system. It is this restriction of reactivity that enables the researcher to discriminate the reference determinants from all other determinants. The combination of the terms specificity of antiserum and antigenic material is often referred to as "specificity of serological reactions". *Fidelity of antiserum*: the degree to which an antiserum reacts with all the different determinants of the reference antigenic material. The fidelity of any one antiserum may be of a high or low magnitude (Moritz, 1960, 1964, 1966). *Determinant groups*: the portion of the antibody molecule that reacts (combines) with a portion of the antigenic molecule. Such portions of the single antigenic molecule, as are "reprinted" by the specific portion of the antibody molecule, are designated "determinant groups", "determinant areas", or "determinants". Correspondingly an antibody as far as it solely bears the serological specificity, may be designated as an "anti-determinant". Thus specific serological reactions occur as a result of the specific affinity between determinant and anti-determinant groups. The reactions become visible (precipitate) because the determinant and anti-determinant groups are attached to molecules and form insoluble complexes under certain conditions of a specific reaction (Moritz, 1965, 1966).

The discussion of terms is presented in detail to make available to the reader some of the principles and concepts of selected serological phenomena. These annotated defintions should also assist the reader in developing an appreciation of these terms in respect to the interpretation of data resulting specifically from systematic serological investigations.

3. *Quantitative Precipitin Methods*

Today, as seventy years ago when Kraus (1897) first reported the precipitin reaction, the basic requirements for such reactions to occur are the presence of antigenic materials, antibodies, and buffered media in which the reaction may be observed. The most important feature of the precipitin reaction is the serological specificity of antibodies, a phenomenon which is dependent upon the ability of the antibodies to combine only with substances possessing certain antigenic determinants. It is this antibody specificity which permits this kind of reaction to be used as a powerful tool for the detection and serological identification of antigenic materials.

However, the precipitin reaction would be of only slight value to science if means for testing it were not available. Perhaps the most common method of evaluating the reaction is the use of serial dilutions of the antigenic material and a constant concentration of antibody. The rates and amount of precipitation are then a function of time and temperature. Since the rate of the reaction increases to approximately 40°C, qualitative evaluation of the reaction is therefore usually conducted at 37°C. It is important that a time-temperature relationship be determined and maintained, in any series of tests, for each system investigated.

(a) *Heidelberger and Kendall Method* These two men (1929, 1935) were the first to quantify the results of the precipitin reaction by actually measuring the amounts of antigen and antibody. They determined the difference between nitrogen content of the antigenic material and of the precipitated complex. This value represented the amount of antibody nitrogen present with a high degree of accuracy. They further discovered that the amount of precipitation varies as to the concentrations and the relative proportions of antigen and antibody. Therefore, the research they later pursued employed the use of varying antigen concentrations while maintaining a constant antibody concentration. The region, with respect to antigen dilutions, in which maximum precipitation was obtained was called the *equivalence zone*; the regions on either side of this zone were designated the *zone of inhibition* (antigen excess in the supernatant) and the *Antibody excess zone*. These same zones are also the ones referred to in the Boyden Precipitin Method using the photronreflectometer. However, it must be indicated that the dilution series employed by the Heidelberger Method was seldom greater than twofold, whereas Boyden employed a doubling dilution series for the antigens. Boyden and co-workers (Boyden, 1942, 1954, 1958; Boyden and DeFalco, 1943; Boyden *et al.*, 1956) have published their rationale for such a dilution series, which is to obtain a more "complete titration curve" for systematic investigations. While the methodology developed by Heidelberger is the foundation upon which other quantitative immunochemical methods are based, the specific technique has been seldom used in serological systematic investigations. This does not mean it is not quantitatively exact or not sound, and it is extensively used in many other areas of serology and immunochemistry. It has not been used extensively in systematic studies because: (1) the range of antigen dilutions is insufficient (lack of whole curve); (2) relatively large amounts of antigen and antiserum are required; (3) a relatively large amount of time is involved in performing a series of tests using micro-Kjeldahl analyses for each antigen and precipitate; (4) knowing the amount of antibody in a unit volume of antiserum does not tell how much or what kinds of antigens those antibodies will precipitate.

(b) *Boyden Procedure* (*BP*) Boyden employs what he has designated "the complete titration curve" or "normal titration curve" as a measure for determining serological correspondence. His technique is based upon the aggregate theory of Heidelberger, but he endeavors to obtain values from maximum precipitation in the equivalence zone to trace or zero precipitation in antigen or antibody excess zone. The relative areas under the curves as determined by turbidity readings with the Libby (1938) photronreflectometer (photron'er), are independent of absolute antigen concentrations so long as the curves are complete. This means the reactions measured are not effected by concentrations of proteins, a fact which is often not considered by critics of the method. Even in mixed (composite) systems every antigen, for which there are any antibodies present in the antiserum used, is present in adequate amounts to satisfy all available antibody. Any deficiency in amounts of antigens present would be detected by the failure of the curves to drop to zero or near zero in the antigen excess zone. This means with the same antiserum, reference and cross reactions are comparable as far as the amounts of the individual antigens are concerned. This same principle applies even in a mixed system (Boyden, 1964, 1966*a*). Thus in the BP the *total* serological reaction is measured and can be expressed as percentage. The data obtained using this procedure (the sum total of the readings from a dilution series) is considered as an *index* of protein similarity or *serological*

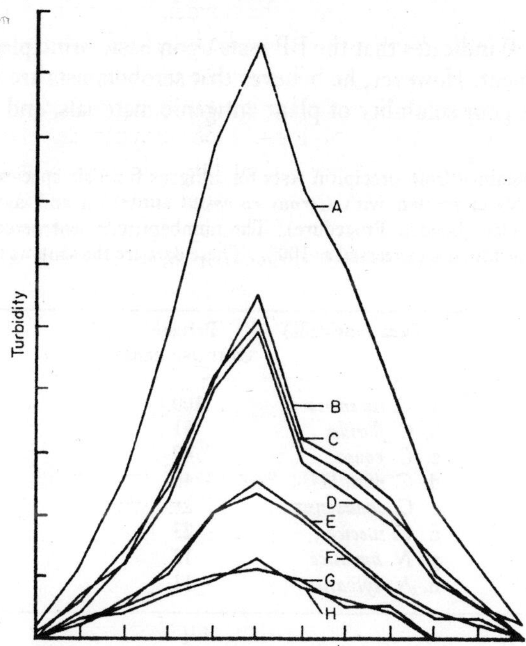

FIG. 1 Curves obtained by the Boyden Procedure showing the precipitin reactions recorded using a photronreflectometer. *Cornus racemosa* antiserum was reacted with antigens of *C. racemosa* (A), *C. florida* (B), *C. kousa* (C), *C. nuttalli* (D), *C. canadensis* (E), *C. suecica* (F), *Nyssa aquatica* (G), and *N. sylvatica* (H). Antigens were prepared in a series of doubling dilutions and mixed with a constant quantity of antiserum. The left slope of the curve represents the *antigen excess zone*, the area around the peak is the *equivalence zone* (optimum proportions), and the right slope represents the *antibody excess zone*.

correspondence. Serological correspondence has been expressed as a percentage of the curve area of the reference reaction (cross-reacting area/reference area) (see Fig. 1 and Table I). Jensen (1962) has also confirmed the validity of the BP with materials tested from the Polycarpicae and Rhoeadales.

Boyden (1964, 1966*a*) has warned of the kinds of problems which can arise if the test systems are not comparable, and of the limitations which result.

This precipitin test makes available relatively objective measures of certain types of biochemical correspondence which are correlated with other diagnostic characteristics of the organism. The qualities of antigens are systematically distributed in organisms and thus provide characteristics of systematic value. Numerous animal and plant systematic serological publications demonstrate the value of the BP (Baum, 1954; Johnson, 1954; Hammond, 1955; Fairbrothers and Johnson, 1964; Johnson and Fairbrothers, 1965; Fairbrothers, 1966*a*, *b*; Kleese and Frey, 1964; Pickering and Fairbrothers, 1966*a*, *b*).

Moritz (1964) indicates that the BP rests upon basic principles with which he is in agreement. However, he believes that serobotanists are handicapped because of the poor solubility of plant antigenic materials, and often by the inability to obtain the "whole curve". Moritz (1966) again discussed these problems, but also included and stressed the importance of complete cross and reference saturation of antibody systems for accurate interpretation of experimental results. He indicated that both the BP and the mutual cross saturation or the saturation placement arrangement (SPA) developed in his laboratory (Kiel, Germany), satisfied this requirement. Both Moritz and Kloz have indicated that the BP requires relatively large amounts of antigenic materials and antisera. I would like to state that in our phytoserology laboratory we have modified the BP so only one half the original amount of antisera and antigenic material is now required for each test.

TABLE I. Data obtained from precipitin tests for antigens from six species of *Cornus* and two species of *Nyssa* reacted with *Cornus racemosa* antiserum and measured by the photronreflectometer (Boyden Procedure). The numbers represent percent area of the reference reaction, which is expressed as 100%. These data are the same as those presented as curves in Fig. 1.

Taxa (antigens)	Percent correspondence
a. *C. racemosa*	100
b. *C. florida*	51
c. *C. kousa*	49
d. *C. nuttalli*	45
e. *C. canadensis*	24
f. *C. suecica*	23
g. *N. aquatica*	12
h. *N. sylvatica*	11

The quantitative measurements of the precipitate systems studied by Boyden and also in our plant investigations were obtained by using the Libby photron'er. This galvanometric instrument measures the light scattered (reflected) from the precipitates in suspension. Baier (1956) has demonstrated that photoelectric light transmission instruments can also be used in systematic serology. This means that standard microdensitometers and spectronic instruments can be adapted for use in quantitative systematic serological investigations.

(c) *Moritz Procedures* (*MP*) The micronephelometry apparatus as developed by Moritz (1960) and the alginate-gel method (Moritz, 1960; Jensen *et al.*, 1964; Frohne *et al.*, 1964) are quantitative methods which have been employed in systematic investigations. The techniques, and advantages and disadvantages are adequately discussed in the above publications. These methods allow less differentiation than the BP; however, less of the antibody and antigenic material are required to obtain a value which is approximately equivalent to the whole curve required in the BP. These publications also show that the alginate-gels have a very high degree of transparency, good adhesion properties (no prior coating necessary) and are easily dissolved, so precipitates or single precipitate layers may be isolated for subsequent investigations without danger of heat denaturation.

More recently the Kiel group has advocated using the saturation placement arrangement (SPA) and have compared this arrangement with the Boyden placement series (BPS). The alginate-gels were mainly used in these presaturation experiments. The reason the presaturation experiments have been stressed by this group was the discovery of "asystematic" or "antisystematic" reactions using plant materials (Jensen *et al.*, 1964; Moritz, 1964, 1965, 1966; Jensen, 1966, 1967). These investigators designate as "antisystematic" reactions those which are surprising or unexpected as a result of convergence. Moritz's introduction of "antisystematic" reactions into the systematic literature caused some biologists to challenge the value of all investigations. Those that still question the value or use of serological data because of possible "antisystematic" reactions, are referred to the publications which indicate this problem can be eliminated in the BPS and the SPA (Moritz, 1965, 1966). Jensen (1966, 1967) has demonstrated how the SPA reveals the distribution of serological characteristics in the Ranunculaceae. Some of his results indicate the close similarity between *Ranunculus-Myosurus*, *Anemone-Clematis*, *Aconitum-Delphinium* and *Actaea-Cimicifuga*. For *Thalictrum*, *Aquilegia*, and *Leptopyrum* a common ancestor could be suggested. The close connection between *Helleborus*, *Ranunculinae*, *Clematidinae* and *Anemoninae* as suggested by the common production of ranunculin (Ruijgrok, 1966) is supported by serological data. The often accepted as self-evident, close relationship between *Trollius* and *Caltha* should be checked since the serological correspondence between the two genera is smaller than expected. The connecting of the genus *Nigella* with *Helleborus* fits with the serological findings. *Helleborus* and *Eranthis* by no means show the high degree of correspondence which other kinds of data have suggested. The genus *Hydrastis* has greater serological correspondence with the Ranunculaceae than the Berberidaceae, as some workers have indicated.

These various methods reported by Moritz and his co-workers once again demonstrate that it is important for botanists using the methods to be aware that they are working with portions of the surface of the antigen molecules. This means the determinant groups are the valuable "serological characteristics". This being so, then the occurrence of similar or convergent characteristics can be expected. This also means the so-called "antisystematic" reactions cease to be non-systematic or unexpected reactions. I do not believe the idea of multispecificity of single antigens restricts the value of serological methods for taxonomic research, because properly designed experiments can now prevent these errors from being included in the interpretation of data.

(d) *Quantitative Ring* (*Layering*) *Precipitation Reaction* (*QRP*) This method has been developed, explained and used by Kloz and co-workers (Kloz, 1960, 1961). It might be said that this method is a "hybrid" of the Heidelberger and Boyden procedures. Kloz used a vertically mounted Libby photronreflectometer to measure turbidity at an antigen-antibody interface. Clarified antiserum was carefully converted with a layer of antigen, thus obtaining a flat meniscus. Readings were made after appropriate time intervals to determine the amount of reaction. This method is different from the "ring" test method used in early animal systematic research and the method reported by Fairbrothers and Johnson (1959, 1961). Evaluation of the QRP has been made comparing it with the BP as a reference point. The Boyden procedure has the advantage of not requiring a precise determination of the original antigen concentration. The theory behind the QRP is that diffusion of the antigen and antibody across the liquid interface will yield the maximum amount of precipitation for that system. Hence the *zones of inhibition*, *equivalence*, *antibody excess*, and the transitional regions between zones are all incorporated into a single determination. This means one photron'er reading incorporates all "points" of Boyden's "normal titration curve". Another advantage with the QRP is that the differentiating power of antisera with low avidity can be substantially increased by raising the concentration of the antiserum to 28% (7% protein in serum is normal). Kloz and his co-workers have shown that using the BP and QRP with the same materials produces the same placement series for the taxa. This research also indicates various organs of the plant have been employed, and these organs gave the same placement of taxa (Kloz *et al.*, 1960, 1966*a*, *b*; Kloz, 1962). Some of these data also prove that such experiments provide positive similarities between protein characteristics and the grafting-affinity, geographical origin and crossability. These and other publications from the Prague group demonstrate the objective validity of data obtained by means of serological methods with plant materials.

Kubo (1964) compared and evaluated the quantitative methods of Heidelberger and Kendall, and those of Boyden. In this publication he developed and explained a mathematical equation for the precipitin reaction system and its application to systematic serology. Kubo concluded that to examine precipitin reaction systems the BP was effective and efficient, and there was no longer a need to insist on the method of Heidelberger and Kendall as the only truly quantitative procedure. Let me indicate at this point that the Boyden view is empirical since no biophysical or biochemical concept has been suggested by him. However, this empirical approach plus one of the concepts or ideas suggested in this discussion on quantitative methods might possibly yield a combination that would help develop a future theoretical explanation of precipitate formation.

It can be seen that the methods of precipitin testing (quantitative or qualitative) vary in their adequacy and in their sources of error. Therefore it is essential to conduct all tests in a comparable procedure. The data presented indicate parts of curves, or single points may be used if they are strictly comparable, though this is not always easily determined. Possible error is still to be found in the present techniques which is often associated with the activity of the antigenic materials. This means it is essential that some investigators constantly pursue experiments which will help uncover these possible errors. The data reported from the various laboratories clearly shows that precipitin methods have been employed to detect similarities and differences between: (1) organs of the same individual, (2) populations, (3) races, (4) varieties, (5) subspecies, (6) species, (7) genera, (8) tribes, (9) families and (10) orders. Such data indicated, to me, that these relatively objective comparisons and placement series are valuable in systematic investigations and additional research will continue to increase their value for systematics in the future.

4. *Qualitative Precipitin Methods*

The principle of antigen-antibody analysis by observing its diffusion in a gel containing reference antigenic material and antiserum was first reported by Bechhold (1905) when he tested extracts obtained from goats.

The diffusion precipitation tests are methods of serological analyses for biological materials which optically reveal complexes of antigens and antibodies formed in various kinds of gel media. These reactions result from the diffusion of prepared materials plus specific immunological reactions. The diffusion of antigens and antibodies in semisolid media (gels) is a migration from a zone of high concentration to lower concentration. Some techniques require the addition of an electric field to help separate the extracts of antigenic materials before the antisera are introduced (immunoelectrophoresis) (see Fig. 2). Antigens have different chemical structures and specific weights, sizes, shapes and electric charges; and consequently various diffusion coefficients (rates).

In the early diffusion experiments gelatin was used as the semisolid medium. Today gelatin has been replaced by various kinds of agar, agarose, alginate and polyacrylamide gels.

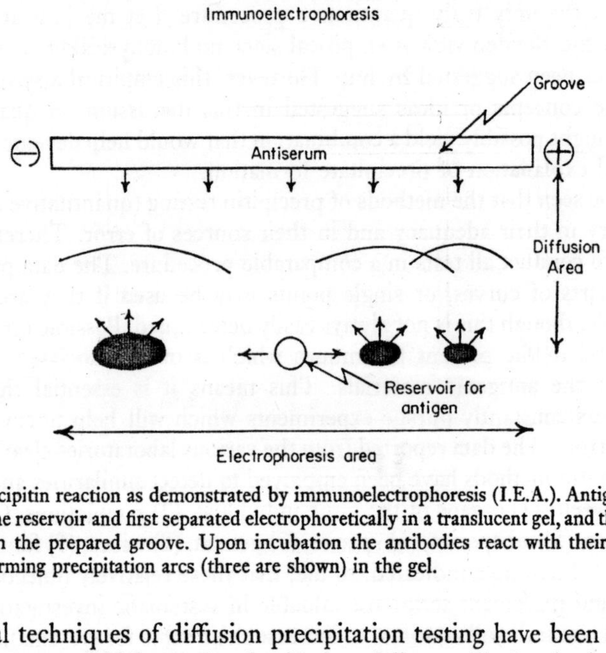

FIG. 2. Precipitin reaction as demonstrated by immunoelectrophoresis (I.E.A.). Antigen extract is placed in the reservoir and first separated electrophoretically in a translucent gel, and the antiserum is placed in the prepared groove. Upon incubation the antibodies react with their appropriate antigens forming precipitation arcs (three are shown) in the gel.

Several techniques of diffusion precipitation testing have been described in the literature which can be classified as follows: (1) the single (simple)-diffusion precipitation method (Oudin, 1946), and (2) the double-diffusion precipitation method. The double-diffusion (D.D.) technique may be performed as a one-dimension method as described by Oakley and Fulthorpe (1953); or as a two-dimension technique as described first by Ouchterlony (1947, 1948) and independently by Elek (1948).

These various techniques have been employed in taxonomic research. However a large portion of the taxonomic research has included the double-diffusion, two-dimension technique, variously modified. In the D.D. two-dimension technique both reactants (antigens and antibodies) migrate through a gel. When, during the diffusion process, an antigen and the precipitating antibody of an active serological system meet in the gel in optimal proportions, immunoprecipitating systems (arcs, lines, bands) are produced forming a precipitation pattern (spectrum) (Ouchterlony, 1964). This is why it is important to experiment with each system to determine the optimal concentrations before interpretations are made. Each arc is the zone of precipitation between one type of antigen from the extract, and one type of antibody from the antiserum. The presence of an arc indicates that there are sufficient molecules of both antigen and antibody to form a precipitate, if certain experimental conditions have been met (Wilson and Pringle, 1954). The arcs can be distinguished from each other because the zones of precipitation occur at different areas in the two-dimensions of the gel medium. An

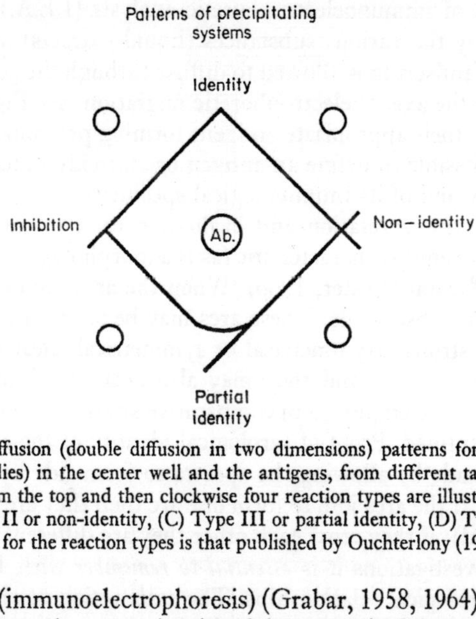

FIG. 3. Immunodiffusion (double diffusion in two dimensions) patterns formed by placing the antiserum (antibodies) in the center well and the antigens, from different taxa, in the four outer wells. Reading from the top and then clockwise four reaction types are illustrated: (A) Type I or identity, (B) Type II or non-identity, (C) Type III or partial identity, (D) Type IV or inhibition. The nomenclature for the reaction types is that published by Ouchterlony (1964).

electrical field (immunoelectrophoresis) (Grabar, 1958, 1964) may sometimes be used to increase the separation (resolving power) of the precipitation zones (arcs, bands, lines).

As a guide to the interpretation of patterns produced in D.D. two-dimension plates, Ouchterlony (1961) originally described three types of reactions; a fourth being subsequently added (Ouchterlony, 1964). These basic patterns are illustrated in Fig. 3. For the first three reactions the names, identity, nonidentity and partial identity were originally suggested. Ouchterlony (1964) later suggested the following revised nomenclature for these patterns: (1) type I or identity, which shows complete fusion of bands; (2) type II or nonidentity, which shows a lack of fusion of bands; (3) type III or partial identity, which shows partial fusion of bands, and (4) type IV or reaction-inhibition, which shows a lack of fusion of bands but differs in appearance from type II. The lack of fusion in type IV is due to the action of an immunospecific barrier. Taxonomists evaluating the D.D. two-dimension technique will find the publication by Ouchterlony (1964) extremely valuable. This method can be employed to compare the number and type of immunoprecipitating systems the various taxa possess.

The double-diffusion two-dimensional technique has been used in taxonomic research to investigate taxonomic categories (rank) from strains to families (Elton and Ewart, 1963; Fairbrothers, 1966*a*, *b*; Fairbrothers and Johnson, 1964; Johnson and Fairbrothers, 1965; Pickering and Fairbrothers, 1967*a*).

The principle of immunoelectrophoretic analysis (I.E.A.) is to separate electrophoretically the various substances (liquid extracts) in a translucent gel and then the antiserum is allowed to diffuse through the gel in a direction perpendicular to the axis of electrophoretic migration (see Fig. 2). The antibodies react with their appropriate antigens forming precipitation arcs in the gel. Thus it is possible to define an antigen or antibody in terms of electrophoretic mobility and of its immunological specificity.

For taxonomic interpretation and evaluation each immunoprecipitating system (arc) is a chemical characteristic (as is a morphological characteristic), specific for each taxon (Lester, 1966). When the arcs are examined several differences may be observable. These arcs may be short or long, straight or curved, weak or strong, asymmetrical or symmetrical, clear or hazy, bluish white or yellowish white, and they may also occur in different positions. These qualities can be employed in comparative studies even though they are chemically unidentified. Proof of serological identity of the arcs in different spectra can be made by allowing the spectra to form side by side using the same antiserum. If the arcs join to form one arc then they are like type I arcs of D.D. (identical); if they cross each other they are different (non-identity). In systematic investigations it is *essential to remember* what I.E.A. helps to explain about serological relationships. The assumption has been made that the larger the number of immunoprecipitating systems (arcs) observed, the closer is the relationship of the taxon extracted (cross reacting antigenic material) to the taxon which produced the antiserum (reference reacting antigenic material). However if two taxa produce the same number of systems (arcs) this does not necessarily infer that they are closely related to each other, but only that they are equally similar to the *reference taxon* used for the antiserum production.

The use of absorbed antiserum has also proven to be a valuable technique in both qualitative and quantitative serological methods. When two or more taxa having certain antigens in common are to be compared it is often helpful to remove the antibodies which combine with the common antigen(s) by absorption. This procedure is done by adding these common antigens to the antiserum being used in the comparative research, and then removing the formed precipitate (common immunoprecipitating systems). If performed correctly, this assures the investigator that the systems now observable in the gel medium (the spectra) are only those systems specific for each taxon (the distinguishing arcs or systems).

Various modifications of I.E.A. have been employed successfully in plant taxonomic research to investigate organisms of various taxonomic rank, including hybrids (Brown and Lester, 1965; Gell *et al.*, 1960; Hall, 1959; Hawkes and Lester, 1966; Kloz, 1962, 1966; Kloz *et al.*, 1966*a*, *b*; Klozova, 1966; Klozova and Kloz, 1964, 1966; Lester, 1965; Vaughan *et al.*, 1966; Vaughan and Waite, 1967*a*, *b*).

ELECTROPHORESIS

Biologists desiring a detailed treatment of gel electrophoresis theory, techniques and clinical, biological and enzymological applications will find the publication *Gel Electrophoresis* (Fredrick, 1964) to be extremely valuable. Taxonomists should find many of the statements helpful in evaluating data obtained from the various gel electrophoretic techniques reported in taxonomic publications.

1. Disc-Electrophoresis

Generally, electrophoresis involves the migration of proteins across a voltage gradient. Disc electrophoresis (D.E.) in particular is a form of "zone electrophoresis" wherein proteins are completely separated within a gel medium. Stated simply, charged molecules will move in a certain direction when subjected to a direct current. This method detects similarities and differences of protein molecules based on two specific properties which affect mobility, ionic charge and molecular weight.

Disc electrophoresis derives its name from the fact that two discontinuous media are placed in conjunction, facilitating the sharp separation of proteins into narrow bands. Acrylamide gels are almost exclusively employed in these tests. The gel's basic characteristic is a "sieving" effect which allows for high resolution separation of proteins. Two gels (differing in pore diameter), the upper or stacking gel having large pore size and the lower or separating gel allowing these protein fractions to spread out, are used. Zone spreading results from a physical (diffusion) and an electrochemical factor (Ornstein, 1964; Morris and Morris, 1964; Nerenberg, 1966). The gel is removed from the voltage gradient, and the separated proteins, enzymes and/or carbohydrates are located by using selected dyes (see Fig. 4).

The same precaution must be used with D.E. as with serological procedures already reported, since different plant organs of the same individual may give different protein patterns. Therefore, it is essential to use comparable organs when comparisons are to be made (Steward *et al.*, 1965). Much of the research of this type which has been applied to systematics has been based upon protein patterns obtained from seed extracts (Boulter *et al.*, 1966; Fox *et al.*, 1964; Thurman *et al.*, 1967; Vaughan *et al.*, 1966, 1967*a*, *b*). We have discovered that seeds, pollen and fern spores can all be used as a

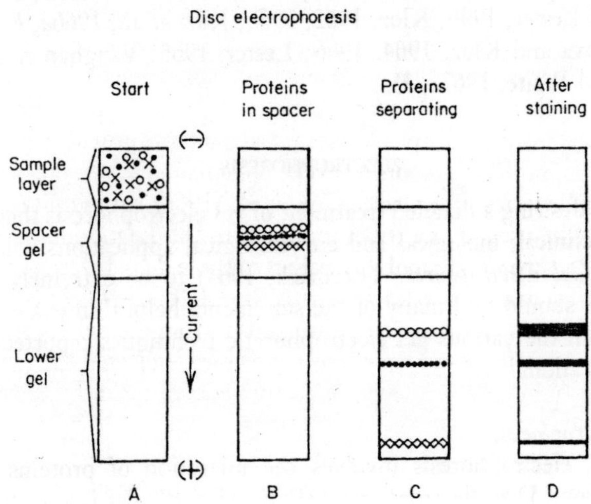

FIG. 4. The separation of a protein mixture into narrow bands within acrylamide gel columns by disc electrophoresis. The test sample is placed on the top of the column containing two gels differing in pore diameter (A). The spacer or upper gel has a large pore size and facilitates the initial sorting of proteins into a closely layered stack of discs (B). The lower or separating gel allows these protein fractions to spread out into bands (discs) (C). These separated protein bands can be observed by applying specific stains (D).

source of extraction material in systematic studies (Lee and Fairbrothers, 1967; Petersen and Fairbrothers, 1967; Pickering and Fairbrothers, 1966*a*, *b*, 1967*a*, *b*). Hagman (1964) has also reported on the use of pollen obtained from taxa of *Betula* in his study of pollen and style relationships as related to genetics and taxonomy.

The D.E. method has been used to compare the total protein band patterns on gel. It has been found that protein band patterns resulting from samples of the same material are highly reproducible. It has also been confirmed by various workers that protein band patterns from different populations of the same species are the same. However, polymorphism has also been reported in some organisms. This indicates the understanding of variation requires additional studies. I would like to refer the reader to the publication by Boulter *et al.* (1966) for consideration and evaluation of the assumptions made when comparing band patterns within our present knowledge.

When bands are compared, a *Rp* value is obtained by measurements; and these values are used as the comparison units to detect similarities and differences. *Rp* expresses the migration ratio between any protein band and a fast-moving stained band accompanying the worker dye in each tube. According to Fox *et al.* (1964), *Rp* is measurement analogous to the *Rf* value of chromatography. Vaughan *et al.* (1966) described the method of calculating *Rp* values. They electrophoresed albumins and globulins separately; however, in our laboratory we have not separated the two fractions prior to migration and have also obtained satisfactory data.

From reported data it appears as if this method provides protein-band data useful at various taxonomic levels. Vaughan *et al.* (1966, 1967*a*) have differentiated species of *Brassica*. Fox *et al.* (1964) have indicated electrophoresis can be used at the tribal grouping in the Leguminosae. Thurman *et al.* (1967) compared the mobility values of three dehydrogenases from taxa of the Leguminosae at species and tribal levels. They concluded that sufficient correlations were observed between data from the dehydrogenases and the globulin data from previous research (which indicated the usefulness of this method and material in systematic studies). Vaughan *et al.* (1966) compared the albumin and globulin fractions of the seeds of three *Brassica* species. They also compared qualitative serological methods using the same material. The data from both methods of protein analysis were in accordance with the placing of the species into a single genus. These data also supported the separation of the three taxa into distinct species. Both methods suggested a closer correspondence between *B. campestris* and *B. oleracea* than between either of these and *B. nigra.*

Lee and Fairbrothers (1967) using pollen collected from four taxa of *Typha* discovered that the serological data and D.E. data were complementary. They also compared results from both methods using pollen and seed material. Both materials gave essentially the same correspondence or similarities and dissimilarities. Pickering and Fairbrothers (1967*a*) have found that the serological and D.E. data supported recent opinions that the cultivated *Magnolia denudata* and *M. liliflora* are mostly self-sterile clones and often produce only hybrid seeds as a result of intercrossing.

The above publications have indicated that this method of protein analysis yields data of value in systematic studies. Some of the publications have also shown that the combination of D.E. and serological data obtained independently can be useful in helping to clarify certain taxonomic questions. That protein band data have a potential use in hybridization studies in which allopolyploids were detected was demonstrated by Johnson and Hall (1965).

2. Disc-Immunodiffusion

Pickering and Fairbrothers (1967*a*) reported a technique which combined D.E. and a serological double diffusion method on the same slide (disc immunodiffusion). This is a modification of the method reported by Seto and Hokama (1964) and Fairbanks *et al.* (1965). Using this combination method to compare the location of the precipitin bands with an identical stained and non-stained D.E. gel column, it was possible to determine which proteins separated by D.E. were antigenic. This method also helped in substantiating band identity when comparing gel columns.

3. Starch-Gel Electrophoresis

Using this technique of electrophoresis the fractionated protein components are detectable as stained bands and yield spectra that have been employed in taxonomic investigations. Coulson and Sim (1964) obtained patterns of wheat endosperm proteins for thirty-four varieties of *Triticum vulgare*. They also compared patterns for eight species of *Triticum* with *Aegilops squarrosa*. The water-soluble endosperm proteins of various grass taxa were also compared. It was reported that the results yielded an accurate identification of different wheat varieties.

Starch electrophoresis has also been used in comparative studies of the seed proteins of amphidiploid species of *Brassica* (see Vaughan and Waite, 1967*a*), and seed proteins of selected species of *Brassica* and *Sinapis* (see Vaughan and Waite, 1967*b*). In these investigations the technique was used for the detection and comparison of β-galactosidase, β-glucosidase and esterase activity. These data were also compared with serological data obtained for the same taxa. The authors reported that results of protein analyses using serological and starch-gel electrophoresis helped them to resolve some taxonomic problems. The data also supported the suggestion of hybridization in the amphidiploid taxa investigation.

Hall and Johnson (1962) reported the value of an electrophoretic analysis of the amphiploid of *Stipa viridula* × *Oryzopsis hymenoidea* and the parental species.

CONCLUSIONS

When it comes to actual construction of any classification, it seems to me, the *sum total* of characteristics from *diverse disciplines* must be interpreted, considered and evaluated. Also required in this construction will be complex decisions regarding the weight (value) to be applied to each characteristic, irrespective of its source. We can be certain that the more data we have for each taxon the more we know about the nature of the taxon.

I have attempted to report on the diversity of chemical characteristics that have proven and potential value as taxonomic characteristics. Now is the proper time for chemical data to be more regularly incorporated with other data in the establishment of taxonomic profiles (the encyclopedic information for each taxon). Compendia can be derived from these profiles for diverse usage in taxonomy.

REFERENCES

ACKROYD, J. F. (ed.). 1964. *Immunological Methods.* F. A. Davis, Philadelphia.

ALSTON, R. E. 1965. The establishment of a joint committee on chemotaxonomy. *Phytochem.*, 4: 779.

ALSTON, R. E. 1967. Biochemical systematics. *In*: T. Dobzhansky, M. K. Hecht, and W. C. Steere (eds). *Evolutionary Biology.* Vol. 1, pp. 197–305. Appleton-Century-Crofts, New York.

ALSTON, R. E. and TURNER, B. L. 1962. New techniques in analysis of complex natural hybridizations. *Proc. natn. Acad. Sci. U.S.A.*, 48: 130–137.

ALSTON, R. E. and TURNER, B. L. 1963. *Biochemical Systematics.* Prentice-Hall, Englewood Cliffs, New Jersey.

ALSTON, R. E. and TURNER, B. L. 1964. Comparative chemistry of *Baptisia*: problems of interspecific hybridization. *In*: C. A. Leone (ed.). *Taxonomic Biochemistry and Serology*, pp. 225–238. Ronald Press, New York.

BAIER, J. G. 1956. Incubation time as a factor in precipitin measurement using photoelectric light-transmission instruments with reference to systematic serology. *Bull. serol. Mus., New Brunsw.*, 17: 3–7.

BAIER, J. G. 1964. An analysis of photoelectric instruments for measurement of turbidity with reference to serology. *Physiol. Zool.*, 20: 172–186.

BATE-SMITH, E. C. and SWAIN, T. 1966. The asperulosides and the aucubins. *In*: T. Swain (ed.). *Comparative Phytochemistry*, pp. 159–174. Academic Press, London.

BAUM, W. C. 1954. Systematic serology of the family Curcubitaceae with special reference to the genus *Curcurbita. Bull. serol. Mus., New Brunsk.*, 13: 5–8.

BECHHOLD, H. 1905. Strukturbildung in Gallerten. *Hoppe-Seyler's Z. physiol. Chem.*, 52: 185–199.

BELL, E. A. 1966. Amino acids and related compounds. *In*: T. Swain (ed.). *Comparative Phytochemistry*, pp. 195–209. Academic Press, London and New York.

BENDICH, A. J. and BOLTON, E. T. 1967. Relatedness among plants as measured by the DNA-agar technique. *Pl. Physiol., Wash.*, 42: 959–967.

BOLTON, E. T. and MCCARTHY, B. J. 1962. A general method for the isolation of RNA complementary to DNA. *Proc. nat. Acad. Sci. U.S.A.*, 48: 1390–1397.

BOULTER, D., THURMAN, D. A. and TURNER, B. L. 1966. The use of disc electrophoresis of plant proteins in systematics. *Taxon*, 15: 135–143.

BOYD, W. C. 1962. *Introduction to Immunochemical Specificity.* Interscience, New York.

BOYDEN, A. 1942. Systematic serology: a critical appreciation. *Physiol. Zoöl.*, 15: 109–145.

BOYDEN, A. 1947. Homology and analogy. A critical review of the meanings and implications of these concepts in biology. *Am. Midl. Nat.*, 37: 648–669.

BOYDEN, A. 1954. The measurement and significance of serological correspondence among proteins. *In*: W. H. Cole (ed.). *Serological Approaches to Studies of Protein Structure and Metabolism*, pp. 74–97. *X A. Conf. Protein Metab. Rutgers Univ.*

BOYDEN, A. 1958. Comparative serology: aims, methods, and results. *In*: W. H. Cole (ed.). *Serological and Biochemical Comparisons of Proteins*, pp. 3–24. *XIV A. Conf. Protein Metab. Rutgers Univ.*

BOYDEN, A. 1964. Perspectives in systematic serology. *In*: C. A. Leone (ed.). *Taxonomic Biochemistry and Serology*, pp. 75–99. Ronald Press, New York.

BOYDEN, A. 1965. A review of the present status of systematic serology. Part I. *Bull. serol. Mus., New Brunsw.*, 33: 5–8.

BOYDEN, A. 1966*a*. A review of the present status of systematic serology. Part II. *Bull. serol. Mus., New Brunsw.*, 34: 1–4.

BOYDEN, A. 1966*b*. "Un-natural" history. *In*: T. R. Buckman (ed.). *Bibliography and Natural History*, pp. 131–139. Kansas Univ. Libraries, Lawrence, Kansas.

BOYDEN, A. and DE FALCO, R. 1943. Report on the use of the photoreflectometer in serological comparisons. *Physiol. Zoöl.*, 16: 229–241.

BOYDEN, A., GEMEROY, D. and DE FALCO, R. 1956. On measuring serological correspondence among antigens. *Bull. serol. Mus., New Brunsw.*, 16: 3–8.

BROWN, R. M. and LESTER, R. N. 1965. Comparative immunology of the algal genera *Tetracystis* and *Chlorococcum*. *J. Phycol.*, 1: 60–65.

BRYSON, V. and VOGEL, H. J. (eds). 1965. *Evolving Genes and Proteins*. Academic Press, New York.

CHESTER, K. S. 1937. A critique of plant serology. *Q. Rev. Biol.*, 12: 19–46; 165–190; 294–321.

COULSON, C. B. and SIM, A. K. 1964. Proteins of various species of wheat and closely related genera and their relationship to genetical characteristics. *Nature, Lond.*, 202: 1305–1308.

CRONQUIST, A. 1965. The status of the general system of classification of flowering plants. *Ann. Missouri Bot. Gard.*, 52: 281–303.

ELEK, S. D. 1948. The recognition of toxicogenic bacterial strains in vitro. *Br. med. J.*, 1: 493–496.

ELTON, G. A. H. and EWART, J. A. D. 1963. Immunological comparison of cereal proteins. *J. Sci. Fd. Agric.*, 14: 750–753.

EMBODEN, W. A., Jr. and LEWIS, H. 1967. Terpenes as taxonomic characters in *Salvia* section *Audibertia*. *Brittonia*, 19: 152–160.

ERDTMAN, H. 1963. Some aspects of chemotaxonomy. *In*: T. Swain (ed.). *Chemical Plant Taxonomy*, pp. 89–125. Academic Press, London and New York.

FAIRBANKS, G., Jr., LEVINTHAL, C. and REEDER, R. H. 1965. Analysis of ^{14}C-labeled proteins by disc electrophoresis. *Biochem. biophys. Res. Commun.*, 20: 393–399.

FAIRBROTHERS, D. E. 1966*a*. Comparative serological studies in plant systematics. *Bull. serol. Mus., New Brunsw.*, 35: 2–6.

FAIRBROTHERS, D. E. 1966*b*. The comparison and interpretation of serological data in plant systematics. *In*: Z. Landa (ed.). *Mechanism of Mutation and Inducing Factors*, pp. 458–464. Proc. Symp. Mut. Process. Topic IV: Evolution of protein characters. *Symposia ČSAV.*

FAIRBROTHERS, D. E. and JOHNSON, M. A. 1959. The precipitin reaction as an indicator of relationship in the subfamily Festucoideae of the family Poaceae (Gramineae). *IX Int. bot. Congr.*, 2: 110–111.

FAIRBROTHERS, D. E. and JOHNSON, M. A. 1961. The precipitin reaction as an indicator of relationships in some grasses. *Recent Advances in Botany*, pp. 116–120. U. of Toronto Press.

FAIRBROTHERS, D. E. and JOHNSON, M. A. 1964. Comparative serological studies within the families Cornaceae (Dogwood) and Nyssaceae (Sour Gum). *In*: C. A. Leone (ed.). *Taxonomic Biochemistry and Serology*, pp. 305–318. Ronald Press, New York.

FOWDEN, L. 1964. The chemistry and metabolism of recently isolated amino acids. *A. Rev. Biochem.*, 33: 173–204.

FOX, D. J., THURMAN, D. A. and BOULTER, D. 1964. Studies on the proteins of seeds of the Leguminosae. I. Albumins. *Phytochem.*, 3: 417–419.

FREDRICK, J. F. (ed.). 1964. Gel electrophoresis. *Ann. N.Y. Acad. Sci.*, 121: 305–650.

FROHNE, D., PFANDER, H. J. and MORITZ, O. 1964. Alginate-gels in immunodiffusion. *Bull. serol. Mus., New Brunsw.*, 31: 5–8.

GARBER, E. D. and STROMMAES, O. 1965. A paper chromatographic study of flower extracts from fourteen species and four interspecific hybrids. *Bot. Gaz.*, 125: 96–101.

GELL, P. G. H., HAWKES, J. G. and WRIGHT, S. T. C. 1960. The application of immunological methods to the taxonomy of species within the genus *Solanum*. *Proc. R. Soc.* (*B*), 151: 364–383.

GHETIE, V. 1966*a*. Evolution of protein characters in ontogenesis of higher plants. *In*: Z. Landa (ed.). *Mechanism of Mutation and Inducing Factors*, pp. 489–503. Proc. Symp. Mut. Process. Topic IV: Evolution of protein characters. *Symposia ČSAV*.

GHETIE, V. 1966*b*. A hypothesis on the mechanism of reserve proteins hydrolysis during seed germination. *Revue roum. Biochimie*, 3: 353–361.

GIBBS, R. D. 1958. Chemical evolution in plants. *J. Linn. Soc.*, 56: 123–166.

GIBBS, R. D. 1965. A classical taxonomist's view of chemistry in taxonomy of higher plants. *Lloydia*, 28: 279–299.

GILIG, E. and SCHURHOFF, P. H. 1926. Die Serodiagnostik in der Verwandtschaftsforschung. *Bot. Jb.*, 60: 439–450.

GILIG, E. and SCHURHOFF, P. H. 1927. Unsere Erfahrungen über die Brauchbarkeit der Serodiagnostik für die botanische Verwandtschaftsforschung. *Berl. dt. Bot. Gesch.*, 45: 315–330.

GRABAR, P. 1958. The use of immunochemical methods in studies on proteins. *Adv. Protein Chem.*, 13: 1–33.

GRABAR, P. 1964. Immunoelectrophoretic analysis. *In*: J. F. Ackroyd (ed.). *Immunological Methods*, pp. 79–91. Blackwell Sci. Pub., Oxford.

GRESHOFF, M. 1893. Gedanken über Pflanzenkräft und phytochemische Verwandtschaft. *Berl. dt. Pharm. Gesch.*, 3: 191–204.

HAGMAN, M. 1964. The use of disc electrophoresis and serological reactions in the study of pollen and style relationships. *In*: H. F. Linskens (ed.). *Pollen Physiology and Fertilization*, pp. 244–250. North-Holland, Amsterdam.

HALL, O. 1959. Immuno-electrophoretic analyses of allopolyploid ryewheat and its parental species. *Hereditas*, 45: 495–504.

HALL, O. and JOHNSON, B. L. 1962. Electrophoretic analysis of the amphiploid of *Stipa viridula* × *Oryzopsis hymenoidea* and its parental species. *Hereditas*, 48: 530–535.

HAMMOND, H. D. 1955. A study of taxonomic relationship within the Solanaceae as revealed by the photroner serological method. *Bull. serol. Mus., New Brunsw.*, 14: 3–5.

HARBORNE, J. B. 1966. Comparative biochemistry of flavonoids—II. 3-desoxyanthocyanins and their systematic distribution in ferns and gesnerads. *Phytochem.*, 5: 589–600.

HARBORNE, J. B. 1967*a*. Phytochemistry comes of age. *Chem. Pl. Taxon. Newsl., Univ. of Liverpool*, 8: 2.

HARBORNE, J. B. 1967*b*. *Comparative Biochemistry of the Flavonoids*. Academic Press, London.

HAWKES, J. G. and LESTER, R. N. 1966. Immunological studies on the tuber-bearing *Solanums*. II. Relationships of the North American species. *Ann. Bot.*, 30: 270–290.

HEGNAUER, R. 1962. *Chemotaxonomie der Pflanzen.* Vol. I. Birkhauser, Basel.

HEGNAUER, R. 1965. Chemotaxonomy—past and present. *Lloydia*, 28: 267–278.

HEIDELBERGER, M. and KENDALL, F. E. 1929. A quantitative study of the precipitin reaction between Type III *Pneumococcus* polysaccharide and purified homologous antibody. *J. exp. Med.*, 50: 809–823.

HEIDELBERGER, M. and KENDALL, F. E. 1935. The precipitin reaction between Type III *Pneumococcus* polysaccharide and homologous antibody. III. A quantitative study and a theory of the reaction mechanism. *J. exp. Med.*, 61: 563–591.

HOYER, B. H., MCCARTHY, B. J. and BOLTON, E. T. 1964. A molecular approach in the systematics of higher organisms. *Science, N.Y.*, 144: 959–967.

JENSEN, U. 1962. Beitrage zur Verwendung der Boydenschen Präzipitationsmethode in der Serobotanik. *Flora, Jena*, 152: 371–398.

JENSEN, U. 1966. On the distribution of serological characters in the family of Ranunculaceae. Preliminary report. *In*: Z. Landa (ed.). *Mechanism of Mutation and Inducing Factors*, pp. 465–470. Proc. Symp. Mut. Process. Topic IV: Evolution of protein characters. *Symposia ČSAV*.

JENSEN, U. 1967. Die Verwandtschaftsverhältnisse innerhalb der Ranunculaceae aus serologischer Sicht. *Ber. dt. bot. Ges.*, 79: 407–412.

JENSEN, U., FROHNE, D. and MORITZ, O. 1964. Serological investigations in the field of Rhoeadales and Ranunculaceae. *Bull. serol. Mus., New Brunsw.*, 32: 3–6.

JOHNSON, B. L. and HALL, O. 1965. Analysis of phylogenetic affinities in the Triticinae by protein electrophoresis. *Am. J. Bot.*, 52: 506–513.

JOHNSON, M. A. 1954. The precipitin reaction as an index of relationship in the Magnoliaceae. *Bull. serol. Mus., New Brunsw.*, 13: 1–5.

JOHNSON, M. A. and FAIRBROTHERS, D. E. 1965. The comparison and interpretation of serological data in the Magnoliaceae. *Bot. Gaz.*, 126: 260–269.

KLEESE, R. A. and FREY, K. J. 1964. Serological predictions of genetic relationships among oat varieties (*Avena sativa* L.) and corn inbreds (*Zea mays* L.). *Crop Sci.*, 4: 379–383.

KLEIN, R. M. and CRONQUIST, A. 1967. A consideration of the evolutionary and taxonomic significance of some biochemical micromorphological, and physiological characters in the thallophytes. *Q. Rev. Biol.*, 42: 105–296.

KLOZ, J. 1960. The quantitative ring (layering) precipitation reaction. *Folia biol., Praha*, 6: 319–325.

KLOZ, J. 1961. The use of the quantitative ring precipitation-reaction for the determination of cross reactions with special reference to comparative serology. *Biologia Pl.*, 3: 217–227.

KLOZ, J. 1962. An investigation of the protein characters of four *Phaseolus* species with special reference to the question of their phylogenesis. *Biologia Pl.*, 4: 85–90.

KLOZ, J. 1966. Protein characters and their genesis in lower taxons. *In*: Z. Landa (ed.). *Mechanism of Mutation and Inducing Factors*, pp. 477–484. Proc. Symp. Mut. Process. Topic IV: Evolution of protein characters. *Symposia ČSAV*.

KLOZ, J., KLOZOVA, E. and TURKOVA, V. 1966*a*. Chemotaxonomy and genesis of protein characters with special reference to the genus *Phaseolus*. *Preslia*, 38: 229–236.

KLOZ, J., KLOZOVA, E. and TURKOVA, V. 1966*b*. Protein characters and relationship between *Phaseolus vulgaris* ssp. *aborigineus* Burk. and related taxa of the genus *Phaseolus*. *Biologia Pl.*, 8: 187–196.

KLOZ, J., TURKOVA, V. and KLOZOVA, E. 1960. Serological investigation of the taxonomic specificity of proteins in various organs in some taxa of the family Viciaceae. *Biologia Pl.*, 2: 126–137.

KLOZOVA, E. 1966. Interrelations among several Asiatic species of the genus *Phaseolus* studied by immunochemical methods. *In*: Z. Landa (ed.). *Mechanism of Mutation and Inducing Factors*, pp. 485–487. Proc. Symp. Mut. Process. Topic IV: Evolution of protein characters. *Symposia ČSAV*.

KLOZOVA, E. and KLOZ, J. 1964. The identification of hybrids of *Phaseolus vulgaris* L. × *Phaseolus coccineus* L. using immunochemical methods. *Biologia Pl.*, 6: 240–241.

KLOZOVA, E. and KLOZ, J. 1966. Protein characters in species hybrids of the genus *Phaseolus*, studied by means of serological methods. *Acta Agric. scand.*, suppl. 16: 225–228.

KRAUS, R. 1897. Uber specifische Reactionen in keimfreien Filtraten aus *Cholera*, *Typhus* und pestbouillon Culturen erzeugt durch homologes Serum. *Wien. klin. Wschr.*, 10: 736–738.

KUBO, K. 1964. An empirical equation for the precipitin reaction system and its application to systematic serology. *Bull. serol. Mus., New Brunsw.*, 31: 1–2.

KWAPINSKI, J. B. 1965. *Methods of Serological Research*. John Wiley and Sons, New York.

LANDA, Z. (ed.). 1966. *Mechanism of Mutation and Inducing Factors*. Proc. Symp. Mut. Process. Topic IV: Evolution of protein characters. *Symposia ČSAV*.

LEE, D. W. and FAIRBROTHERS, D. E. 1967. Serological and disc-electrophoretic studies of North American *Typha*. *Am. J. Bot.*, 54: 660.

LEONE, C. A. (ed.). 1964. *Taxonomic Biochemistry and Serology*. Ronald Press, New York.

LESTER, R. N. 1965. Immunological studies on the tuber-bearing *Solanums*. I. Techniques and South American species. *Ann. Bot.*, 29: 609–624.

LESTER, R. N. 1966. Interpretation of the spectra of double diffusion and immunoelectrophoresis for systematic purposes. *In*: Z. Landa (ed.). *Mechanism of Mutation and Inducing Factors*, pp. 471–476. Proc. Symp. Mut. Process. Topic IV: Evolution of protein characters. *Symposia ČSAV*.

LIBBY, R. L. 1938. The photronreflectometer—an instrument for the measurement of turbid systems. *J. Immun.*, 34: 71–73.

McCARTHY, B. J. and BOLTON, E. T. 1963. An approach to the measurement of genetic relatedness among organisms. *Proc. nat. Acad. Sci. U.S.A.*, 50: 156–164.

McNAIR, J. B. 1965. *Studies in Plant Chemistry including Chemical Taxonomy, Ontogeny and Phylogeny*. Pub. by author, California.

MABRY, T. J. 1964. The betacyanins, a new class of red-violet pigments, and their phylogenetic significance. *In*: C. A. Leone (ed.). *Taxonomic Biochemistry and Serology*, pp. 239–254. Ronald Press, New York.

MABRY, T. J. 1966. The betacyanins and betaxanthins. *In*: T. Swain (ed.). *Comparative Phytochemistry*, pp. 231–244. Academic Press, London and New York.

MABRY, T. J., TAYLOR, F. A. and TURNER, B. L. 1963. The betacyanins and their distribution. *Phytochem.*, 2: 61–64.

MENTZER, C. 1966. Biogenetic classification of plant constituents. *In*: T. Swain (ed.). *Comparative Phytochemistry*, pp. 21–31. Academic Press, London and New York.

MEZ, C. 1922. Anleitung zu serodiagnostischen Untersuchungen für Botaniker. *Bot. Arch.*, 1: 177–200.

MIROV, N. T. 1963. Chemistry and plant taxonomy. *Lloydia*, 26: 117–124.

MIROV, N. T. 1967. *The Genus Pinus*. Ronald Press, New York.

MORITZ, O. 1960. Some variants of serological technique developed in sero-botanical work. *Bull. serol. Mus., New Brunsw.*, 24: 1–8.

MORITZ, O. 1964. Some special features of sero-botanical work. *In*: C. A. Leone (ed.). *Taxonomic Biochemistry and Serology*, pp. 275–290. Ronald Press, New York.

MORITZ, O. 1965. Uber die homologe Restreaktivität von Immunopräzipitaten aus heterologen Reaktionen und ihre Bedeutung. *Enzymologia*, 29: 155–168.

MORITZ, O. 1966. Revealing systematical distribution of protein characters by serological methods. *In*: Z. Landa (ed.). *Mechanism of Mutation and Inducing Factors*, pp. 443–457. Proc. Symp. Mut. Process. Topic IV: Evolution of protein characters. *Symposia ČSAV*.

MORRIS, C. J. O. R. and MORRIS, P. 1964. *Separation Methods in Biochemistry*, pp. 620–663; 712–715. Intersci. Pub., New York.

NERENBERG, S. T. 1966. *Electrophoresis, a Practical Laboratory Manual*, pp. 5–13; 110–115; 218–232. F. A. Davis, Philadelphia.

NUTTALL, G. H. F. 1901. The new biological test for blood in relation to zoological classification. *Proc. R. Soc.*, 69: 150–153.

OAKLEY, C. L. and FULTHORPE, A. J. 1953. Antigenic analysis by diffusion. *J. Path. Bact.*, 65: 49–60.

ORNSTEIN, L. 1964. Disc electrophoresis: I. Background and theory. *Ann. N.Y. Acad. Sci.*, 121: 321–349.

OUCHTERLONY, O. 1947. In vitro method for testing the toxin-producing capacity of diphtheria bacteria. *Scand. Path. Cong.*, Abs. 8.

OUCHTERLONY, O. 1948. In vitro method for testing the toxin-producing capacity of diphtheria bacteria. *Acta path. microbiol. scand.*, 25: 189–191.

OUCHTERLONY, O. 1961. Interpretation of comparative immune precipitation patterns obtained by diffusion-in-gel techniques. *In*: M. Heidelberger, and O. J. Plescia (eds). *Immunochemical Approaches to Problems in Microbiology*, pp. 5–19. Rutgers Univ. Press, New Brunsw.

OUCHTERLONY, O. 1964. Gel-diffusion techniques. *In*: J. F. Ackroyd (ed.). *Immunological Methods*, pp. 55–78. Blackwell Sci. Pub., Oxford.

OUDIN, J. 1946. Méthode d'analyse immunochimique par précipitation spécifique in milieu gélifié. *C.r. Acad. Sci., Paris*, 222: 115–116.

OWEN, R. 1847. Report on the archetype and homologies of the vertebrate skeleton, pp. 169–340. *XVI Rep. Br. Ass. Advmt. Sci.*

OWEN, R. 1848. *On the Archetype and Homologies of the Vertebrate Skeleton*. (Reprint, revision and extension of 1847 publication.) John van Voorst, London.

PETERSEN, R. L. and FAIRBROTHERS, D. E. 1967. Investigations of the fern genus *Osmunda* employing serological and disc-electrophoretic techniques. *Am. J. Bot.*, 64: 660.

PETIVER, J. 1699. Some attempts made to prove that herbs of the same make or class for the generality, have the like virtue and tendency to work the same effects. *Phil. Trans. R. Soc.*, 21B: 289–294.

PICKERING, J. L. and FAIRBROTHERS, D. E. 1966*a*. A serological investigation of *Magnolia* hybrids and parental species. *Bull. serol. Mus., New Brunsw.*, 36: 4–7.

PICKERING, J. L. and FAIRBROTHERS, D. E. 1966*b*. Taxonomic investigation of *Magnolia denudata*, *M. liliflora*, *M.* × *Soulangiana*, and *M.* × *Soulangiana* "*Lennei*" using serological and acrylamide gel electrophoretic methods. *Am. J. Bot.*, 53: 636–637.

PICKERING, J. L. and FAIRBROTHERS, D. E. 1967*a*. A serological and disc electrophoretic investigation of *Magnolia* taxa. *Bull. Torrey bot. Club.*, 94: 468–479.

PICKERING, J. L. and FAIRBROTHERS, D. E. 1967*b*. Taxonomic investigation of selected Umbelliferae using serological and acrylamide gel electrophoretic methods. *Am. J. Bot.*, 54: 660.

POLLARD, C. J. 1964. The specificity of ribosomal ribonucleic acids of plants. *Biochem. biophys. Res. Commun.*, 17: 171–176.

PRIDHAM, J. B. (ed.). 1967. *Terpenoids in Plants*. Academic Press, London.

ROGERS, D. J., FLEMING, H. S. and ESTABROOK, G. F. 1967. Use of computers in studies of taxonomy and evolution. *In*: T. Dobzhansky, M. K. Hecht, and W. C. Steere (eds). *Evolutionary Biology*. Vol. 1, pp. 169–196. Appleton-Century-Crofts, New York.

RUIJGROK, H. W. L. 1966. The distribution of ranunculin and cyanogenetic compounds in the Ranunculaceae. *In*: T. Swain (ed.). *Comparative Phytochemistry*, pp. 175–186. Academic Press, London and New York.

SAKAGUCHI, S. 1965. Immunological techniques for diagnosis of varieties relationship in plant. *SABCO J. Japan*, **1**: 188–194.

SAKAGUCHI, S. and ARAI, S. 1965. Application of adjuvant method in phyto-immunological studies. *Jap. J. Breed.*, 15: 7–12.

SETO, J. T. and HOKAMA, Y. 1964. Disc electrophoretic analysis of sialidase from influenza virus. *Ann. N.Y. Acad. Sci.*, 121: 640–650.

SMITH, D. M. and LEVIN, D. A. 1963. A chromatographic study of reticulate evolution in the Appalachian *Asplenium* complex. *Am. J. Bot.*, 50: 952–958.

STEWARD, F. C., LYNDON, R. F. and BARBER, J. T. 1965. Acrylamide gel electrophoresis of soluble plant proteins: a study on pea seedlings in relation to development. *Am. J. Bot.*, 52: 155–164.

SWAIN, T. (ed.) 1963. *Chemical Plant Taxonomy*. Academic Press, London and New York.

SWAIN, T. (ed.). 1966. *Comparative Phytochemistry*. Academic Press, London and New York.

TÉTÉNYI, P. 1967. From "chemo"—to chemotaxonomist. *Taxon*, **16**: 179–181.

THURMAN, D. A., BOULTER, D., DERBYSHIRE, E. and TURNER, B. L. 1967. Electrophoretic mobilities of formic and glutamic dehydrogenoses in the Fabaceae: A systematic survey. *New Phytol.*, **66**: 37–45.

TORRES, A. M. and LEVIN, D. A. 1964. A chromatographic study of cespitose zinnias. *Am. J. Bot.*, **51**: 639–643.

TURNER, B. L. and ALSTON, R. E. 1959. Segregation and recombination of chemical constituents in a hybrid swarm of *Baptisia laevicaulis* × *B. viridis* and their taxonomic implications. *Am. J. Bot.*, **46**: 678–686.

VAUGHAN, J. G. and WAITE, A. 1967*a*. Comparative electrophoretic studies of the seed proteins of certain amphidiploid species of *Brassica*. *J. exp. Bot.*, 18: 269–276.

VAUGHAN, J. G. and WAITE, A. 1967*b*. Comparative electrophoretic studies of the seed proteins of certain species of *Brassica* and *Sinapis*. *J. exp. Bot.*, **18**: 100–109.

VAUGHAN, J. G., WAITE, A., BOULTER, D. and WAITERS, S. 1966. Comparative studies of the seed proteins of *Brassica campestris*, *B. oleracea* and *B. nigra*. *J. exp. Bot.*, 17: 332–343.

WILLIAMS, A. H. 1966. Dihydrochalcones. *In*: T. Swain (ed.). *Comparative Phytochemistry*, pp. 297–307. Academic Press, London and New York.

WILLIAMS, C. A., Jr. 1964. Immunochemical analysis of serum proteins of the primates: a study in molecular evolution. *In*: J. Buettner-Janusch (ed.). *Evolutionary and Genetic Biology of the Primates*, pp. 25–74. Vol. II. Academic Press, London and New York.

WILSON, M. W. and PRINGLE, B. H. 1954. Experimental studies of the agar-plate precipitin test of Ouchterlony. *J. Immun.*, 73: 232–243.

POSTSCRIPT

The publication by Klein and Cronquist (1967) became available to me after the completion of my review. Since it is of such general interest to botanists and reviews an area of chemotaxonomy, I wanted to have it included as a portion of my own review of the subject.

This publication is an evaluation of the phylogeny and evolution of the plants classified as bacteria, algae and fungi. The evolution of chemical characteristics (porphyrins, carotenoids, carbohydrates, lipids and sterols), micromorphological characteristics (cell wall, genetical apparatus, flagella apparatus, photosynthetic apparatus, mitochondria and Golgi apparatus), and functional characteristics (metabolism of hydrogen, nitrogen, sulfur and carbon, including respiration, photosynthetic energy transformation, and photosynthetic carbon fixation) were discussed. Phylogenetic interrelations among thallophytes were evaluated on the bases of each of these characteristics.

11 | Botanical Problems of Numerical Taxonomy

J. CULLEN
University of Liverpool Botanic Gardens, Ness, Neston, Cheshire, England

INTRODUCTION

To attempt, at the present time, to give a balanced and up-to-date review of Numerical Taxonomy is not really feasible, at least from the point of view of an Angiosperm taxonomist—the subject is still in a developing state and has not, as yet, provided us with sufficient results for broad comparisons to be made. Therefore, I have chosen to discuss three topics which emphasize conceptual rather than practical problems, and which illustrate, to some extent, the differences in approach between Numerical and Orthodox taxonomic methodologies.

CHARACTER CONCEPTS

Even the most cursory survey of the literatures of Numerical and Orthodox taxonomy will reveal a most striking difference between them—a difference relating to the amount of discussion and controversy about the theoretical bases of classification. The rise of Numerical taxonomy in recent years has been marked by a (still continuing) controversy about the theory and methodology of the subject. The progress of Orthodox taxonomy, on the other hand, has been brought about largely by *ad hoc*, empirical methods, without much controversy about the theoretical bases underlying the practical techniques involved. Such discussions of the theoretical aspects as have been published (e.g. Bentham, 1874; Hooker, 1855; Wettstein, 1935; Van Steenis, 1957; Gilmour, 1961; Davis and Heywood, 1963—giving further references) have not, on the whole, aroused much controversy, and the practical business of classification has gone on, largely unaffected by the different theoretical interpretations that have been offered.

Because of this, the concepts employed by Numerical taxonomists are explicit and easily available: those employed by Orthodox taxonomists are usually implicit, unstated, and have to be deduced. This involves not only a study of the theoretical works mentioned above, but the study of the results of taxonomic activity—published monographs, revisions and Floras.

This contrast is very marked as far as the concept of characters is concerned. Numerical taxonomists have devoted considerable thought to what they mean by "characters" and to the use of the concept in practice. Orthodox taxonomists, however, while using the term frequently, seldom provide a definition or explanation of what they mean by it (but *cf.*, for a notable exception, Hedberg, 1957). In fact, it seems that most orthodox taxonomists regard the meaning of the term as self-evident—a proposition reinforced by such definitions as have been published, which all approximate to the form given by Davis and Heywood (1963: 113): "any attribute (or descriptive phrase) referring to form, structure or behaviour which the taxonomist separates from the whole organism for a particular purpose such as comparison or interpretation". Reduced to its simplest form, this becomes: "any attribute of an organism which can, in some way, for some purpose, be regarded as separate from the rest". In a taxonomic context this can be reduced still further, to: "any attribute of an organism which shows variation within the group being studied".

This definition is so wide and all-embracing as to be largely meaningless. If we consider an Oak tree, for example, and look at it in terms of this definition, then we find that there must be literally millions of characters, from the shape of the whole tree to the angle formed by a hair with the surface of one of its leaves. The definition does not, however, tell us how or why we are to select some of these potential millions for use.

To arrive at a better *working* definition, it is necessary to consider how an Orthodox taxonomist actually operates. To do this, let us assume that he is starting to make a revision of a group, with fully adequate material, untrammelled by any preconceptions derived from previously published classifications of the group concerned—two conditions hardly ever satisfied in practice. Our taxonomist will begin by *looking* over the material, noting (subconsciously) striking features of the variation, including similarities and differences. Then he will begin to sort the specimens into piles, each pile containing specimens that, to his eye, show morphological continuity and homogeneity. This process will continue (and it often takes a long time), and specimens will be transferred from one pile to another, new piles will be formed, old piles merged, until he is satisfied that he can make no further improvements. While doing this, possible hierarchical arrangements of the taxa represented by his piles, will have suggested themselves.

What should be pointed out here is that all this is done without any concept of character being used at all; what the taxonomist is doing is comparing mental images of whole specimens as wholes; he is comparing a series of what have been called *Gestalten**. Discussion of these images would take us into very alien fields, notably those of the psychology of perception; all that it is necessary to point out here is that they are flexible, allowing for variation, and able to cope with somewhat inadequate or incomplete material.

The idea of "characters" only arises when it is necessary to put the already formed mental classification of images into a readily communicable form, i.e. words. The necessity for communication by means of words is a limitation imposed by our present technological situation—it is possible that eventually the need for words will be eliminated. At present, however, words are necessary, and to put his mental classification into verbal form, the taxonomist chooses, from the many available potential characters (*cf.* definition above), those which will most effectively and accurately communicate his mental classification to the reader. To do this, he will choose features which he thinks are stable, and which can least ambiguously be put into verbal form. In some cases this is easy, in others difficult—in difficult cases the taxonomist may have to search for a long time to find features which are efficient and accurate, and this may lead him to a search for microcharacters (*cf.* Gilg's classification of the Gentianaceae, 1895), to use as markers (or symbols) for overall differences which cannot be expressed in words. In such cases a first-class illustration may well be much more useful than any other form of communication, as it avoids the need for verbal expression. What I am concerned to stress is that the use of, say, stipule venation, arises from the need to use a marker (symbol) for some larger difference; the taxonomist does not say to himself: "stipule venation looks an interesting character—let's see how the group divides up on that". Nor does the taxonomist look at features like stipule venation as a matter of course in every group he studies; such characters are only sought when a larger difference needs to be marked by some convenient feature.

Of course, the description of taxonomic activity given above deals with an ideal case. Mostly, taxonomists have to deal with something less than adequate material, and, naturally, they take note of earlier classifications of the group under consideration. This often leads to the "inheritance", as it were, of previously used characters, which, by long continued use acquire a spurious reality as *the* characters which are used in particular groups. However, this does not invalidate the theoretical picture I have tried to draw.

Two important points emerge from this discussion. First, the characters used in a classification, as published, arise out of the need to communicate a pre-existing mental classification; and they are given context by that mental classification. The characters used in keys and descriptions are only a sample

* The use of this term does not imply acceptance of the theories put forward by the school of "idealistic" morphology.

of those potentially available; they are not *the* classification itself, but an abstract of it, expressed in the most helpful way. Second, for communication to be most effective, the characters chosen will involve, as far as possible, large, easily observable and expressed features.

On the basis of this discussion, the following definition of the term character, in an Orthodox taxonomic context, may be presented: a character is any attribute of an organism chosen to communicate effectively a pre-formed mental classification of a group which includes the organism.

When we turn to Numerical taxonomy we find that the concept of characters is much more explicit. Sneath and Sokal (1963) have given a widely accepted definition of what they call "unit characters" as follows [a unit character is] "a taxonomic character of two or more states, which, within the study at hand, cannot be further subdivided logically". The use of "character" in this definition refers, I think, to some such definition as I quoted above (p. 176).

Most methods of Numerical taxonomy seem to call for the employment of a large number of such unit characters (*cf.* Sneath and Sokal, 1963), and so the emphasis is definitely on the "logical subdivision" mentioned in Sneath and Sokal's definition. This large number of characters seems to be necessary in order to provide the computer with something approximating to the *Gestalt* mentioned above. The important point to note is that these unit characters must be chosen first, and then the classification is computed from them. This means that the characters must be chosen with no apparent context (but see below) to give them relevance and point. It seems to me that there is a danger in this: Numerical taxonomists may find themselves using features of unknown overall variability and environmental plasticity, which are difficult to observe and to score. If this is combined with the use of "average individuals" or some other form of selective sampling to represent the basic taxa (OTU's) of the group under consideration, there is a real danger that random and unassessed variations will play a large part in the classification produced. The characters used by orthodox taxonomists have often had over 200 years of observation under all sorts of environmental conditions, and their variation-potential is reasonably well-known: this is not the case with the micro- and chemical characters which Numerical taxonomists will tend to use in their studies. It seems to me that only in the context of a previously existing, non-numerical classification can the variability and plasticity of these new characters be assessed.

I mentioned above that in a numerical study the unit characters must be chosen outside the context of a pre-formed classification. It is possible, however, that the search for the characters may produce in the taxonomist an implicit, unformulated classification, precisely in the manner suggested above in the case of the Orthodox taxonomist (p. 176). It is also quite possible that, if this subconscious classification is formed relatively early on during the search for characters, all characters chosen later will, in fact, be chosen because they reinforce that subconscious classification. If this is the case, then clearly, the resulting Numerical classification will correspond fairly closely to the subconscious one: this really means that the actual computing process is gratuitous.

Such unconscious distortion (or bias) in the data used for computation is obviously objectionable to the Numerical taxonomist, as it renders impossible the achievement of one of his frequently stated aims: the production of a *more objective* classification. It is difficult to see, however, how such distortion can be avoided, and its effect may well be increased in projects in which the search for characters and the actual computing are done by different groups of people, neither of which fully understands the techniques of the other.

On the basis of this discussion, then, we may view with some scepticism all Numerical classifications that correspond more or less closely with previously existing orthodox classifications. Such scepticism must remain until we have machines which can do the character selection and scoring for us, because it seems unlikely that any method of correcting for the distortion can be developed while human beings are directly involved at any stage in the classificatory procedure. This is a point on which some discussion with workers trained in the psychology of perception might be helpful.

Thus we have a clear contrast between Orthodox and Numerical techniques as regards their use of the concept of characters. In Orthodox techniques the characters used arise out of the classification, and are used in their largest possible form. With Numerical techniques the characters are chosen *a priori*, used in their smallest possible (logically subdivided) form, and the classification is constructed from them.

WEIGHTING

This is a subject which has caused a considerable amount of controversy among taxonomists, both orthodox and numerical. The controversy about weighting *versus* not weighting in Numerical Taxonomy is still going on (*cf.* Kendrick, 1965; Long, 1966), and no generally agreed decision appears to have been reached. Of course, it has frequently been pointed out that the Numerical Taxonomist's choice of particular characters to use is in itself a form of weighting; but this may be extended, because, as pointed out above, the characters chosen lack a context, and therefore their selection and subdivision may well involve a great deal of unconscious weighting. However, until we have many more examples of Numerical classification to judge from, we will not be in a position to assess the effects of such weighting.

Orthodox taxonomists have frequently been accused of weighting individual characters in their classifications, but, it seems to me that, apart from some obvious, and usually explicit cases (e.g. Linnaeus' Sexual System, which was avowedly artificial, and cases involving much stress on particular characters because of their assumed phylogenetic importance), most apparent cases of weighting in Angiosperm classification are due to an entirely different cause, relating to the evolutionary state of the group in question—I refer, of course, to the existence of reticulate variation in the groups concerned.

An example of this sort of variation is provided by a group with which I have recently been concerned (*cf.* Davis, 1967: 99–244)—the subfamily *Silenoideae* of the family Caryophyllaceae. Here the genera are known to be difficult to define and distinguish, and plants of very similar overall appearance are placed in different genera because of the occurrence or non-occurrence of certain individual features.

The most striking example of this is provided by the species *Gypsophila confertifolia* Huber-Morath, which is endemic to a small area of southwest Turkey. Plants of this species are not very like any other species of *Gypsophila* in overall appearance, but they do resemble closely some of the species of the genus *Velezia*. However, the technical distinction between these genera is to be found in the seeds—those of *Gypsophila* having a lateral hilum, those of *Velezia* a facial hilum. On this basis there is no doubt that *G. confertifolia* should be placed in *Gypsophila*. Here we have a case that, on the surface, appears to involve *a priori* weighting of the seed character as against overall similarity. But if we look further into the subfamily, we find it riddled with similar cases—*Petrorhagia* and *Dianthus*, *Saponaria* and *Gypsophila*, *Silene* and *Melandrium* are separated from each other by equally (apparently) weighted characters. The reason for this is the existence of reticulate variation. Certain "areas" of the variation pattern are, as it were, crystallized out forming the "cores" of the major genera. But between these cores is an intergrading mass of species in which the correlations distinguishing the major genera are broken down, and their diagnostic characters are recombined in a kaleidoscopic manner. In order to deal taxonomically with a situation of this type, in which interrelationships flow in all directions, the taxonomist must adopt somewhat arbitrary methods. A classification must not only describe the variation patterns in a group—it must also be workable. Therefore, the taxonomist has to define the "cores" of the major genera as best he can, and then assign the intervening species to one side or other of artificial lines drawn on the basis of the distinctions between the major genera. This produces a classification which, although it is not very efficient as a picture of the variation patterns is, at least, useful in terms of recognition, identification and information storage. The apparent weighting is introduced to allow of the production of a classification that is useful in these terms. From the evolutionary point of view, anyway, the existence of such a reticulum is probably of greater importance than the details of its subdivision.

Similar reticulate variation patterns occur throughout the Angiospermae, at all levels (e.g. the species *Anthyllis vulneraria* L., the genus *Erysimum*, the family Cruciferae). It would be interesting to see the results of a numerical study on some of these groups.

ASSESSMENT OF A NEW CLASSIFICATION

The second topic I want to discuss is, how are we to assess a new classification produced by Numerical techniques? Of course, the relatively simple tests which we apply to any Angiosperm classification can be used, such as: will the new classification accommodate new material without strain?; will it accommodate new information about the old material?; can we identify our specimens by means of it?; can we use it to predict the overall variation pattern in some feature newly investigated? (this last a test more frequently talked about than applied). I do not need to stress here how successful orthodox techniques have been in producing classifications which pass these tests (*cf.* Blackwelder, 1964).

However, one of the most frequently used tests of a Numerical classification is to see how it compares with the previous Orthodox classification(s) of the group in question. I have mentioned above some objections to such comparisons (see p. 179). There are two main possibilities which may result from this comparison: the new classification may broadly confirm the old; or it may diverge sharply from it. If the new classification agrees with the old, then, as mentioned above, we cannot rule out the possibility that the Numerical classification has been produced from distorted data. If, on the other hand, the new classification differs sharply from the old, then we are in a difficult position. All we can do is apply the criteria mentioned above, and, if the new classification turns out to be superior to the old, then, I am sure, all taxonomists will be glad to accept it as a better classification. However, we do not yet know how often this is likely to happen, and we must wait for more Numerical classifications to be produced before we can go very far in this matter.

PHYLOGENY

The problem of phylogeny, which, for my purposes here, refers to the evolutionary relationships of groups above the species level, is a thorny one. In spite of about one hundred years of study and effort, there is little agreement about the phylogeny of the Angiosperms. The hey-day of Phylogenetic System-building has passed, and most taxonomists, in Britain at least, appear to have relegated phylogeny to the limbo of lost causes. This is a pity; but it is not my purpose here to enquire into why phylogeny has declined, but to consider whether the advent of Numerical classification is likely to advance it. In my view this is not the case; if the limiting factor in the failure of phylogeny was the inadequacy or poor quality of the classification on which it had to work, the matter might have been different. But the main cause preventing the advance of phylogenetic speculations seems to me to reside in the lack of agreement about the answers to such questions as: how are we to decide what is homologous with what?; what criteria can we use to decide whether a group is mono- or polyphyletic, and precisely what is implied by these terms?; what criteria can we use to decide what is primitive and what is advanced? These are questions to which a classification, however good as a classification, cannot provide an answer. The study of phylogeny can only be advanced by finding answers to these questions, and then applying those answers in an interpretation of a classification. Of course, the better the classification to be interpreted in this way, the better the phylogeny will be.

CONCLUSIONS

It is very difficult to provide a summary relating to the botanical problems of Numerical Taxonomy. As I mentioned at the beginning, Numerical Taxonomy is in a rapid state of development, and today's dogma may well be tomorrow's heresy; new classifications produced by numerical means are still too few for us to be able to decide precisely what the problems are. However, we orthodox taxonomists should welcome the advance of Numerical techniques, which may well provide means of checking and improving our classifications. We should realize that Numerical classifications are not likely to supplant orthodox ones—they will either confirm them, or, if very different, exist side by side with them. We must also welcome a number of "fringe benefits" that will derive from numerical projects. The search for large numbers of unit characters will lead to the discovery of many new facts about plants which, whether taxonomically useful or not, will be important biologically. Also, Numerical techniques offer a method of dealing with material that is fragmentary—here one thinks particularly of fossils, where one needs to use the maximum amount of information to link individual fossils together, e.g., the linking of fossil stems and leaves, inflorescences and stems, etc., and even the linking of fossils with present day plants. If this can be done more effectively, it may well have an enormous effect on phylogenetic studies. We should also welcome the use of Numerical techniques in groups in which a "visual" classification is proving unsatisfactory, e.g. bacteria and some other micro-organisms.

Finally, a study of the history of taxonomy shows us the truth behind Heiser's comment (in Sharp *et al.*, 1962: 122; quoted by Constance, 1964): "Biochemical Taxonomy will not cure all. Neither will Numerical Taxonomy". One can replace "Biochemical Taxonomy" (or "Numerical Taxonomy") with "Cytology" or "Biosystematics" or "The new Morphology", etc., without for one moment altering the basic truth. All these approaches have been put forward in the past with grandiose and over-emphatic claims as to their ability to "cure" taxonomic ills. But taxonomy has absorbed their contributions without any great overall strain, and it seems likely that the same will happen to Numerical Taxonomy, which will then take its place as one of the many available tools which can be used to study the fascinating diversity of living organisms.

ACKNOWLEDGEMENTS

I would like to record here my thanks to Professor V. H. Heywood (Liverpool), Dr P. H. Davis (University of Edinburgh) and Mr B. L. Burtt (Royal Botanic Garden, Edinburgh), with whom I have had many stimulating discussions about the topics discussed above. They should not, however, be regarded as being necessarily in agreement with the views put forward here.

REFERENCES

BENTHAM, G. 1874. On the recent progress and present state of knowledge of systematic botany. *Rep. Br. Ass. Adv. Sci.* for 1874: 27–54.

BLACKWELDER, R. E. 1964. Phyletic and phenetic *versus* omnispective classification. *In*: V. H. Heywood and J. McNeill, (eds). *Phenetic and Phylogenetic Classification*, pp. 17–28. Systematics Association, London.

CONSTANCE, L. 1964. Systematic botany—an unending synthesis. *Taxon*, **13**: 257–273.

DAVIS, P. H. 1967. *Flora of Turkey*, 2. Edinburgh University Press.

DAVIS, P. H. and HEYWOOD, V. H. 1963. *Principles of Angiosperm Taxonomy*. Oliver and Boyd, Edinburgh and London.

GILG, E. 1895. Gentianaceae. *In*: A. Engler and K. Prantl, (eds). *Die Natürlichen Pflanzenfamilien*, 4(2): 50–108.

GILMOUR, J. S. L. 1961. Taxonomy. *In*: A. McLeod and L. S. Cobley. (eds). *Contemporary Botanical Thought*, pp. 27–45. Oliver and Boyd, Edinburgh and London.

HEDBERG, O. 1957. Afroalpine Vascular Plants. A taxonomic revision. *Symb. Bot. Upsal.*, 15(1).

HOOKER, J. D. 1855. Introductory essay to J. D. Hooker and T. Thomson, *Flora Indica*, pp. 1–44. London.

KENDRICK, W. B. 1965. Complexity and dependance in computer taxonomy. *Taxon*, **14**: 141–154.

LONG, C. A. 1966. Dependence in Taxonomy. *Taxon*, **15**: 49–51.

SHARP, A. J., THOMSON, J. W., FULFORD, M., ANDERSON, L. E. and HEISER, C. B., Jr. 1962. Modern species concept: a symposium. *The Bryologist*, **66**: 93–124.

SNEATH, P. H. A. and SOKAL, R. R. 1963. *Principles of Numerical Taxonomy*. Freeman and Co., San Francisco and London.

VAN STEENIS, C. G. G. J. 1957. Specific and infraspecific delimitation. *Flora Malesiana* ser. 1, 5(3): clxvii–ccxxix. Djakarta.

WETTSTEIN, F. 1935. *Handbuch der Systematischen Botanik*. Franz Denticke, Leipzig and Wien.

12 | On Property Structure, Numerical Taxonomy and Data Handling

M. B. DALE

*Department of Botany, The University, Hull, England**

INTRODUCTION

The object of this paper is to present the basic procedures which underlie Numerical Taxonomic (N.T.) methods, and to outline a more general structure which allows greater freedom to the taxonomist. A detailed survey of the various methods which have been proposed is not intended, nor is any method claimed to have universal applicability, since, at the present stage of development, such a method is certainly not available. For reasons of space, no consideration will be given to other methods of organizing data such as "ordination" (Goodall, 1954), "curve seeking" methods (Sneath, 1966) or "predictive" classification (Williams and Dale, 1965: 39).

Numerical taxonomy will be here restricted to heuristic methods of grouping entities or operational taxonomic units (O.T.U.) into classes, by means of properties or characters whose values describe the entities. This grouping is carried out by maximizing or minimizing some function of the properties. As with any heuristic process, there is no formal proof of the validity of the methods, a statement equally true of traditional taxonomic practice. The hope is that useful results will be obtained and two advantages result from using numerical methods for such a purpose.

(1) The decisions made and the rules used are overt and should be explicitly stated.

(2) The rules and decisions are uniformly and consistently applied throughout the analysis.

One covert advantage of numerical methods is that the collection of data is itself made more orderly, consistent observations and complete records being desirable. This reduction of errors is equally apparent in traditional methods, when these are performed well, but the use of electronic computation re-emphasizes its importance. It is worth noting here that numerical

* Now at Division of Plant Industry, C.S.I.R.O., Canberra, Australia.

methods can be, and have been, carried out without a computer, though this is laborious. Such a practice does have the advantage, however, that the procedure is not defined by the action of a particular programme for a particular computer, but by the original definition. Not all programmes are well documented and some do not follow their own documentation.

THE PROCEDURE OF NUMERICAL TAXONOMY

Any Numerical Taxonomic method has five basic stages, with two further optional stages, when used in isolation. When incorporated in the context of an information storage and retrieval project, further complications can occur but these need not concern us here.

The procedures are outlined here and then discussed in greater detail.

(1) Selection of the O.T.U. or entities to be classified.

(2) Selection of properties or characteristics whose values provide a description of each O.T.U.

(a) Optionally the property values may be subjected to transformations in order to establish some desirable feature. For example, normalizing so that all properties have zero mean and unit variance, or multidimensional scaling (Shepard, 1962; Kruskal, 1964*a*, *b*) to introduce metric properties.

(3) The definition of some measure of likeness between the O.T.U. and/or previously erected groups of O.T.U. The definition is usually between pairs although this need not be so.

(4) The definition of rules which, using the measure of likeness, are applied to organize the O.T.U. into groups, that is to sort them.

(4a) Optionally, if the rules used initially for sorting are inefficient, a second process may be desirable to optimize the structure of the groups resulting from the first sort.

(5) The derivation of rules of allocation permitting new O.T.U. to be allocated to an appropriate group. This may be complicated as in some of the two-parameter methods which have been proposed (Macnaughton-Smith, 1965).

FURTHER DISCUSSION OF THE STAGES

The first stage is of obvious importance, but basically it must remain a function of the user. Whether a machine could be programmed to "want" data for analysis is a debatable point, philosophically and semantically, and no effort has yet been made to obtain machine performance of this kind. The succeeding stages may be restrictive, notably in the definition of the single individual but these restrictions will become apparent in the later discussion.

Of the other stages, most study has been given to stage (5), under the name of pattern recognition. The literature here is vast in extent and little of it is biological with the notable exception of Fishers discriminant function (see Kendall, 1957). There has been much more work than this however (see Sebestyen, 1962) and much of this may be relevant to taxonomists.

Returning to Numerical Taxonomic methods proper, most study has been devoted to stages (3) and (4). The former stage has generated a variety of coefficients (see Goodman and Kruskal, 1954, 1959; Dagnelie, 1960; Sokal and Sneath, 1964, for most of the common coefficients). The latter stage has perhaps caused less controversy but again there is relevant work outside taxonomy in the general field of "maximisation under constraint" (see Ruben, 1967: 107 footnote), and in the social sciences (Ball, 1965).

By far the least studied is stage (2) and it is on this stage that this paper will concentrate. The optional stages can be temporarily neglected since they are of little importance for the present.

PROPERTIES, STRUCTURE AND REPRESENTATION

Almost all Numerical Taxonomic procedures have made use of a list of properties, with their associated values, to describe each O.T.U. By fixing the sequence of properties these can be implied by order and the result represented by a vector (Fig. 1). By grouping the vectors for each O.T.U. the familiar table or matrix of data is obtained. Vector representation (using rectangular Cartesian coordinates) leads easily to the concept of "distance" between O.T.U. being used as a measure of similarity. It also leads to sorting rules such as nearest neighbour and furthest neighbour.

Such a representation may be complicated by weighting of properties, which leads to a non-Euclidean space. Other systems of representation based on the Shannon-Wiener Information Theory (Hyvarinen, 1962; Macnaughton-Smith, 1965) can be related to quasimetric spatial systems (Williams and Dale, 1965).

The vector concept is very simple, but more critical inspection suggests that it does not match the actual data, neglecting some important relationships between properties.

Consider the following property, "Hairs on leaf branched" with values *true* or *false*. If the value is *true*, no particular problem arises, but the answer *false* may mean one of three things:

(1) There are no leaves.

(2) There are leaves but these are not hairy.

(3) There are hairy leaves but the hairs are unbranched. The property of "branchedness" is dependent on the existence of leaves and of hairs on the leaves, and is irrelevant if hairs or leaves are non-existent. Using three properties, leaves present, hairs present, branching present, all with values

true or *false* does not solve the problem since this introduces logically necessary correlations between these properties which may bias the analysis.

A similar problem occurs with properties having quantitative values. The length of a petal is not measurable if the plant is apetalous.

The first method proposed for solving this problem was the partitioning of correlation (Williams and Dale, 1962). This, while useful in ecology where only the quantitative problem form is important, is not well suited to the complexities of taxonomic data. It is of more use in a second problem, that of missing data. Here some separation must be maintained between properties

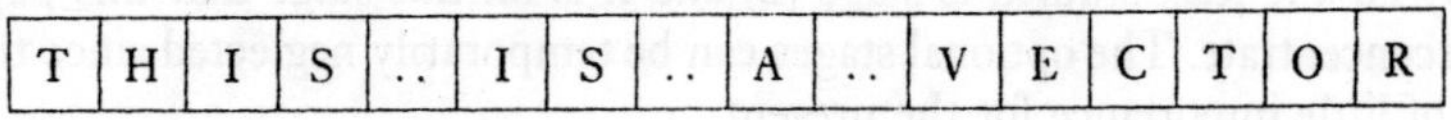

(a) A vector

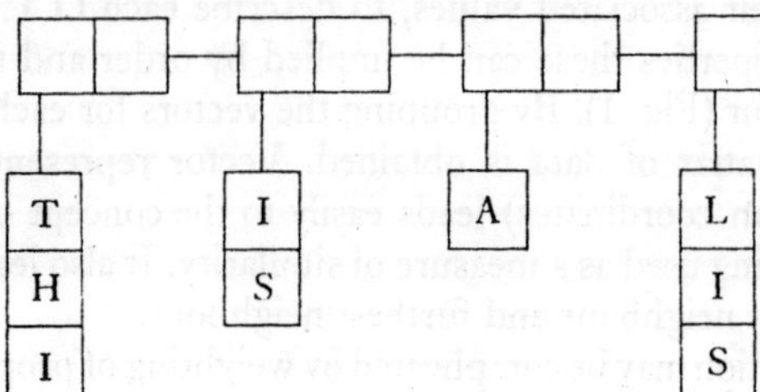

(b) A simple list structure

FIG. 1. Representation of data.

examined and properties not examined and therefore unknown. Missing data may be unavoidable if, for example, fossil material is to be included in an analysis, which would be essential for almost all phylogenetic studies (see Lange *et al.*, 1965: 1203).

Still further complications may be introduced if the O.T.U. may vary in space or time. Examples of this are developmental stages, which could and perhaps should be included in a single O.T.U. for many purposes, and geographical variation. This problem is rather intractable unless some common scale of change can be established (Rayner, 1966; Lance and Williams, 1967, *b*). Thus pollen diagrams have proved very difficult to analyse.

The complexity introduced by dependency, by missing values and by spatio-temporal variation suggests that the simple vector may not be the most suitable representation. The problem is to find others, and to determine methods of manipulating them. Two such alternative representations are available, the *record* and the *list*.

Records have a fixed form and provide space for all possible alternatives, indicated by a positional notation. They are much used in business data processing but in taxonomic studies, where complexity *may* be great but often is not, they are very wasteful. They will not be considered further here.

LISTS, ATOMS AND LIST STRUCTURES

While a little more complex initially, lists and developments from them provide a very flexible representation, and conceptually seem preferable to both records and vectors. They were introduced during work on artificial intelligence (McCarthy, 1960; Newell and Tonge, 1960) and have since found a variety of other uses (Sussenguth, 1963; Prywes and Gray, 1966). A good introduction can be found in Woodward and Jenkins (1961).

A list is made up of *elements* each of which consists of two parts at least. The first indicates the successor, if any, to the present element (sometimes the predecessor is also noted). The second may be either an *atom*, a single indivisible unit of information, or another list. A list, amongst whose elements is another list, is termed a *list structure* (Fig. 1). The definition of a list structure involves itself, a situation called recursive.

This may seem a little far from the complexity of taxonomic data. However, a list structure holds two kinds of information. One is derived from the values given to the atoms, which may be numerical, logical or alphabetical (alphabetical values would include perhaps colour—red, green, blue etc. and texture—rough, smooth etc.). The other source of information is the structure of the list itself, which can be used to represent dependency, missing values and some forms of spatio-temporal variation. An example is shown in Fig. 2 where some of the properties of individuals in the genus *Salix* have been represented. Dependency is essentially the replacement of an atom by a list, missing values occur as null lists, which could, conceptually, have entries.

It is interesting to note here that studies of computer techniques for handling graphic data, both diagrams and words, and for handling spoken language, almost always involve list structures or their derivatives. It might be possible, therefore, to use drawings or formal descriptions directly without the error producing stages of selecting and coding properties to describe such information. To the author's knowledge, no such system is available, but most of the techniques necessary have been studied.

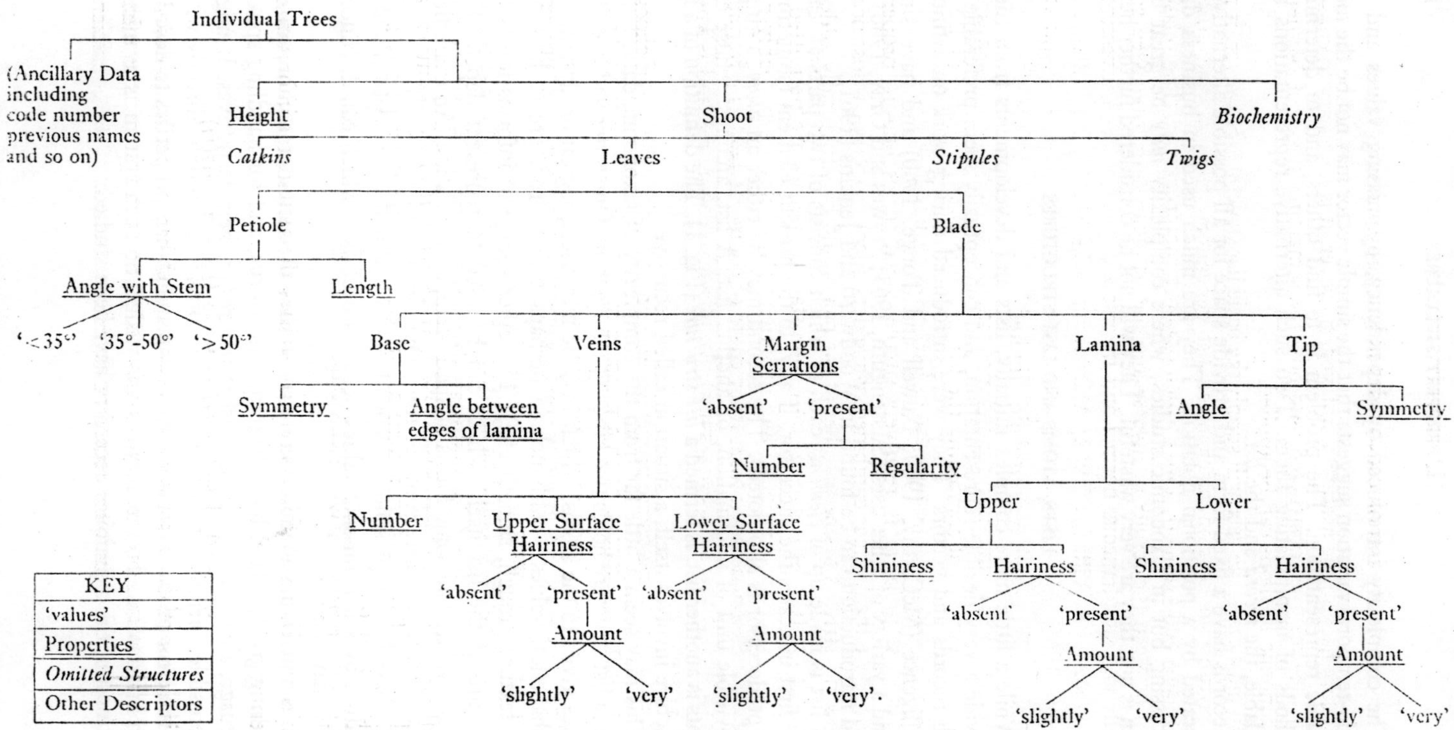

FIG. 2. Portion of list description for *Salix* data.

PROBLEMS OF USING LISTS IN NUMERICAL TAXONOMY

Before lists can be usefully employed, however, it is necessary to consider how they may be constructed, and how the remaining stages of Numerical Taxonomic methods can be applied.

In most cases, the user will supply the basic list structure. However, it may be possible to develop mechanized list generators using inverse analysis (Williams and Lambert, 1961) which organize the properties. The relationship between the results of inverse and normal analysis has received some study and simultaneous normal and inverse analysis has been investigated (Lambert and Williams, 1962; Macnaughton-Smith, 1965; Williams and Tharu, 1966). No taxonomic applications have been published and their usefulness in taxonomy is probably limited.

To return to Numerical Taxonomic methods, two problems are apparent. First some means of combining the list structure descriptions of O.T.U. is necessary. Second some measure of likeness is required.

The first problem is relatively simple, since corresponding properties would have their values combined, while dependent properties whose occurrence is restricted to one member of the pair would be structurally added, a simple operation with lists. If only multistate properties are permitted the problem is further simplified and quantitative variables are probably most easily handled in a coded form (they could of course be held as a frequency distribution if this seems necessary). An objective coding method is known but unpublished.

Restriction to multistate properties also simplifies the problem of choosing the measure of likeness. Such a measure must be defined recursively since the list is recursive, and would take the form:

The likeness of this property combined with the likeness of all properties dependent on it, if any.

The combination should be additive for simplicity and while many coefficients could be defined, for multistate data, a useful statistic is an information coefficient (Macnaughton-Smith, 1965; Williams *et al.*, 1966; Lambert and Williams, 1966). This coefficient has several convenient properties and is defined as follows.

Consider a property with two values:

If n O.T.U. have the first value and m have the second, the information is given by

$$(n + m) \cdot \log(n + m) - n \cdot \log(n) - m \cdot \log(m)$$

This is essentially a measure of disorder and can be easily extended to properties with more than two states including "not known" for missing values (there is a problem regarding degrees of freedom if the coefficient is used probabilistically). Extensions to permit quantitative data have been proposed but none has yet been accepted for general use, and another coefficient would be necessary if coding of quantitative data was considered undesirable, though since any number of states is permissible, this seems an unlikely situation.

Before leaving lists, it is necessary to consider whether such complex structures can be manipulated easily. Fortunately list handling facilities are becoming available on many computers and such facilities have been incorporated into general language such as ALGOL and FORTRAN.

The final problem with properties is one of scale of measure. Properties may have values which are multistate (nominal scale), quantitative (interval or ratio scales) ranked (ordinal scale) or some other esoteric scale such as the Domin scale much used in ecology in Europe (an ordered interval scale). A good measure of likeness should permit as many scales as possible to be incorporated but only one such coefficient, to be discussed in the next section, is known. Scaling (Shepard, 1962) can avoid some troubles and simple scaling methods are becoming available (Gower, 1966).

WEIGHTING AND LIKENESS MEASURES

Only general consideration will be given here since a complete survey is not proposed. Weighting is a delicate subject, since practising taxonomists feel it to be undesirable. An objective weighting method has been proposed (Williams *et al.*, 1963) but is little used. The major difficulty is the inherent circularity of weighting, the results supporting the weights and the weights producing the results. The best advice is still—when in doubt don't!

Comparative studies of coefficients of likeness have started to appear (Williams *et al.*, 1966; Lambert and Williams, 1966; Lange *et al.*, 1965; Beers *et al.*, 1962) and the results suggest that major groups are stable in most instances. There is a growing interest in information coefficients (see above, Rescigno and Maccaccaro, 1960; Kullback, 1959). Some work has also been published on a probabilistic coefficient (Goodall, 1964), which allows properties in any scale of measure but in the absence of published reports it is difficult to assess the utility of the results. The only available, unpublished, evidence seen suggests some undesirable properties but this can only be a tentative assessment.

SORTING

Almost all Numerical Taxonomy methods have been dominated by hierarchical sorting routes, although this is not true of work outside biology, where non-hierarchical methods have been used (Ball, 1965; Bonner, 1964;

ancey, 1966; Ruben, 1967). Whether this reflects the usefulness of hierarchies or the conservatism of taxonomists is not to be decided here.

Only three types of sorting need be considered here:

(1) Similarity polythetic;
(2) Dissimilarity monothetic;
(3) Dissimilarity polythetic.

Polythetic similarity is the most commonly used route in taxonomy, and was apparently first introduced into biology by Tuomikoski (1942). Earlier psychological work was mostly non-hierarchical. The method is relatively simple and therefore fast in operation. First the pair of O.T.U. most alike are found, these then being fused and the similarities of this group with all other O.T.U. are calculated. The next most similar pair is then found and the process repeated. Using information coefficient similarities the process is unambiguous (Lance and Williams, 1967*a*); but using the commoner distance coefficients, the various sorting methods—nearest neighbour, furthest neighbour, median, centroid and group average (Lance and Williams, 1967*a*, 1966*a*, *b*) have been shown to be variants of a general method of recalculating the similarities of the last fused pair and the remaining O.T.U., or groups. By altering the parameters of a single equation, all the methods could be obtained. The hierarchies produced by the methods are not identical, and this indeterminacy does not seem desirable. A further problem is that if weighting is employed, the space is not Euclidean and the distance function should be adjusted to allow for this. For similarity methods only the coordinate space of the entire population is available, whereas locally there can be great variation. While not of critical importance this does suggest that dissimilarity will in general be a more powerful approach.

Dissimilarity monothetic analysis is found in one widely used method, association analysis (Williams and Lambert, 1959, 1960; Lance and Williams, 1965; Lange, 1966), and in several proposed methods (Macnaughton-Smith, 1965; Goodall, 1953; Crawford and Wishart, 1967). Monothetic indicates the use of single character divisions as in a dichotomous key and it was originally restricted to qualitative data. An extension to permit quantitative data is known (Dale, unpublished), for association analysis. The fundamental basis of these methods is derived structuring (Williams and Dale, 1965) and they have proved useful ecologically. Taxonomically they have been less useful because of the complexities discussed earlier, which make the underlying basis less relevant.

Dissimilarity polythetic methods have been proposed by Macnaughton-Smith *et al.* (1964), Edwards (1963), Rescigno and Maccaccaro (1960), and Macnaughton-Smith (1965). Of these only the first and last can be employed except for very small studies, and even these are complex computationally. These methods attempt to divide the O.T.U. into the two most different groups. Weights here can be changed after each division so that they apply to the population under study. Unfortunately n objects can be divided into two groups in $2^{n-1} - 1$ ways. For example 39 O.T.U. would present more than 2,000,000,000 possible divisions. Some form of directed search is therefore essential, and this immediately allows the possibility of misdirected search due to local maxima, so that the division made is not the best possible.

Besides choice of route, the other major sorting problem is the provision of a "level of importance" for any division or fusion. All methods except three use the coefficient of likeness which determines the route as a level indicator. The exceptions are association analysis, which uses a crude indicator solely on the grounds of expediency, the probabilistic coefficient which uses the probability level itself, and the information coefficient. The last uses the information content of the group as a level, but uses change of information due to dividing or fusing as the route finder. In other words this is using a measure of disorder to indicate level, and change in amount of disorder to indicate the route, a procedure which is intuitively reasonable.

Finally on sorting, a few methods based on graph theory have been proposed but these all use a fixed level of likeness to determine connections. While repeated applications would allow variation of this level, the complexity is such that these methods are not likely to prove useful.

PRESENTATION OF RESULTS

The organization of data is the primary object of Numerical Taxonomy methods, but the results are often the only contact which the taxonomist has with the procedures. At present much needs to be done to produce output in the most useful and acceptable form. This will probably mean storing information not used in the analysis, such as previous names, authorities, dates and bibliographic references, and presenting this information on request to the user. Much more use of graphic output seems desirable to avoid the cumbersome and bulky form of the present output. It would be possible to display diagrams and dichotomous keys to aid identification. But the necessity of convenient output does presuppose that taxonomists know in what form the results should be presented.

TESTS OF RESULTS

While the application of two or more methods does test correspondence, it does not show the usefulness of any results. Lange *et al.* (1965) have used random data and have shown that this will be classified into groups. Since they do not specify how the random data were prepared it is difficult to assess the importance of this effect which is, in any case, expected. Goodall (1966) has suggested that utility can be measured by predicting random variables and obtaining a frequency distribution of prediction probabilities. This, of course, neglects the obvious feature of many analyses, to predict one variable only and Goodall's procedure does not provide an objective method of measuring utility except in a highly abstract sense.

As an example of a study where some properties were not used in the analysis, and could therefore be used to reflect power, a preliminary study of the anatomy of genera of the Gramineae suggested that groups were anatomically distinguishable. These groups did not coincide with the present tribes however, but did show similarities in embryo structure, chromosome number and some other features. Another worker has attempted a more detailed analysis and since his results are as yet unpublished, no comment can be made on the final conclusions. It is noteworthy, however, that the data took several years to collect, and were still marked by many omissions. The analysis took a few days only, a meagre expenditure of time and effort. It is clear that the major taxonomic problems are not in the organization of data, but in its collection and recording.

CONCLUSION

This brief survey has attempted to outline the fundamental procedures of numerical taxonomy, emphasizing the difficulties remaining. Any taxonomist proposing to use such methods must still be careful in his choice and be wary of what may seem to be unimportant details.

REFERENCES

BALL, G. H. 1965. Data analysis in the social sciences, p. 533. *A. F. I. P. S. Proc. Fall Joint Computer Conf.*

BEERS, R. J., FISHER, J., MEGRAW, S. and LOCKHART, W. R. 1962. A comparison of methods for computer taxonomy. *J. gen. Microbiol.*, 28: 641–652.

BONNER, R. E. 1964. On some clustering techniques. *I. B. M. J. Res. Dev.*, 8: 22.

CRAWFORD, R. M. M. and WISHART, D. 1967. A rapid multivariate method for the detection and classification of groups. *J. Ecol.*, 55: 505–524.

DAGNELIE, P. 1960. Contribution à l'étude des communautés végétales par l'analyse factorielle. *Bull. Serv. Carte Phytogeogr. B.*, 5: 7.

EDWARDS, A. W. F. 1963. *Proc. 5th Internatl. Biometric Conf. Cambridge.*

GOODALL, D. W. 1953. Objective methods for the classification of vegetation. I. The use of positive interspecific correlations. *Aust. J. Bot.*, 1: 39–63.

GOODALL, D. W. 1954. Objective methods in the classification of vegetation. III. An essay in the use of factor analysis. *Aust. J. Bot.*, 2: 304–324.

GOODALL, D. W. 1964. A probabilistic similarity index. *Nature, Lond.*, 203: 1098.

GOODALL, D. W. 1966. Classification, probability and utility. *Nature, Lond.*, 211: 53.

GOODMAN, L. A. and KRUSKAL, W. H. 1954, 1959. Measures of association in cross-classification. I and II. *J. Amer. Statist. Ass.*, 49: 732; 54: 123.

GOWER, J. C. 1966. *Biometrics*, 22: (in press).

HYVARINEN, L. 1962. Classification of qualitative data. *Br. Info. Theory J.*, 83.

JANCEY, R. C. 1966. Multidimensional group analysis. *Aust. J. Bot.*, 14: 127.

KENDALL, M. G. 1957. *A course in multivariate analysis*. Griffin, London.

KRUSKAL, J. B. 1964*a*. Multidimensional scaling by optimizing goodness of fit to a non-metric hypothesis. *Psychometrika*, 29: 1–27.

KRUSKAL, J. B. 1964*b*. Nonmetric multidimensional scaling—A numerical method. *Psychometrika*, 29: 115–129.

KULLBACK, S. 1959. *Information Theory and Statistics*. J. Wiley, New York.

LAMBERT, J. M. and WILLIAMS, W. T. 1962. Multivariate methods in plant ecology. IV. Nodal analysis. *J. Ecol.*, 50: 775–802.

LAMBERT, J. M. and WILLIAMS, W. T. 1966. Multivariate methods in plant ecology. VI. Comparison of information analysis and Association analysis. *J. Ecol.*, 54: 635–664.

LANCE, G. N. and WILLIAMS, W. T. 1965. Computer programs for monothetic classification. *Comp. Jour.*, 8: 246.

LANCE, G. N. and WILLIAMS, W. T. 1966*a*. Computer programs for hierarchical polythetic classification. *Comp. Jour.*, 9: 60–64.

LANCE, G. N. and WILLIAMS, W. T. 1966*b*. A generalized sorting strategy for computer classification. *Nature, Lond.*, 212: 218.

LANCE, G. N. and WILLIAMS, W. T. 1967*a* and 1967*b*. *Comp. Jour.*, 10: (in press).

LANGE, R. H. 1966. Sampling for Association analysis. *Aust. J. Bot.*, 14: 373–378.

LANGE, R. H., STENHOUSE, N. S. and OFFLER, C. E. 1965. Experimental appraisal of certain procedures for the classification of data. *Aust. J. biol. Sci.*, 18: 1189–1205.

MACNAUGHTON-SMITH, P. 1965. Some statistical and other numerical techniques for classifying individuals. *H.M.S.O. Home Office Research Unit Report No. 6.*

MACNAUGHTON-SMITH, P., WILLIAMS, W. T., DALE, M. B. and MOCKETT, L. G. 1964 Dissimilarity analysis. *Nature, Lond.*, 202: 1034.

MCCARTHY, J. 1960. Recursive functions of symbolic expressions. *Comm. A. C. M.*, 3: 184.

NEWELL, A. and TONGE, F. M. 1960. An introduction to information processing language. V. *Comm. A. C. M.*, 3: 205–211.

PRYWES, N. S. and GRAY, H. J. 1966. Multilist system for real-time storage and retrieval. *Proc. 21st Ann. Conf. A. C. M.*

RAYNER, J. H. 1966. Classification of soils by numerical methods. *J. Soil. Sci.*, 17: 79.

RESCIGNO, A. and MACCACCARO, W. B. 1960. The information content of biological classification. *In*: *Symposium on Information Theory*. London.

RUBEN, J. 1967. Optimal classification into groups. *J. Theoret. Biol.*, 15: 103–144.

SEBESTYEN, G. S. 1962. *Decision Making Processes in Pattern Recognition*. Macmillan, New York.

SHEPARD, R. N. 1962. The analyses of proximities. Multidimensional scaling with an unknown distance function. I and II. *Psychometrika*, 27: 125–140 and 219–246.

SNEATH, P. H. A. 1966. A method for curve seeking from scattered points. *Comp. Jour.*, 9: 383–391.

SOKAL, R. R. and SNEATH, P. H. A. 1964. *Numerical Taxonomy*. W. H. Freeman, San Francisco and London.

SUSSENGUTH, E. H. 1963. Use of tree structures for processing files. *Comm. A. C. M.*, 6: 272.

TUOMIKOSKI, R. 1942. Untersuchungen über die Vegetation der Bruchmoore in Ostfinnland 1. *Ann. bot. Soc. Zool.-bot. fenn. Vanamo*, 17(1): 1–203.

WILLIAMS, W. T. and DALE, M. B. 1962. Partition correlation matrices for heterogeneous quantitative data. *Nature, Lond.*, 196: 602.

WILLIAMS, W. T. and DALE, M. B. 1965. Fundamental problems in numerical taxonomy. *Adv. bot. Res.*, 2: 35–68.

WILLIAMS, W. T., DALE, M. B. and MACNAUGHTON-SMITH, P. 1963. An objective method of weighting in similarity analysis. *Nature, Lond.*, 201: 426.

WILLIAMS, W. T. and Lambert, J. M. 1959. Multivariate methods in plant ecology. I. Association analysis. *J. Ecol.*, 47: 83–101.

WILLIAMS, W. T. and LAMBERT, J. M. 1960. Multivariate methods in plant ecology. II. The use of an electronic digital computer. *J. Ecol.*, 48: 689–710.

WILLIAMS, W. T. and LAMBERT, J. M. 1961. Multivariate methods in plant ecology. III. Inverse analysis. *J. Ecol.*, 49: 717–729.

WILLIAMS, W. T., LAMBERT, J. M. and LANCE, G. N. 1966. Multivariate methods in plant ecology. V. Similarity analysis and information analysis. *J. Ecol.*, 54: 427–445.

WILLIAMS, W. T. and THARU, J. 1966. Concentration of entries in binary arrays. *Nature, Lond.*, 210: 549.

WOODWARD, P. M. and JENKINS, D. P. 1961. Atoms and lists. *Comp. Jour.*, 4: 47.

13 | Numerical Taxonomic Studies in the Genus *Sarcostemma* R. Br. (Asclepiadaceae)

MICHAEL P. JOHNSON

Department of Biological Sciences, Kent State University, Kent, Ohio, U.S.A.

and

RICHARD W. HOLM

Department of Biological Sciences, Stanford University, Stanford California, U.S.A.

INTRODUCTION

Although numerical taxonomy has a relatively long history (Sokal and Sneath, 1963), it must be considered essentially a development of this decade. The techniques to be used are still being developed; for example, a number of measures of similarity are being tested. The three measures most commonly used are product-moment correlation coefficients, taxonomic distance (Sokal, 1961), and similarity coefficients (Sneath, 1957, and modified by Rogers and Tanimoto, 1960). Each measure has advantages and disadvantages, and each gives different information about the relationships among the objects being compared. A more thorough discussion of similarity measures is given in Sokal and Sneath (1963). Comparisons among the similarity measures for taxonomic studies have been made on objective or subjective bases by Sokal and Rohlf (1962), Rohlf (1963), Sokal and Sneath (1963), Heiser *et al.* (1965), Moss (1966), Sokal and Michener (1967), Breedlove (1968), Ehrlich and Ehrlich (1967), and Hickman and Johnson (1968). Which of these methods will give the information most appropriate to the goals of a particular taxonomic study will be known only after these methods have received more thorough testing both practically and theoretically. So far the detailed comparisons have been with animal groups and general agreement has not been reached as to the best measure of similarity. Rohlf (1963) preferred taxonomic distance for classification of mosquitos of the genus *Aedes*. Sokal and Michener (1967), working with bees, prefer correlation coefficients. Moss (1966) suggested that some combination of the two measures would produce the most acceptable classification in the mite genus, *Dermanyssus*. The criteria for selection of the best measure have been somewhat different in each case.

When the characters selected for study are considerably affected by size Sokal and Sneath (1963) suggest that correlation coefficients are preferable to taxonomic distance as a similarity measure for classification. Animals of similar size tend to cluster together using taxonomic distance (Sokal and Camin, 1965, and Rohlf and Sokal, 1965). This effect has also been observed in studies of plants by Breedlove (1968), Hickman and Johnson (1968), and Johnson and Raven (manuscript in preparation). The similarity coefficient (Rogers and Tanimoto, 1960; Rogers and Fleming, 1964) is only now being tested extensively. Studies using this measure are given in Katz and Torres (1965) and Estabrook and Rogers (1966). Heiser *et al.* (1965) have used both correlation coefficients and similarity coefficients and consider them to give "a slightly different arrangement". Comparisons of standardized and unstandardized character states have been made by Rohlf (1962) and Sokal and Michener (1967). The latter also includes comparisons of various cluster techniques. It is clear from these studies that each of the numerical methods gives a different classification, and therefore different information about the entities being compared.

A second aspect of numerical taxonomy which has received much attention is the selection of characters, and in particular a testing of the hypothesis of nonspecificity (Sokal and Sneath, 1963). This hypothesis may be stated simply as follows. Because of the pleiotropic effects of genes, and since there are no large independent classes of genes, any sampling of characters will reflect the same degree of similarity as any other sample. Character sampling in taxonomic studies may be largely from one or a few organs (for example floral morphology) or from one life history stage (for example adult insects). If phylogenies are to be determined from these character sets, it is important that nonspecificity exists. As pointed out by Sokal and Sneath (1963) it is likely that the nonspecificity hypothesis holds only in part. For example, each organ has some function or set of functions. Selection for changes in one organ may not affect any other organ because each organ has different functions and thus rates of divergence between organs could be largely independent. Interest in the nonspecificity hypothesis is not restricted to numerical taxonomic studies, but this hypothesis is best tested numerically.

To examine this hypothesis numerically, matrices of correlation or distance between species (Q-matrices) based on different sets of characters may be compared using a correlation coefficient. This has been done for several groups of animals. Sneath (in Sokal and Sneath, 1963) compared Q-matrices of similarity coefficients of galliform birds, based on foot *versus* leg characters, which gave a correlation of $r = 0{\cdot}720$, and extensor *versus* flexor muscles which gave a correlation of $r = 0{\cdot}785$. Rohlf (1963) compared adult and larval character sets in mosquitos which gave a correlation of $r = 0{\cdot}30$ for Q-matrices of correlation coefficients, and a correlation of $r = 0{\cdot}60$ for Q-matrices of distance. Sokal and Michener (1967) in their studies on bees compared male and female character sets which gave correlations of $r = 0{\cdot}71$ and $r = 0{\cdot}34$ for correlation and distance matrices respectively; and for head and nonhead characters correlations of $r = 0{\cdot}61$ and $r = 0{\cdot}34$ for correlation and distance matrices respectively. The correlation coefficients given above are all significant at $P < 0{\cdot}001$. While there

TABLE I. The distribution of correlation coefficients between randomly selected character sets. Data from Rohlf (1965). See text for explanation.

Reference	Similarity measure	Correlation coefficients 0·50–0·59	0·60–0·69	0·70–0·79	0·80–0·89
Rohlf (1962)	*r*	8	16		
Adult–larvae	*d*		3	15	6
Ehrlich and	*r*			14	10
Ehrlich (1967)	*d*		1	12	11
Internal–external					
Sokal and	*r*			3	21
Michener (1967)	*d*	3	17	4	
Head–nonhead					
Sokal and	*r*			4	20
Michener (1967)	*d*		18	6	
Male–female					

are significant correlations between classifications based on different character sets, there are obvious and varying degrees of disparity between classifications based upon them.

Rohlf (1965) has tested several of these correlations using a randomization test. For each of the data matrices of the bee and mosquito studies and from the data matrix of Ehrlich and Ehrlich (1967), 24 random partitions were made of the characters into sets of sizes equal to the sets of the original study. Correlation coefficients were computed between these random sets. Distributions of the number of random partitions into frequency classes of correlation coefficients between them are given in Table I. All randomly partitioned character set pairs gave significant correlations ($P < 0{\cdot}001$). However, in seven of the eight cases the character sets used in the original studies fell considerably outside the distribution of correlations between random sets. Because randomly selected character sets showed better correlations than the nonrandom character sets the hypothesis of nonspecificity was rejected for those seven comparisons (Rohlf, 1965). In other words, in these seven, there is lower correlation between nonrandom sets (head and nonhead) than between randomly chosen sets.

In a more comprehensive study of the correlations between character sets made by Ehrlich and Ehrlich (1968) with butterfly taxa, 196 characters were divided into eight character sets including one set made up of all characters. Q-matrices were computed for each character set and correlations between Q-matrices were computed. The results are given in Table II (with permission of the authors). As would be predicted, none of the classifications correlates perfectly with any other. Again all correlation coefficients between Q-matrices were significant at $P < 0{\cdot}001$. The Q-matrix from each character

TABLE II. Product moment correlation coefficients among Q-matrices from Ehrlich and Ehrlich (1968). Left of diagonal: correlation matrices; right of diagonal: distance matrices. Number of characters in parentheses.

Character sets	All	Ext.	Int.	Mus.	Hd.	Tho.	T.W.A.	Abd.
All (196)	1·000	0·975	0·882	0·879	0·800	0·954	0·923	0·514
External (100)	0·957	1·000	0·759	0·759	0·751	0·940	0·879	0·468
Internal (96)	0·866	0·692	1·000	0·994	0·771	0·820	0·867	0·534
Musculature (75)	0·867	0·700	0·992	1·000	0·772	0·822	0·873	0·536
Head (38)	0·858	0·787	0·820	0·827	1·000	0·684	0·761	0·278
Thorax (105)	0·944	0·933	0·781	0·790	0·787	1·000	0·941	0·367
Thorax without appendages (70)	0·926	0·902	0·793	0·800	0·788	0·977	1·000	0·356
Abdomen (27)	0·543	0·455	0·555	0·559	0·368	0·349	0·354	1·000

set constitutes a special classification. Each of these special classifications is correlated with each of the other subset classifications and with the classification based on all characters. In addition, differing numerical techniques give special classifications from the same character sets (Sokal and Michener, 1967). It has been argued (Ehrlich, 1964) that a general classification, phenetic or phylogenetic, is an ideal that is not likely to be achieved for any group with certainty. Furthermore, the information content of such a classification is considered to be low (Ehrlich, 1964) in terms of our understanding of the processes of evolution in that group. Examination of characters, character sets, and relationships among these will give more information with the same data. Character sets consistently give classifications which are correlated. Insofar as there are "true" phenetic relationships among the entities being classified, any set of characters would be expected to reflect them.

It is the purpose of this study to begin to examine some of the questions discussed above in plants using the tools of numerical taxonomy. The American species of *Sarcostemma* R. Br. were selected because data were already available. Floral morphology is known to be specialized for insect pollinators and varies accordingly (Holm, 1950). The species occupy sufficiently different habitats that the vegetative morphology may be expected to vary extensively. We will explore to what extent these two character sets give similar phenetic classifications and how this type of analysis might affect considerations regarding the taxonomic classification of these species.

METHODS AND MATERIALS

The 24 taxa in Holm (1950) were used as the operational taxonomic units (O.T.U.s). These are listed in Table III with code numbers which will be used for identification hereafter. The data matrix was constructed from the species descriptions. The character list is therefore previously published (Holm, 1950) and will not be repeated here. Qualitative characters were divided into character states and quantitative characters were scored as their listed values. The midpoint was used when ranges were given. Ninety-three characters were used. Because of the way in which the species descriptions were made, there were very few missing values in the data matrix. The problems of applying numerical procedures to monographic studies are discussed by Johnson and Raven (manuscript in preparation).

The characters were divided into two sets, floral and vegetative. The 61 floral characters included all characters of the flower including the calyx, but did not include inflorescence characters. The latter were grouped with the 32 vegetative characters. Q-matrices of product moment correlation coëfficients and taxonomic distances were computed using floral, vegetative, and all characters. Each of these six Q-matrices was clustered using the weighted pair group method with arithmetic averages. New similarity values (cophenetic values) were computed by the cluster analysis. Phenograms were constructed from these cophenetic values. Correlation coefficients were computed between the original Q-matrices and the cophenetic values computed from them, among Q-matrices of different character sets, and between Q-matrices of the two similarity measures for each character set.

As a test for dissimilarity of character sets between species pairs, the methods of Rohlf (1963) were used. Each element of the three Q-matrices was compared to the equivalent element of the other two. We will use species

TABLE III. Taxa included in the present study.

Taxon code no.	Taxa
1	*S. pannosum* Decne.
2	*S. glaucum* HBK.
3	*S. clausum* (Jacq.) Roem. & Schult.
4	*S. bilobum* Hook. & Arn. subsp. *bilobum* (Hook. & Arn.) R. Holm
5	*S. bilobum* Hook. & Arn. subsp. *lindenianum* (Decne.) R. Holm
6	*S. odoratum* (Hemsl.) R. Holm
7	*S. gracile* Decne.
8	*S. crispum* Benth.
9	*S. elegans* Decne.
10	*S. torreyi* (Gray) Woods.
11	*S. cynanchoides* Decne. subsp. *cynanchoides* (Decne.) R. Holm
12	*S. cynanchoides* Decne. subsp. *hartwegii* (Vail) R. Holm
13	*S. hirtellum* (Gray) R. Holm
14	*S. arenarium* Decne.
15	*S. angustissimum* (Anderss.) R. Holm
16	*S. flavum* Decne.
17	*S. campanulatum* Lindl.
18	*S. gilliesii* (Hook. & Arn.) Decne.
19	*S. vailiae* (Rusby) R. Holm
20	*S. solanoides* (HBK.) Decne.
21	*S. stipitatum* (Lillo) R. Holm
22	*S. lysimachioides* (Wedd.) R. Holm
23	*S. andinum* (Ball) R. Holm
24	*S. lehmannii* (Schltr.) R. Holm

1 and 2 as examples. Taxonomic distance was compared as follows. The distance between species 1 and 2 for each character set was converted to t by —

$$t_{1,2} = \frac{\left(\frac{d^2_{1,2}}{2}\right)^{\frac{1}{3}} + \frac{2}{9n_{1,2}} - 1}{\left(\frac{2}{9n_{1,2}}\right)^{\frac{1}{2}}}$$

where $n_{1,2}$ is the number of characters used in the computation of $d_{1,2}$. The absolute difference in t between two character sets for $d_{1,2}$ is then compared to the values 2·77 for $P < 0{\cdot}05$, 3·64 for $P < 0{\cdot}01$, and for 4·65 $P < 0{\cdot}001$ (degrees of freedom $=\infty$). We have thus statistically compared two values of distance obtained from different samples of characters. Similarly, correlation coefficients between species 1 and 2 ($r_{1,2}$) may be compared for two sets of characters by converting $r_{1,2}$ to $z_{1,2}$ by—

$$z_{1,2} = \frac{\log_e(1 + r_{1,2}) - \log_e(1 - r_{1,2})}{2}$$

The absolute differences between the z's of two matrices are then divided by a function of the number of characters used to compute each $r_{1,2}$—

$$\frac{1}{\left(\frac{1}{n_{1,2,k} - 3} + \frac{1}{n_{1,2,k} - 3}\right)^{\frac{1}{2}}}$$

where k is the index of the character set. This gives a value of t which can be compared to 1·960 for $P < 0{\cdot}05$, 2·576 for $P < 0{\cdot}01$, and 3·291 for $P < 0{\cdot}001$.

RESULTS

1. Comparisons of Q- and Cophenetic Matrices (Table IV)

Ehrlich and Holm (1962, 1963) have pointed out that much information may be lost in converting Q-matrices into a phenogram. This loss of information is an unfortunate consequence of the cluster analysis. This loss of information must be weighed against the value of expressing the Q-matrix in a form in which the similarities among O.T.U.s can readily be seen.

TABLE IV. Correlation coefficients between Q-matrices of correlation coefficients (r), and taxonomic distance (d), and the Q-matrices of cophenetic values (c) computed from them. $n = 276$.

Q-matrix Comparison	Character sets: Floral	Vegetative	Total
$d \times c_d$	0·893	0·827	0·838
$r \times c_r$	0·691	0·631	0·676

Rohlf (1967) has demonstrated methods of expressing phenograms in several dimensions which reduce this information loss. Two dimensional displays of the cophenetic matrix are given in this study. The correlation coefficients computed between the various Q-matrices and their derived cophenetic matrices are consistently higher for Q-matrices of distance as compared to Q-matrices of correlation coefficients. By this criterion distance is the better measure of similarity.

2. Comparisons Between Q-Matrices Based on Different Character Sets (Table V)

Correlation coefficients between Q-matrices based on different character sets are slightly higher for distance Q-matrices than correlation Q-matrices. They are all significant at $P < 0{\cdot}01$, but those observed between floral and vegetative are lower than previously observed between any two character

TABLE V. Correlation coefficients among Q-matrices of various character sets. Right of diagonal: correlation matrices (r); left of diagonal: distance matrices (d); diagonal: $d \times r$. $n = 276$. Number of characters is given in parentheses.

Character sets	Floral	Vegetative	All
Floral (61)	−0·545	0·169	0·803
Vegetative (32)	0·232	−0·515	0·670
All (93)	0·832	0·698	−0·542

sets in any group of organisms thus far studied. Correlation of floral and vegetative character sets to all characters was high, as would be expected since each set contributes greatly to the total. Correlations between the Q-matrices of the two measures of similarity are about —0·5 for all character sets.

3. Comparisons of Phenograms (Figs 1–7)

A summary of the relationships among the species included in this study based on classical taxonomic approaches is given by Holm (1950). This summary has been converted to a dendrogram (Fig. 1) for purposes of comparison with the numerically computed phenograms based on the same characters. The phenon lines of this dendrogram are labelled by taxonomic categories rather than computed cophenetic values. The similarities between *Sarcostemma pannosum* (1) and *S. flavum* (16) will be used as examples to compare the methods of classification. These species are each considered members of a monotypic series in the subgenus *Ceramanthus*. The species are quite distinct in floral morphology and to a much lesser degree in vegetative morphology, except that the overall size of vegetative parts of *S. flavum* is greater. In the phenogram constructed from correlation coefficients and vegetative characters (Fig. 3), species 1 and 16 cluster first with each other. They are reasonably close in terms of the maximum degree of phenetic similarity computed in this analysis. The two species do not show the same degree of similarity on the distance phenogram of vegetative characters (Fig. 2). Species 16 clusters first with 24 (*S. lehmannii*), but not at all close to

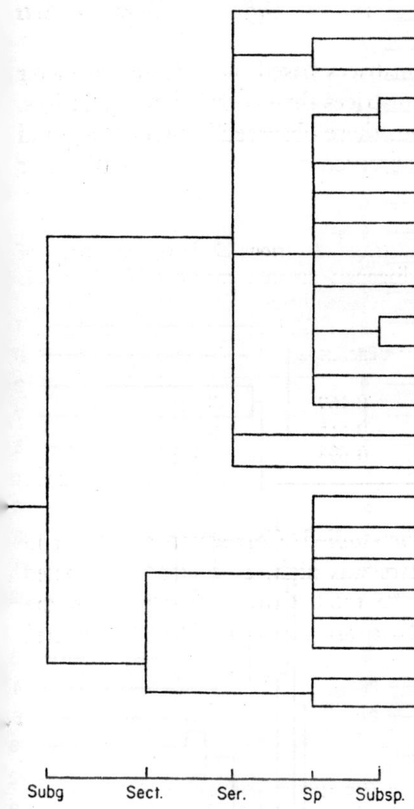

Figure 1.

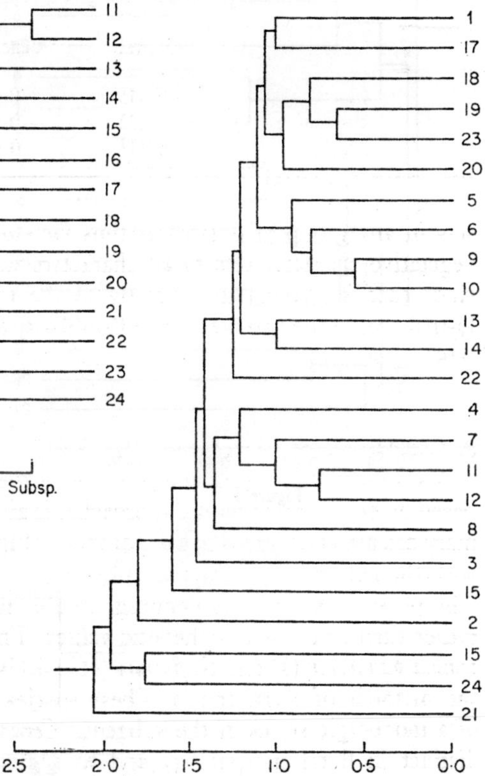

Figure 2.

FIG. 1. Dendrogram drawn from the relationships suggested by Holm (1950) for the 24 taxa included in this study.

FIG. 2. Phenogram constructed from the cluster analysis of the Q-matrix of taxonomic distance based on vegetative characters. Taxonomic distances are given below.

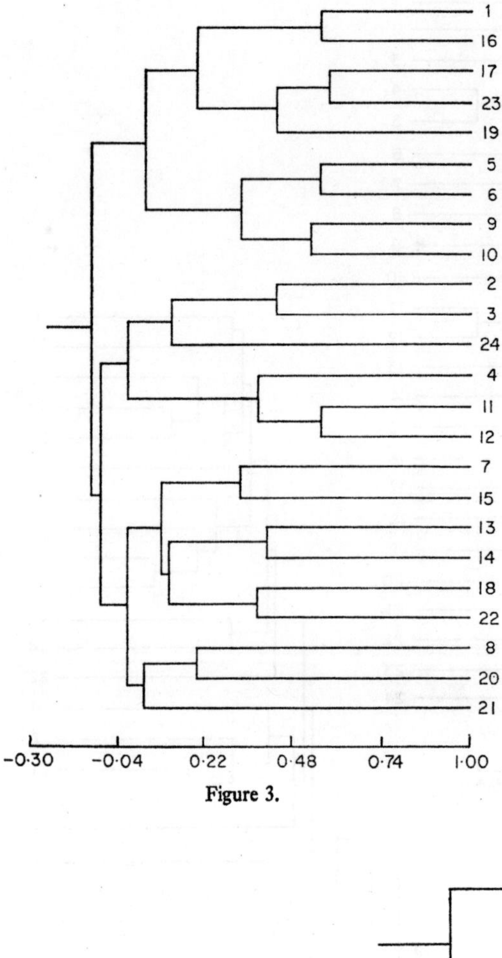

Figure 3.

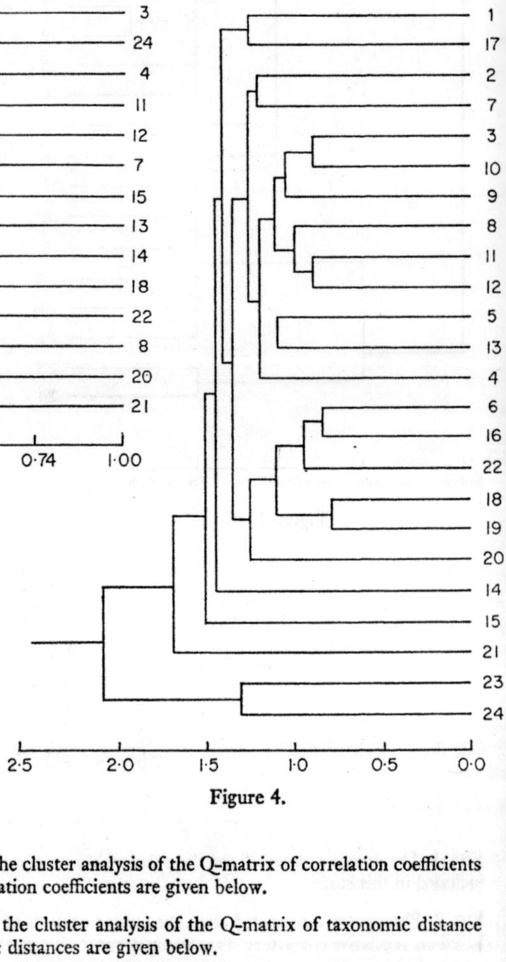

Figure 4.

FIG. 3. Phenogram constructed from the cluster analysis of the Q-matrix of correlation coefficients based on vegetative characters. Correlation coefficients are given below.

FIG. 4. Phenogram constructed from the cluster analysis of the Q-matrix of taxonomic distance based on floral characters. Taxonomic distances are given below.

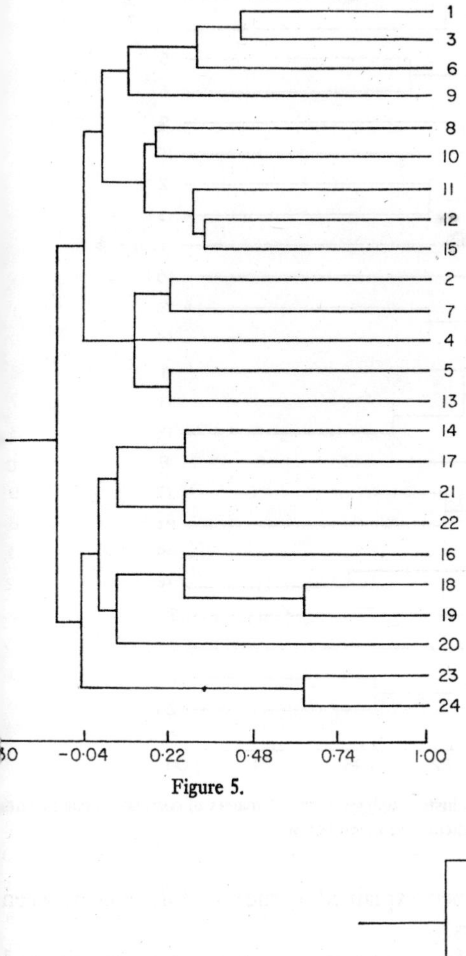

Figure 5.

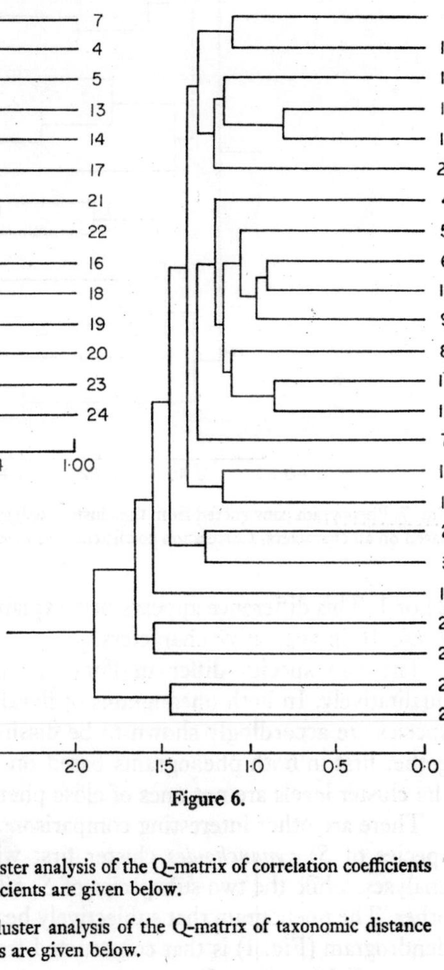

Figure 6.

FIG. 5. Phenogram constructed from the cluster analysis of the Q-matrix of correlation coefficients based on floral characters. Correlation coefficients are given below.

FIG. 6. Phenogram constructed from the cluster analysis of the Q-matrix of taxonomic distance based on all characters. Taxonomic distances are given below.

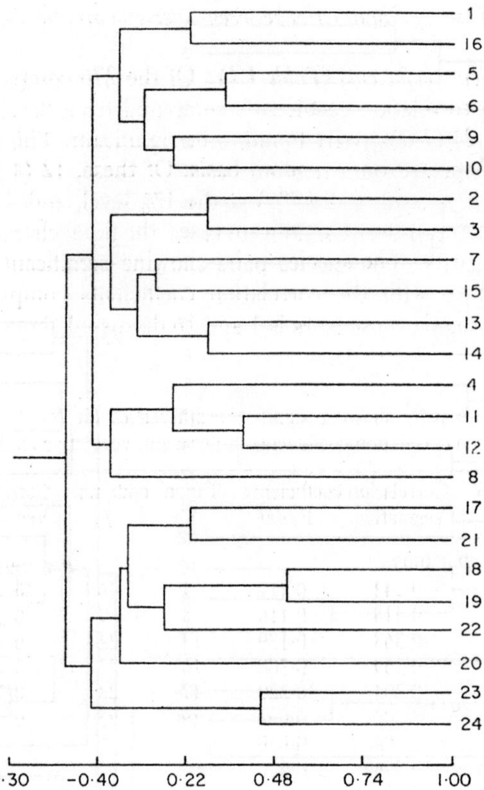

FIG. 7. Phenogram constructed from the cluster analysis of the Q-matrix of correlation coefficients based on all characters. Correlation coefficients are given below.

24 or 1. This difference appears best explained by the size difference between 1 and 16 in vegetative characters.

The two species differ in floral morphology both quantitatively and qualitatively. In both phenograms of floral characters (Figs 4 and 5) the two species are accordingly shown to be dissimilar. The two species cluster together first in both phenograms based on all characters (Figs 6 and 7), but the cluster levels are not ones of close phenetic similarity in either case.

There are other interesting comparisons that can be made. The two subspecies of *S. cynanchoides* cluster first with each other in five of the six analyses, while the two subspecies of *S. bilobum* never cluster first with each other. The phenogram that subjectively bears the greatest resemblance to the dendrogram (Fig. 1) is that constructed from the floral Q-matrix of correlation coefficients (Fig. 5).

4. *Statistical Tests of Similarities between Species-pairs for Different Character-sets*

(a) *Correlation Coefficients* (*Table VI*) Of the 276 comparisons of differences between correlation coefficients computed from floral and vegetative character-sets, 21 (7·6%) were found to be significant. This is slightly more than the 5% expected on a random basis. Of these, 12 (4·35%) were significant at the 5% level, 6 (2·17%) at the 1% level, and 3 (1·09%) at the 0·1% level. In nine of the 21 significant cases, the floral character set gave the greater dissimilarity. The species pairs showing significant differences are given in Table 6 with the correlation coefficients computed from each character set. Among those pairs is 1 and 16 discussed above.

TABLE VI. Species pairs showing significant differences (at $P < 0{\cdot}05$, $P < 0{\cdot}01$, and $P < 0{\cdot}001$) between correlation coefficients of floral and vegetative characters.

Taxon code no. *i*	*j*	Correlation coefficients Vegetative	Floral	Taxon code no. *i*	*j*	Correlation coefficients Vegetative	Floral
		$P < 0{\cdot}05$				$P < 0{\cdot}01$	
1	3	−0·142	0·417	1	16	0·565	−0·079
1	19	0·418	−0·116	5	7	−0·534	0·131
2	10	−0·368	0·139	13	22	0·344	−0·359
3	6	−0·269	0·308	16	22	−0·468	0·221
3	10	−0·204	0·329	17	23	0·581	−0·010
4	11	0·500	0·092	19	23	0·504	−0·243
4	13	−0·404	0·050				
14	17	−0·274	0·280			$P < 0{\cdot}001$	
14	19	0·308	−0·282	6	7	−0·535	0·197
16	18	−0·174	0·353	10	17	0·351	−0·405
18	19	0·176	0·631	23	24	−0·166	0·649
18	23	0·414	−0·129				

(b) *Taxonomic Distance* (*Tables VII and VIII*) Significant differences between taxonomic distances were observed for 58 (21·01%) of the species pairs. This number is significantly higher than the expected 5% (X^2 test, $P < 0{\cdot}001$). Of these, 28 (10·14%) were significant at the 5% level, 13 (4·71%) at the 1% level, and 17 (6·16%) at the 0·1% level. Twelve of the 21 pairs showing significant differences in correlation coefficients also show significant differences for taxonomic distances. Three species, 2, 16, and 23 are involved respectively in 14, 13, and 20 instances of species pairs showing significant differences. In 26 of the significantly different pairs, the floral characters were less similar. Those species pairs significant at the 5% and 1% levels are given in Table VII with the taxonomic distances computed from each character set. The species pairs showing significance at the 0·1% level are given in Table VIII with habitats and gynostegium height for each species. The gynostegium height is a floral character considered important in determining insect pollinators (for a description of insect pollination in Asclepiadaceae, see Holm, 1950). For other members of this family, differences in gynostegium height between species of a species pair indicate differences in insect pollinators.

TABLE VII. Species pairs showing significant differences (at $P < 0{\cdot}05$ and $P < 0{\cdot}01$) between taxonomic distances of floral and vegative characters.

Taxon *i*	code no. *j*	Taxonomic distances Vegetative	Floral	Taxon *i*	code no. *j*	Taxonomic distances Vegetative	Floral
		$P < 0{\cdot}05$		10	17	0·841	1·405
1	3	1·361	0·980	11	15	1·572	1·122
1	19	0·778	1·290	13	16	1·735	1·216
1	20	1·094	1·558	14	24	1·662	2·094
2	4	1·871	1·259	15	23	1·615	2·131
2	7	1·677	1·186	16	19	1·501	1·052
2	9	1·711	1·230	21	23	1·604	2·156
2	11	1·790	1·242				
2	12	1·875	1·251			$P < 0{\cdot}01$	
2	15	1·920	1·340	2	13	1·952	1·223
2	19	1·888	1·362	3	6	1·488	0·923
2	21	2·246	1·582	3	16	1·807	1·150
2	22	2·088	1·450	5	23	1·268	1·864
3	5	1·546	1·067	6	16	1·400	0·851
3	10	1·309	0·917	8	23	1·296	1·955
3	23	1·392	1·964	10	23	0·966	1·707
4	23	1·380	1·963	11	16	1·625	1·007
5	16	1·655	1·182	12	16	1·690	0·957
6	7	1·470	1·063	13	23	1·306	2·093
7	16	1·761	1·185	14	19	0·975	1·583
8	16	1·670	1·146	16	18	1·624	1·016
9	24	1·189	1·720	22	23	1·150	1·894

It is much more difficult, on the other hand, to evaluate degrees of differences between the habitats of the species. Nevertheless, it is reasonable to expect that selection for different vegetative morphology will exist between widely divergent habitats. Where gynostegium height between species differs greatly and the habitats are similar, phenetic similarity for the floral character set is significantly less than that of the vegetative character set and the converse (Table VIII) is also true.

TABLE VIII. Species pairs showing significant differences (at $P < 0{\cdot}001$) between taxonomic distances of floral and vegetative characters. A floral character, considered important in determining pollinators, and habitats are given for each species.

Taxon code no.		Taxonomic distances		Gynostegium height (in mm)		Habitat (elevation in feet in parentheses)	
i	*j*	Vegetative	Floral	*i*	*j*	*i*	*j*
1	23	1·085	1·904	4	7	grassy rocky slopes (500–6000)	shady or open slopes (5500–13000)
2	5	2·080	1·162	2	4	marshy or arid thickets (0–2500)	desert (50–80)
2	6	1·991	1·099	2	2·5	marshy or arid thickets (0–2500)	hillsides (4000–5000)
2	10	1·854	1·049	2	3	marshy or arid thickets (0–2500)	dry hillsides (4500–7200)
2	16	2·212	1·237	2	2·5	marshy or arid thickets (0–2500)	lowland areas
6	23	1·113	1·961	2·5	7	hillsides (4000–5000)	shady or open slopes (5500–13000)
7	23	1·171	2·117	2	7	halophyte, grassy plains	shady or open slopes (5500–13000)
9	23	0·919	1·843	3·5	7	rocky slopes–lava fields (5000–8000)	shady or open slopes (5500–13000)
11	23	1·088	2·001	2	7	sandy soils–rocky canyons (1600–6400)	shady or open slopes (5500–13000)
12	23	1·040	2·017	2·5	7	rocky soils, plains and arroyos (100–5000)	shady or open slopes (5500–13000)
14	23	1·342	2·201	2·8	7	dry coastal plain	shady or open slopes (5500–13000)
15	16	2·218	1·211	2	3	lava beds	lowland areas
16	22	2·124	0·878	2	3	lowland areas	alpine (9000–14000)
17	23	0·908	1·868	5·5	7	moist ground (6000–11000)	shady or open slopes (5500–13000)
18	23	0·789	1·992	3	7	dry rocky soil (450–3000)	shady or open slopes (5500–13000)
19	23	0·652	1·996	3	7	open woods (3500–6000)	shady or open slopes (5500–13000)
20	23	0·979	1·954	2·5	7	dry rocky soils–talus slope (6500–16000)	shady or open slopes (5500–13000)

DISCUSSION

The six analyses of phenetic similarity used in this study (two similarity measures times three character sets) must all be considered special classifications. The selection of the appropriate classification in any study should be based on requirements set by the goals of that study. Each of these classifications reveals different aspects of the similarities between the organisms studied. Each classification may add information about the biology of these organisms. That the different methods of analysis do not give the same classification may be one of the strong points of numerical taxonomy rather than a weak point (as implied by Heiser *et al.*, 1965). That different numerical methods give different systems cannot be used to argue that numerical methods are not objective. Nor is one justified in criticizing a classification because it does not show the same relationship among the entities as that system to which it is being compared. Systems of classification should be judged by the amount of information that they contain and on their utility with respect to the goals of the classification.

In the study by Heiser *et al.*, the phylogenetic relationships are known between many of the O.T.U.s because their origin was manipulated. The computed phenetic similarities did not agree with the known phylogenies. Nor is it likely that any method of analysis, numerical or otherwise, would give the actual phylogenetic relationships in this case if they were not known to the investigator. On the other hand, the data matrix and the Q-matrix computed from it have a wealth of information about the variation within the group which can not be construed from simply knowing the phylogenetic relationships among the organisms. Once in a data matrix, the characters and O.T.U.s may be analysed using a variety of approaches and the data are readily available for analysis using new techniques as they are developed. We will discuss the results of this study in terms of the information content of the various analyses.

The two Q-matrices based on phenetic similarities of pooled floral and vegetative characters give phenograms which are correlated, but nevertheless quite distinct. Increasing the number of characters by pooling the character sets in *Sarcostemma* does not improve the correlation between the Q-matrices and the cophenetic matrices derived from them. However, the large number of characters employed does make it difficult to understand why a species clusters differently in the two phenograms. Based on the analyses of the subset phenograms and on similar studies mentioned above, we presume that assumptions about size inherent in the two statistics account for much of this lack of correlation. Without further analysis, the information contained in these Q-matrices is not useful. Certainly, we do not consider any of these classifications to represent phylogenetic relationships closely.

Useful and interesting information regarding ecology, evolution, and taxonomy is obtained when the characters are partitioned into character sets and similar analyses made. When the Q-matrices of vegetative and floral characters are compared by correlation coefficients, the correlations are observed to be lower than in any character partition study made for animal groups. These correlations reflect the correlations among the individual characters. The degree to which the character sets are different reflects the degree to which these character sets are evolving independently. Since the floral morphology of species 1 has had the same length of time to diverge from the floral morphology of species 2 as have the vegetative parts of the two to diverge, significant differences between measures of similarity of the two species should indicate differing rates of divergence in the two character sets for those species. If species 1 and 2 have similar pollinators, but exist in different habitats, they will exhibit floral similarity and vegetative dissimilarity and the converse. The differences between species pairs do appear to be correlated in this fashion (Table VIII). The number of pairs exhibiting increased floral divergence is about equal to the number with increased vegetative divergence. This casts doubt on the validity of using floral characters alone, or with heavy weighting, for determining phylogenetic relationships in this group.

This study of the genus *Sarcostemma* has led us to the following view. Thorough analyses of character sets by taxonomists will lead to a better understanding of the process of evolution and the role of the environment in determining patterns of variation. Since statistically significant correlations are always observed between classifications based on character sets, any well done character analysis will give a classification which approximates to some degree to any other (probably including a phylogenetic one). Such classifications may also be useful for identification. We would conclude that the nonspecificity hypothesis holds only in part for plants as in animals. The analysis of the reasons for character and character set divergence should be of great interest to the evolutionary biologist and the ecologist. Furthermore, different methods of measuring similarity may contribute separate sets of useful information. In this study, for example, the size of floral parts may be very important in determining insect pollinators. In addition, reduction in the size of vegetative organs may be associated with such changes in habitat as coastal to alpine, shaded to open, or mesic to xeric. The more appropriate measure for analyzing these factors may be distance which tends to weight for size. On the other hand, the numerical classification based on correlation coefficients bears closer resemblance to the classical taxonomic classification.

SUMMARY

The phenetic relationships between the 24 species and subspecies defined and described by Holm (1950) were analysed numerically. Q-matrices of *r* (correlation coefficients) and of *d* (taxonomic distance) were computed for all characters, for vegetative characters only, and for floral characters only. Phenograms were produced from each. Correlation coefficients between Q-matrices were computed. Floral Q-matrices and vegetative Q-matrices both correlate well with the Q-matrices of all characters but not well with each other. The elements of the floral Q-matrices are compared statistically with the equivalent elements of the vegetative Q-matrices to determine those species pairs having significant phenetic divergence for these two sets of characters. In general greater dissimilarity between species pairs is shown for vegetative characters when the habitats vary while the pollinators are similar, and the converse. More information on the biology of the organisms being studied is revealed when the data matrix is broken into subsets of characters.

ACKNOWLEDGEMENTS

We appreciate the assistance of Mr Peter Brussard and Miss Valerie Chase in scoring the characters. The Q-matrices and phenograms were computed at the University of Kansas Computation Center using NT-SYS Version 1. The assistance of Dr F. James Rohlf and Mr Ronald Bartcher with these computations is greatly appreciated. The species pair analyses were computed at the Stanford University Computation Center using programs written by the senior author. Dr Peter H. Raven, Dr Tom S. Cooperrider, Dr Shirley Graham, Mr James Hickman and Mr C. David White kindly read and commented on the manuscript. We are grateful to Dr Paul Ehrlich for his discussions during the course of the study and for permitting the use of his unpublished data. This study was supported by U.S.P.H.S.-N.I.H. Grant 2G-365-R1.

REFERENCES

BREEDLOVE, D. E. 1968. A biosystematic study of *Fuchsia* sect. *Encliandra* (Onagraceae). Ph.D. thesis, Stanford Univ.

EHRLICH, P. R. 1964. Some axioms of taxonomy. *Syst. Zool.*, 13: 109–123.

EHRLICH, P. R. and EHRLICH, A. H. 1967. Phenetic relationships of butterflies. I. Adult taxonomy and the nonspecificity hypothesis. *Syst. Zool.*, 16: 301–317

EHRLICH, P. R. and HOLM, R. W. 1962. Patterns and populations. *Science, N.Y.*, 137: 652–657.

EHRLICH, P. R. and HOLM, R. W. 1963. *The Process of Evolution*. McGraw-Hill, New York. 347 pp.

ESTABROOK, G. F. and ROGERS, D. J. 1966. A general method of taxonomic description for a computed similarity measure. *Bioscience*, 16: 789–793.

HEISER, C. B., SORIA, J. and BURTON, D. L. 1965. A numerical taxonomic study of *Solanum* species and hybrids. *Am. Nat.*, 99: 471–488.

HICKMAN, J. C. and JOHNSON, M. P. 1968. An analysis of geographic variation in *Menziesia* Sm. (Ericaceae). *Madroño*, in press.

HOLM, R. W. 1950. The American species of *Sarcostemma* R. Br. (Asclepiadaceae). *Ann. Missouri Bot. Garden*, **37**: 477–560.

JOHNSON, M. P. and RAVEN, P. H. Studies in *Ranunuculus* sect. *Flammula*. I. Numerical analyses of classical data. In preparation.

KATZ, M. W. and TORRES, A. M. 1965. Numerical analyses of cespitose zinnias. *Brittonia*, 17: 335–349.

MOSS, W. W. 1966. The biological and systematic relationships of the martin mite, *Dermanyssus prognephilus* Ewing (Acari: Mesostigmata: Dermanyssidae) Ph.D. thesis, University of Kansas.

ROGERS, D. J. and FLEMING, H. S. 1964. A computer program for classifying plants. II. A numerical handling of non-numerical data. *Bioscience*, **14**: 15–28.

ROGERS, D. J. and TANIMOTO, T. T. 1960. A computer program for classifying plants. *Science, N.Y.*, **132**: 1115–1118.

ROHLF, F. J. 1962. A numerical taxonomic study of the genus *Aedes* (Diptera: Culicidae) with emphasis on the congruence of larval and adult classifications. Ph.D. thesis, Univ. of Kansas.

ROHLF, F. J. 1963. Congruence of larval and adult classification in *Aedes* (Diptera: Culicidae). *Syst. Zool.*, **12**: 97–117.

ROHLF, F. J. 1965. A randomization test of the nonspecificity hypothesis in numerical taxonomy. *Taxon*, 14: 262–267.

ROHLF, F. J. 1967. Correlated characters in numerical taxonomy. *Syst. Zool.*, **16**: 109–126.

ROHLF, F. J. and SOKAL, R. R. 1965. Coefficients of correlation and distance in numerical taxonomy. *Univ. Kansas Sci. Bull.*, **45**: 3–27.

SNEATH, P. H. A. 1957. Application of computers to taxonomy. *J. gen. Microbiol.*, **17**: 201–226.

SOKAL, R. R. 1961. Distance as a measure of taxonomic similarity. *Syst. Zool.*, **10**: 70–79.

SOKAL, R. R. and CAMIN, J. H. 1965. The two taxonomies: areas of agreement and conflict. *Syst. Zool.*, **14**: 176–195.

SOKAL, R. R. and MICHENER, C. D. 1967. The effects of different numerical techniques on the phenetic classification of bees of the *Hoplitis* complex (Megachilidae). *Proc. Linn. Soc. London*, **178**: 59–74.

SOKAL, R. R. and ROHLF, F. J. 1962. The comparison of dendrograms by objective methods. *Taxon*, **11**: 33–40.

SOKAL, R. R. and SNEATH, P. H. A. 1963. *Principles of Numerical Taxonomy*. W. H. Freeman, San Francisco.

14 | Observations on a Computer-aided Survey of the Jamaican species of *Columnea* and *Alloplectus*

WILLIAM T. STEARN

British Museum (Natural History), London, England

INTRODUCTION

Experience has shown that, contrary to optimistic early expectations of the objectivity and infallibility of computer-aided techniques, the same body of facts may yield different results when treated by different taxonometric procedures as Sokal and Michener (1967), for example, have demonstrated; these should accordingly be regarded as decision-indicating rather than decision-making when used in preparing or assessing taxonomic revisions of groups. A major difficulty for the uninitiated is to choose the best procedure for his purpose. Another difficulty concerns the number of characters needed in order to obtain by these mechanical aids satisfying results, i.e., results as good as or better than those derived from the customary procedures and thinking of experienced taxonomists. Obviously the smaller the number of characters taken, the greater becomes the risk of distortion by abnormalities, since they will have a greater proportionate weight, but in many groups of organisms the range of ascertainable characters falls much below the 40–100 characters which have been postulated for statistically valid results. Taxonomists usually manage with far less than that. Consequently many tend to view taxonometric procedures with caution or extreme scepticism, particularly on account of the time-consuming and laborious collecting, tabulating and encoding of data involved, and they also lack confidence in the results obtained. Before taxonometric procedures can be recommended with assurance for the elucidation of obscure or controversial groups, they have to be tested empirically by application to well-understood groups since a technique which makes nonsense of the known is unlikely to do better with the unknown. Such groups can provide standards for assessing performance. It is desirable, when satisfactory results have come from the use of a large number of characters, to ascertain whether a smaller number would have sufficed or would have yielded results which, even if not so detailed and meaningful, would nevertheless contribute to understanding the group by associating organisms in a useful manner according to their degree of overall resemblance and by keeping separate those which diverge from them. It is furthermore desirable to ascertain the extent to which a reduced set of selected characters and an equally reduced set of randomly taken characters agree in their results with each other and with a random set. Many such testing investigations need to be made. The following remarks outline experience in applying taxonometric methods to the Jamaican species of *Columnea* and *Alloplectus* and are a by-product of work on *Gesneriaceae* for "The Flora of Jamaica"; much information having been collected, it seemed a pity not to make further use of this, particularly in view of some earlier remarks on numerical taxonomy (cf. Stearn, 1964).

CHARACTERS USED

During my visit to Jamaica in 1955–56 I paid special attention to the native species of *Columnea* and *Alloplectus* and was privileged to collect and study almost all in a living state; since then they have been studied cytologically by Mr Brian Morley. The species as now defined are well distinguished and have distinctive ranges within the island (cf. Stearn, 1959: 141). They furnish enough characters for taxonometric survey. Thus, by including cytological, biochemical and anatomical data along with macroscopic morphological data, it was possible to tabulate 51 characters varying from species to species, i.e. constant within a species but not for the whole group. Twenty-seven of these had been used in making a key for identification purposes; they thus formed a reduced set weighted in favour of readily observed characters. For comparison 27 characters were taken at random out of the total of 51; 13 of these were found to coincide with those of the other reduced set; the 14 different ones included some biochemical, cytological and anatomical characters; this random set was thus somewhat more representative of the whole organization of the plants concerned. For coding purposes the characters fell into three groups: two-state or dichotomous; multistate; quantitative (cf. Watson *et al.*, 1967). They provided material for three computations, using Gower and Ross's program CLASP, which were made by Mr Gavin Ross on an Orion electronic computer at the Rothamsted Experimental Station, Harpenden, Herts. The "median sort" results were expressed as three dendrograms. Later three scatter diagrams were made from the similarity matrices, by means of Gower's method of Principal Coordinates Analysis (cf. Gower, 1966), using the scaled first and second vectors of the transformed matrix as coordinate axes; this is the procedure used by Sims (1966).

ANALYSIS OF RESULTS FROM COMPUTATIONS

Details of this study are being published elsewhere (Stearn, 1968). It must suffice here simply to indicate some results. Revision of the group had shown that some species are manifestly close in morphological characters, e.g. (1) *Columnea hirsuta* and *C. fawcettii*; (2) *C. subcordata* and *C. proctorii*; (3) *C. argentea*, *C. brevipila* and *C. harrisii*; *C. rutilans*, although possessing many distinctive features, links on to *C. hirsuta* and *C. fawcettii*. Some other species, e.g. *C. hispida* and *C. jamaicensis*, are relatively isolated. Hence it was postulated that, unless the dendrograms and scatter diagrams linked and separated such species accordingly, they would have little or no relevance to the actual situation within the group. In fact all three computations satisfied

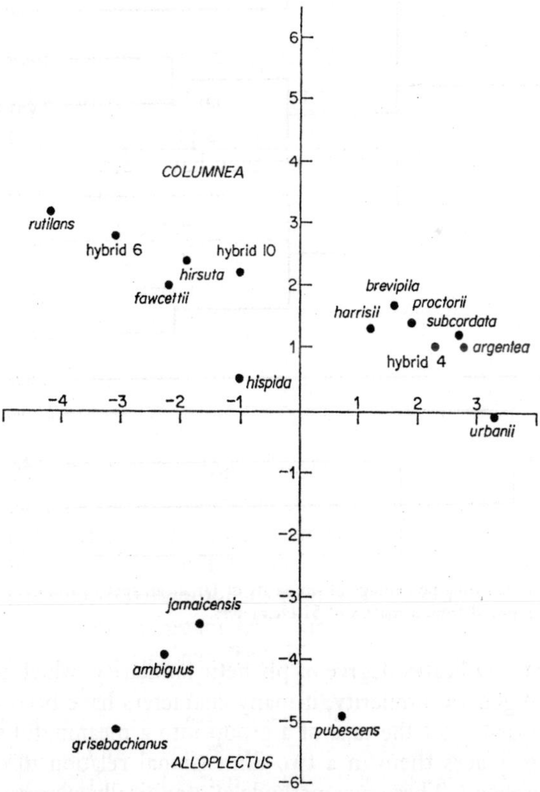

FIG. 1. Vector diagram showing relative similarity of Jamaican species and hybrids of *Columnea* and *Alloplectus* computed from a matrix of 51 characters, the vertical coordinates derived from the first latent roots, the horizontal coordinates from the second latent roots of the transformed matrix.

these basic requirements. The dendrogram from the full matrix of 51 characters made the most satisfactory differentiations and groupings; next came that from the reduced matrix of 27 unselected characters, then that from the 27 selected characters. Although varying in detail, each gave a reasonable classification not greatly different from one devised by traditional methods. They associated the nothomorphs of *C. rutilans* × *C. urbanii* (Hybrids 4, 6 and 10) with the one parent species they most resembled phenotypically.

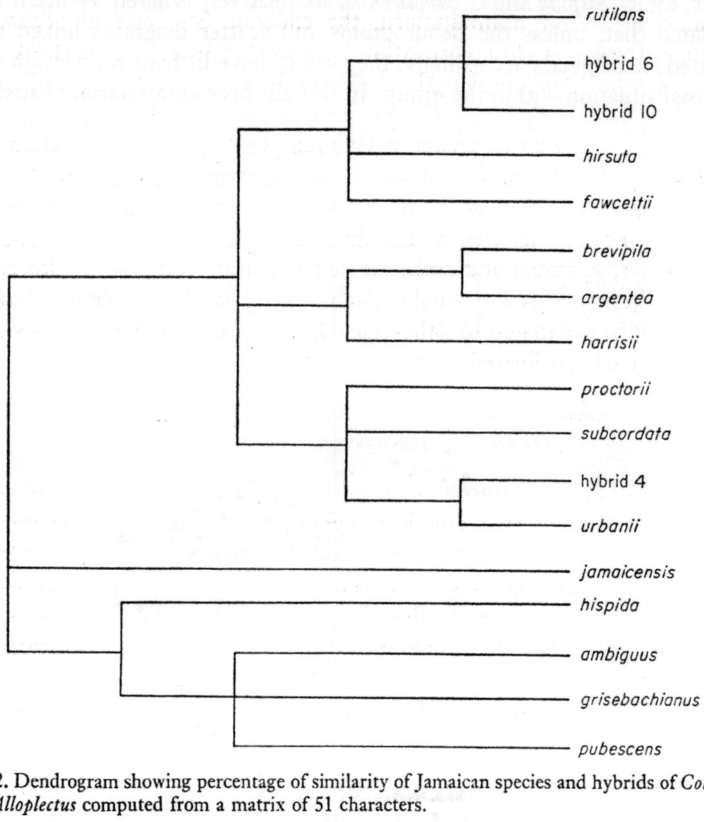

FIG. 2. Dendrogram showing percentage of similarity of Jamaican species and hybrids of *Columnea* and *Alloplectus* computed from a matrix of 51 characters.

A dendrogram indicates degree of phenetic similarity, which is also likely to be degree of genetic similarity, if many characters have been used, but it does not necessarily sort the taxa of a group into a meaningful sequence; a scatter diagram places them in a two-dimensional relation to one another easier to comprehend. These two methods of graphically expressing relationships can be combined. Thus, by keeping together the species of *Columnea* and *Alloplectus* associated at a given level of phenetic similarity on the dendrogram obtained from the full matrix but rearranging them within the groups to correspond more or less with the sequence of the scatter diagram obtained also from the full matrix (Fig. 1), a classification resulted which coincided almost exactly with the one made by traditional methods after much study and indicated the course of development of the group. Thus the rearranged dendrogram (Fig. 2) separates the species into three major divisions. It places together the typical species of *Columnea* with fairly large strongly zygomorphic flowers which in Jamaica seem all to be pollinated by the same species of hummingbird, the endemic Streamertail (*Trochilus polytmus*); there is a close correlation between the length of the beak and head of this bird and the length of the corolla-tube and stamens of these Jamaican *Columnea* species. It then separates these into three acceptable groups, beginning with *C. urbanii* having relatively primitive characters and ending with *C. rutilans* having relatively advanced characters. It emphasizes the isolated nature of *C. jamaicensis* and also of *C. hispida* and recognizes the general resemblance as well as the distinctness of the *Alloplectus* species *grisebachianus*, *pubescens* and *ambiguus* (not Jamaican, but inserted for comparison). These species with smaller and more regular flowers than *Columnea* proper must be pollinated by other agents; indeed the species of *Alloplectus* are probably self-pollinated.

CONCLUSION

This survey thus demonstrated, as others have done, the capacity of computer-aided taxonometric methods to build from an assemblage of characters a grouping of species comparable in validity to one made by conscious taxonomic effort. It also indicated that the number of characters used is less important than their range, i.e. they should be diverse enough to epitomize the whole organization of the organisms concerned. Taxonomy is concerned essentially with making diagrammatic pictures of complex states of affairs in nature; taxonometric methods provide some new ways of drawing them.

ACKNOWLEDGEMENTS

Grateful acknowledgements are made to Miss Barbara Heywood, Mr Brian Morley, Mr G. J. S. Ross and Dr J. C. Sheals for their cooperation.

REFERENCES

GOWER, J. C. 1966. Some distance properties of latent root and vector methods used in multivariate analysis. *Biometrika*, 53: 325–338.

SIMS, R. 1966. The classification of the Megascolecoid earthworms: an investigation of Oligochaete systematics by computer techniques. *Proc. Linn. Soc. Lond.*, 177: 125–141.

SOKAL, R. R. and MICHENER, C. D. 1967. The effects of different numerical techniques on the phenetic classification of bees of the *Hoplitis* complex. *Proc. Linn. Soc. Lond.*, 178: 59–74.

SOKAL, R. R. and SNEATH, P. H. A. 1963. *Principles of Numerical Taxonomy*. W. H. Freeman, San Francisco and London.

STEARN, W. T. 1959. A botanist's random impressions of Jamaica. *Proc. Linn. Soc. Lond.*, 170: 134–147.

STEARN, W. T. 1964. Problems of character selection and weighting: introduction. *In:* Heywood, V. H. and McNeill, J. (eds). *Phenetic and Phylogenetic Classification*, 83–86.

STEARN, W. T. 1968. The Jamaican species of *Columnea* and *Alloplectus* (Gesneriaceae). *Bull. Br. Mus. nat. Hist. Bot.* (in the press).

WATSON, L., WILLIAMS, W. T. and LANCE, G. N. 1967. A mixed-data numerical approach to angiosperm taxonomy: the classification of *Ericales*. *Proc. Linn. Soc. Lond.*, 178: 25–35.

Geography and Ecology

15 | The Scale of Genecological Differentiation

D. A. WILKINS

Department of Botany, University of Birmingham, England

INTRODUCTION

Taxonomic insight consists largely of the ability to choose the right characters for attention. We were once enjoined to avoid adaptive characters, but the statisticians now urge us to use all characters, and even to treat them as all of equal importance until the computer decides otherwise. Whether or not we are prepared to go as far as that we are certainly faced with the need to use adaptive characters in taxonomy, even if for no better reason than that we cannot rely on recognizing them when they occur. It is a serious weakness of the whole discussion that we can never say that a character is non-adaptive: experiment only allows us to say that so far we have found no use for it. We thus have to accept that any character we handle may one day be shown to be of vital importance to the plant. The selectionists are in an invulnerable position. This paper is concerned with the consequences of using adaptive characters at the lower levels of taxonomy: at those levels where gene flow is still an active possibility, and our taxa are not yet free to evolve in the safety of genetic isolation. This is the region where evolution is still reversible and taxonomy correspondingly provisional.

When gene flow is extensive there are good reasons to expect that most differences found will be of adaptive importance. There must be selection pressure to balance gene migration. The characters distinguishing two major taxa, such as families, between which there has been a complete sterility barrier for a long time, may be of many kinds. Some may be directly adaptive to present conditions, others may have been selected by past environments and not now be relevant, while others may possibly have arisen by chance "drift". In the case of two adjacent populations of an outbreeding species only the first of these possibilities is normally open. Any differences must be due to selection, and Jain and Bradshaw (1966) have recently discussed the degree of selection pressure necessary in these particular cases. This is not the place for a detailed survey of the experimental study of adaptation. A defect of much theoretical discussion of the subject is that the inevitably comparative nature of experimental results tends to be ignored. We can compare the fitness in a chosen environment of two plants differing in a particular character or we can test the same plant in two different environments. In either case the result is a relative one. When we talk about a prostrate plant being adapted to a mountain environment it is important to bear in mind what other habits and habitats are being implied. It would seem prudent to retain the language of comparison even in those cases where experimental tests cannot be so easily envisaged.

It has often been pointed out that many adaptive differences are physiological, and do not involve the kind of characters used in morphological taxonomy at all. It does not follow from this that they cannot be treated in much the same way, once the technical difficulties of assessing them have been overcome. There is every sign that the genetic bases of these two types of character are much the same, and in any case morphological differences must presumably have their origin in biochemical differences which influence development. Morphological and physiological characters are often genetically linked, and Clausen's evidence for the adaptive value of these linkage groups in *Potentilla* will be discussed later. Hutchinson (1966) made the interesting observation that a mutation which happened to have a beneficial physiological effect would only in very rare cases have an equally beneficial morphological effect as well. He argued from this that pleiotropy might be less common than expected in adaptive characters. This argument hardly applies at the taxonomic level, however, where we deal with big groups of linked genes rather than with the various manifestations of a single one.

Linkage and pleiotropy are both mechanisms which lead to character correlation. This topic is central to all modern discussion of taxonomic procedure, but it has been adequately reviewed elsewhere and the intention here is to extend the discussion slightly to take in the special properties of adaptive characters. We have to consider not only the mutual relations of the observed character differences but also their correlation with the environment. We may even in the end have to examine the correlations between the habitat factors themselves.

SIMPLE POLYMORPHISM

There are many examples of polymorphisms due to major genes, but in most cases there is little information about the adaptive advantages of the two or more forms in the wild environments in which they occur. Sometimes the relative frequency of the morphs varies in a suggestive way, however, and shows a correlation with the habitat which demands an explanation. Spotted individuals of *Arum maculatum*, for instance, increase in frequency from north to south in Britain (Prime, 1955). The correlation is strictly speaking with distance rather than with the habitat, but so many factors vary in the same direction that a selective explanation seems called for. The alternative is to suppose some historical accident such as a northward spread of spots which has not yet reached saturation. There seems to be no conclusive evidence about this yet available. It is important to note in this kind of study that we cannot say how many genes or characters are involved in the north–south topocline altogether. The work was concerned with a chosen character only, and not with a taxonomist's hypothetical list of "all characters", or with adaptation to climate. The same applies to most studies of polymorphism. A particular character is studied in detail, and the adaptive properties of the morphs perhaps related to their distribution, but correlations with other characters are not sought unless they are clearly related to the primary investigation.

An example of a geographical cline which did prove to have an ecological basis is provided by cyanogenesis in *Trifolium repens* (Daday, 1958). The crude correlations first observed with altitude and latitude were eventually shown to be regulated by adaptation to temperature and probably also to the activities of grazing animals (Jones, 1966). Experimental evidence for the adaptive value of the two morphs under contrasting conditions in the laboratory made the deductions about selection in the wild reasonably certain, at any rate in outline.

The case of glaucousness in *Eucalyptus* spp. (Barber, 1955) is interesting because of the circumstantial argument put forward for its adaptive value. The frequency of glaucous trees increased with altitude, and this was found in parallel in a number of intersterile species. The argument is that only selection could have induced the same character change under roughly the same conditions in several species independently. The presence of gene flow against the cline was made evident by differing ratios of glaucous to green individuals in seedlings and in adults. Experiment showed the selective factor to be winter temperature, but as in the case of *Arum* it cannot be assumed that there were no other differences involved. Very many climatic factors are correlated up an altitude gradient, and one would expect a correspondingly complex set of character differences to be selected along its length. It is no criticism of the original study to say that these were not looked for, but the lack of information about them means that the taxonomic position is uncertain. If each species of *Eucalyptus* has merely two forms, based on a major gene, classification is easy but trivial. If each shows a complex cline of many characters its subdivision is more difficult but possibly of more interest because of its predictive value.

POLYGENIC VARIATION

The distinction between this category and the last is one of convenience and cannot be justified by rigorous genetical evidence in many of the examples chosen, but continuous variation poses acute problems of sampling and measurement which markedly affect the results obtained. Striking polymorphisms are not common, but when they occur their study is reasonably straightforward and often permits direct presentation of the results in terms of gene frequency. When variation is effectively continuous, on the other hand, it is much more difficult to penetrate below the phenotype to its genetic basis. In addition there are dozens of varying characters which might be studied in any chosen species, and the investigator usually starts either with adaptation to some particular habitat difference in mind, or with the deliberate intention of examining as many characters as possible to get a picture of the total pattern of variation. Examples of both will be given.

Most studies of adaptive physiology come into the first category, where it is the habitat as much as the species which is sampled. If we are interested in heavy-metal tolerance (Wilkins, 1960; Gregory and Bradshaw, 1965), we sample from soils of a wide range of metal content and measure the physiological characters involved in tolerance. If we choose sites which differ only in lead content we find differences in lead tolerance and little else. If we sampled from calcareous and acidic mine tips we might well find a differential calcium requirement superimposed.

McMillan (1959) sampled prairie grasses from a wide latitude range and studied their photoperiodic responses. Björkman and Holmgren (1963) sampled *Solidago virgaurea* from shade and sun habitats and studied their photosynthetic efficiencies at different light intensities. In each of these cases a particular adaptive system was under investigation: a particular character correlated with a habitat factor, and in each case the results suggested how the selective mechanism had operated to produce the observed correlation. In the case of metal tolerance associated morphological differences were only looked for rather casually, and not on the whole found. In *Solidago* the adaptive differences turned out to involve a complex of anatomical and physiological features. In none of these cases was it possible to be neutral about the choice of characters, because the original sampling had been done with chosen factors and characters in mind.

It is not strictly accurate, on the other hand, to treat the above examples as if each were concerned with one character only. Overall photosynthetic efficiency may be measured as a single character but it has many components. The case of lead tolerance may be fairly simple in that only a single metal is being investigated, but the tips vary in the amounts of associated metals they contain and so even this assumption may sometimes be misleading. In the related example of serpentine tolerance (Kruckeberg, 1954) the complex of factors may include high concentrations of magnesium, and sometimes of nickel and chromium, coupled with unusually low amounts of calcium. The complexity of the adaptive mechanism depends directly on the degree of correlation shown by the habitat factors. One reason for the existence of the calcicole/calcifuge "problem" is the large number of correlated factors involved, such as pH, calcium concentration, phosphorus availability, and the solubility of iron and aluminium. So far a number of studies have been made of the adaptive differences between species in this situation, but much remains to be done at the infraspecific level. Snaydon (1961) has shown that in *Trifolium repens* calcicole and calcifuge populations differ in their response to both calcium and phosphorus. It seems likely that many more characters will be found to be correlated with these.

Probably the most clearcut example of an environment within which most of the habitat variables are correlated is the intertidal zone on the shore, where the tidal movements lead to a gradient in salinity, temperature, humidity and light intensity. As before there is a certain amount of information about physiological differences between species of algae which grow at different levels on the shore, but none about variation in the properties of a single species over its vertical range.

The second type of study involving continuously varying characters takes a large sample of a species from a chosen region and aims to build up a picture of the major variational trends. Much of the early experimental taxonomy was of this type, but concerned with morphological differences only. Once physiological measurements are included in such a programme the number of characters available for study becomes very large indeed, and the time required even greater because of the complexity of the techniques. This is the basic snag to the "all characters" approach to higher plants. They have too many characters, and ideas of random sampling of characters seem difficult to clarify. Some of the morphological features analysed by the pioneers of genecology were clearly of adaptive importance in themselves, such as the prostrate habit selected by intensive grazing or extreme exposure. Bradshaw's study of *Agrostis tenuis* (1959) showed a mosaic of morphological variation apparently related to various habitat factors. The recent work of McNaughton (1966*a*) on three species of *Typha* considered both morphological and physiological characters, and is especially interesting in that detailed studies were made of individual enzymes, whose properties were shown to vary in an adaptive manner according to the habitat of origin (1966*b*). Again many correlations were found between particular characters and habitat factors, and some of these were shown by tests in controlled environments to be of adaptive importance. Similar work has been in progress for some time on *Mimulus* in California (Milner and Hiesey, 1964).

The interest of any one of these examples lies largely in our ability to extrapolate from it. Taxonomic results are assumed to have a very high degree of predictive power, on the grounds that they are based on a wide sample of the taxa concerned, and that the characters studied are stable over reasonable distance and time intervals. This may be less true of intraspecific studies of adaptive differences for a number of reasons. The sampling may be limited in scope. The correlation of *Arum* spots with latitude in Britain cannot be assumed to be repeated in other parts of Europe without further study. At a more subtle level it could be that lead tolerance in a species could have quite different genetic bases in two continents, which raises the issue of what we mean by the "same" character.

We always feel safest with characters whose adaptive function seems obvious. Björkman's results with *Solidago* were based on three carefully selected plants from each of four sites, two shaded and two exposed, and yet so thorough was the experimental technique and so convincing the results that we do not dismiss this as a hopelessly inadequate sample. We just have to accept the fact that no physiologist can handle the number of plants which the statistician recommends. All we can ask is that the sampling should be as carefully planned as the experiments. The test is always whether we can safely extrapolate from the sample to the whole population of plants occupying the same habitats. In this case it seems likely that other shade and sun populations would show similar differences. In less clear examples we should ask for bigger samples, not of the particular populations but of the habitat types involved. This point has been analysed in detail elsewhere (Harberd, 1961; Wilkins, 1959).

The taxonomic treatment of the type of variation discussed in this section has long been a matter of argument. If we maintain that classification becomes possible when discontinuities are present, and interesting when characters are correlated, we can quickly outline the possibilities. Where adaptive characters are concerned these are seen to depend largely on the habitat, in that variation which is habitat-correlated can be classified when the habitat can be classified. Taking the correlation requirement first—some habitat factors are highly correlated together in groups, as in the comparison of acid and calcareous soils already mentioned. This means that information about any one factor allows predictions to be made about the probable values of others. It also means that any species which occurs over this range of soil types is likely to show adaptive variation in many characters together. This can be discussed in terms of clines. The original topocline was a gradient on a map: a change of genetics with distance as in *Arum*. The later concept of a subjective ecocline relating a plant character to a habitat factor need not be amenable to mapping at all, and certainly need not involve straight lines. Both variables may be distributed in a mosaic, as in the case of lead tolerance, and the only possible map then has one set of "contours" showing lead concentration and another set showing the degree of tolerance. The calcicole/calcifuge gradient may be thought of as a group of parallel ecoclines of this type, where the clines cross the "contours" at right angles. In this situation classification would be interesting because of its predictive value, but it is only possible if there are discontinuities, unless of course we are prepared to manufacture arbitrary ones by breaking up the gradient into sections.

Much has been said in the past about the causes of taxonomic discontinuity. Breeding systems, sterility barriers, geographical barriers and several other mechanisms may be involved. At the genecological level we have a different primary cause of discontinuity: a break in the environmental gradients. This has to be a genuine gap and not just a local pecularity. In a small area there may often be only two contrasting soil types, but all the intermediates can be found in a wider survey. This raises the many problems of the classification of habitats and of vegetation, where real discontinuities do not seem to be very frequent. One of the very few examples which may be relevant here is to some extent parochial, in that it draws on British material only, but it is perhaps the best so far available. *Festuca ovina* has been shown by Snaydon and Bradshaw (1961) to be differentiated in its response to the calcium concentration in culture solutions. They sampled four populations from soils within the pH range 7·8–8·9, and four from the range 4·3–4·6, and found marked differences between the two groups in the expected direction. This situation has so far only been analysed in terms of one character: response to calcium, but doubtless involves many other characters in parallel. It is of interest here because the discontinuity shown in the authors' sampling probably reflects the wild situation fairly accurately. From the point of view of this species the edaphic environment is markedly discontinuous. Intermediate soils abound, but they do not carry *F. ovina*, which cannot compete with the pasture and meadow grasses of neutral soils. This looks very much like a case of a discontinuity in the plant characters being imposed by a discontinuity in the available habitats. It deserves further study.

It should be emphasized that there is a wide gap between the theoretical analysis here and the experimental evidence. Even in a case which has been carefully investigated, such as that of *F. ovina*, the authors do not explicitly say whether they think the discontinuity is an artefact or not. The deduction that it might be genuine is based on independent knowledge of the ecology of the species. In most other cases there is no information available at all. Sampling is normally done in a highly discontinuous manner, especially in physiological work where the object is to establish how large a difference exists between extreme environments, and the question of intermediates remains one for the future. It is probably safest to assume, from what detailed evidence is available, that adaptive variation is usually continuous, and to ask for evidence of discontinuity in those cases where it is claimed. If this analysis is correct the place to look for distinct ecotypes is a species with an ecologically disjunct distribution.

ECOLOGICAL RACES

It has been implied throughout the discussion so far that the variation studied could not easily be ordered in terms of taxonomy. Most of it showed neither a high enough degree of correlation nor sufficiently objective discontinuities for the erection of satisfactory taxa. It should be noted that this judgement does not rest directly on the adaptive nature of the differences discussed, although this may affect both the correlation and the continuity observed, and neither does it depend on the fact that many of the characters were physiological in nature. We can find adaptive physiological differences at all taxonomic levels. At the higher infraspecific levels to be discussed here these are more closely linked with morphological differences, which in many cases have already been the subject of taxonomic attention.

There are many intermediate examples. Mooney and Billings' work on *Oxyria digyna* (1961) reveals some interesting ambiguities. They refer to the arctic and alpine "races" in their material as if these could in general be recognized on morphological grounds, and some of the measurements they obtained under controlled culture conditions do suggest a moderate degree of discontinuity. On the other hand the characters involved do not all show a break in the same place. They sampled 16 populations over a big latitude gradient and of these the central four were in some way ambiguous. When they turned to physiological characters they again found some with discontinuities in more or less the same place on the gradient, but unfortunately in choosing a smaller number of samples for intensive study they did not include any from the group showing intermediate morphology. The effect of this was to exaggerate the discontinuity. Other characters showed clinal variation correlated closely with latitude across the whole series. Most of the differences found were discussed in terms of adaptation to climate. This example is particularly instructive because it is on the verge of taxonomy. Two races could plausibly have been distinguished. On the other hand many plants could not be placed confidently in either race, and most of those intermediate plants came from geographically and climatically intermediate sites. Each race showed gradients in several characters which were continued across the boundary into the other race. These facts must be seen in the light of the rather peculiar nature of the material sampled. All the sites were virtually in a straight line transect covering 34° of latitude, which resulted in a very high degree of correlation between climatic factors. Attention was then concentrated on the effects of those climatic factors. It is impossible to say what other patterns of differentiation would have been found if, for instance, adaptation to soils had been studied. This is not intended as a criticism of one of the most interesting papers on the subject yet produced, but rather as a reminder of how difficult it is to extract answers to your own questions from work which was designed to answer others. This is particularly acute in a taxonomy which claims to be synthetic, and which aims to use the results of many specialized investigations as the source of its raw material. *Oxyria* affords an excellent example in this context of the difficulty already touched on of defining a random sample of characters. The authors were concerned with the adaptation of populations of the species to contrasting alpine and arctic environments, about which they discovered a great deal. We are asking in retrospect how the characters they chose to study relate to the total genetic variation in their material, and our answer has to be very tentative indeed. One of the great strengths of herbarium taxonomy is that the plants themselves are preserved, so that future workers can relate their characters to ours in the same material. There seems no way of doing this with physiology. To investigate the edaphic relations of Mooney and Billings' material we should have to attempt to repeat their original sampling from the wild.

A great deal of work on large scale ecological races was discussed by Clausen and Hiesey (1958), and it remains to place some of their examples in the present context. Their own work on *Potentilla glandulosa* needs little introduction. In California this species comprises at least four morphologically distinct subspecies, with different habitat preferences and geographical distributions. Most of their work was on morphological characters, but it was clear that the differential performance they so frequently encountered between their three contrasting experimental gardens was a reflection of adaptive physiological differences related to climate. *P. glandulosa* was sampled in such a way that its geographical pattern was emphasized. The transect across California was so placed that it crossed the four subspecies in succession, at a point where they were clearly allopatric, but the authors remark that the situation was less clear in other parts of the range and that the subspecies could in fact overlap. Although few illustrative figures were quoted it is clear from the discussion that many scales of variation were present. The subspecies were merely the major groupings, and within each there was a great deal of variation which was probably adaptive and by implication often continuous.

The particular interest of the *Potentilla* example lies in the attempt made to discover the genetic bases of the subspecies differences studied. Many crosses were made between plants from different subspecies, and one or two of these analysed in great detail. The aim was to estimate the number of genes controlling the difference between the subspecies in each of the many morphological characters measured in the transplant experiments. The technique used, which involved picking out and counting the numbers of parental forms segregating in the F_2, is recognized to be a highly approximate one, but if the results are taken to represent the minimum possible number of genes concerned they are probably not misleading. Most character differences appeared to depend on a minimum of three to six loci. Some were single gene effects, while a few could have involved up to 20. The interesting point which Clausen makes is that even if his rather low numbers of genes per character are accepted there is still plenty of genetic variation available amongst the various recombinants to produce an enormous range of forms adapted to an equally wide range of habitats.

The linkage between morphology and fitness in different environments was established by transplanting samples of the segregating progeny to the experimental gardens, which were located within the natural areas of occurrence of the two races concerned. There was much differential mortality, and the most successful plants in each garden were those which carried the largest proportion of characters of the parent from the local race. Thus not only were the parental combinations of morphological characters linked together but they in turn were linked to physiological characters of high adaptive value. This linkage is the basis of introgression in Anderson's sense, but here the adaptive significance of the recombination spindle is emphasized.

There seems every reason to expect the differences between many other close pairs of ecologically distinct subspecies to be controlled in a similar way. Examples such as *Geum urbanum/rivale* and *Silene vulgaris/maritima* come readily to mind. In these cases the correlation between characters is high, as is that between plant characters and habitat factors, and there is a clear discontinuity which does not seem to depend on a discontinuity in the habitat. The assumption must surely be that these pairs of taxa differentiated under conditions of reasonable isolation, and that any intermediacy found today is the result of secondary intergradation. Their taxonomy affords no worse a problem than a decision about the treatment of hybrids.

TAXONOMIC CONSEQUENCES

The physiologist and the ecologist rarely choose their characters with taxonomy in mind. In spite of one or two disclaimers, such as that of Burtt (1964) in which he maintains that the two genera *Astragalus* and *Oxtropis* are separated by a single character which does not correlate with anything else, there is fairly wide agreement that a good taxonomic character must be highly correlated with a large number of others. This is clearly the best perspective from which to view key characters, if they are to have the predictive properties required of them. Once we go beyond individual judgement in this matter, and ask for actual lists of characters and tests of correlation, we are conducting an exercise in methodology which has already forced taxonomists to examine their assumptions and procedures in a highly salutary way. Whether we can actually apply such methods in routine practice is another matter, on which the hard-pressed herbarium worker is apt to have strong views. It can be argued that when correlations are very high almost any character will give the same result. We can then choose to emphasize those which are convenient for other reasons.

Assuming that our intention is simply to allocate each individual to its appropriate taxon, and that we are not concerned with the relationships between our taxa, this picture of high correlation throughout is probably true for the *Potentilla glandulosa* example. We can classify each plant with fair confidence that its morphology allows us to predict its habitat tolerance. In the case of *Oxyria digyna* this argument would be much too crude. Even the morphological characters do not correlate highly enough for that, and when the physiological ones are considered as well it is clear that any division into two taxa would be highly arbitrary. The other danger here is that the reason for the study was not taxonomic but physiological, and the characters chosen for study reflect a specialized concern with adaptation to climate. They cannot therefore be taken over uncritically and used for taxonomic purposes if that involves extrapolating from this group of characters to the sum total of genetic differences amongst the populations of the species.

At this level and below we can no longer expect that classifications based on different groups of characters will resemble each other. The idea of a general-purpose classification is then illusory, unless we are content with a very vague scheme of low predictive value based on as wide a sample of characters as we can obtain. The classification of *Oxyria digyna* on the hypothetical basis of characters adaptive to soil differences would doubtless give a very different picture from that based on climatic adaptations, and there would be little prospect of any combination of the two being satisfactory. This is related to the multidimensional nature of habitat variation. To take a simple example four soil types might be recognized: dry calcareous, wet calcareous, dry acidic and wet acidic respectively. These can be classified in two dimensions, but not in a hierarchy, as neither factor can be chosen as the primary one. It has been suggested that climatic ecotypes should be regarded as primary and edaphic ones as secondary, so that at least the beginnings of a hierarchy can be constructed, but this does not accord at all closely with the ecologist's view of habitat classification. Although there is no fundamental objection to a multidimensional classification of plants it cannot easily be grafted on at the bottom of the conventional system, and in practice it is apt to degenerate into a one character/one dimension arrangement which is neither convenient nor informative.

The lack of correspondence between different character systems at lower levels has further implications when we consider the effect of absorbing completely new characters. The general experience of biochemical and serological taxonomy at the level of species, genera, and families has been that it gives a picture very similar to that given by morphology. The new characters correlate highly enough with the old to be accepted as a useful source of compatible information. At lower levels there is no reason to expect this to happen. We know very little yet about the adaptive significance of biochemical differences, but the analogy of human blood groups, whose distribution it seems can only be explained in terms of persistent selection, should lead us to expect that adaptive functions will probably be found for them in time. Their correlation with other characters will depend to a large extent on what those functions are.

REFERENCES

BARBER, H. N. 1955. Gene substitutions in Eucalypts. *Evolution*, 9: 1–14.

BJÖRKMAN, O. and HOLMGREN, P. 1963. Adaptability of the photosynthetic apparatus to light intensity in ecotypes from exposed and shaded habitats. *Physiol. Plant.*, 16: 889–914.

BRADSHAW, A, D. 1959. Population differentiation in *Agrostis tenuis*. I. Morphological differentiation. *New Phytol.*, 58: 208–227.

BURTT, B. L. 1964. Angiosperm taxonomy in practice. *Phenetic and Phylogenetic Classification*, p. 5. *Syst. Ass. Publ.* 6. London.

CLAUSEN, J. and HIESEY, W. M. 1958. Experimental studies in the nature of species. IV. Genetic structure of ecological races. *Publ. Carneg. Instn.* No. 615.

DADAY, H. 1958. Gene frequencies in wild populations of *Trifolium repens*. III. World distribution. *Heredity*, 12: 169–184.

GREGORY, R. P. G. and BRADSHAW, A. D. 1965. Heavy metal tolerance in populations of *Agrostis tenuis* and other grasses. *New Phytol.*, 64: 131–143.

HARBERD, D. J. 1961. The case for extensive rather than intensive sampling in genecology. *New Phytol.*, 60: 325–338.

HUTCHINSON, E. G. 1966. Sensory aspects of taxonomy, pleiotropism, and the kinds of manifest evolution. *Am. Nat.*, 100: 553.

JAIN, S. K. and BRADSHAW, A. D. 1966. Evolution in closely adjacent plant populations. I. The evidence and its theoretical analysis. *Heredity*, 22: 407–441.

JONES, D. A. 1966. On the polymorphism of cyanogenesis in *Lotus corniculatus*. Selection by animals. *Can. J. Genet. Cytol.*, 8: 556–567.

KRUCKEBERG, A. R. 1954. The ecology of serpentine soils. III. Plant species in relation to serpentine soils. *Ecology*, 35: 267–274.

McMILLAN, C. 1959. The role of ecotypic variation in the distribution of the central grassland of North America. *Ecol. Monogr.*, 29: 285–307.

McNAUGHTON, S. J. 1966*a*. Ecotype function in the *Typha* community-type. *Ecol. Monogr.*, 36: 297–325.

McNAUGHTON, S. J. 1966*b*. Thermal inactivation properties of enzymes from *Typha latifolia* ecotypes. *Plant Physiol.*, 41: 1736–1738.

MILNER, H. W. and HIESEY, W. M. 1964. Photosynthesis in climatic races of *Mimulus*. I. Effect of light intensity and temperature on rate. *Plant Physiol.* 39: 208–213.

MOONEY, H. A. and BILLINGS, W. D. 1961. Comparative physiological ecology of arctic and alpine populations of *Oxyria digyna*. *Ecol. Monogr.*, **31**: 1–29.

PRIME, C. T. 1955. Variation in *Arum maculatum*. *Watsonia*, **3**: 181.

SNAYDON, R. W. 1961. A study of differentiation within the species *Trifolium repens*, with special reference to mineral nutrition. Ph.D. thesis. University of Wales.

SNAYDON, R. W. and BRADSHAW, A. D. 1961. Differential response to calcium in *Festuca ovina*. *New Phytol.*, **62**: 219–234.

WILKINS, D. A. 1959. Sampling for genecology. *Rep. Scottish Plant Breed. Sta.* **1959**: 92–96.

WILKINS, D. A. 1960. Measurement and genetic analysis of lead tolerance in *Festuca ovina*. *Rep. Scottish Plant Breed. Sta.* **1960**: 85–98.

For a comprehensive review of genecology in general see:

HESLOP-HARRISON, J. 1964. Forty years of genecology. *Adv. Ecol. Res.*, **2**: 159–247.

16 | The Role of Geographical and Ecological Studies in Taxonomy

F. J. F. FISHER

Simon Fraser University, British Columbia, Canada

INTRODUCTION

Because taxonomy, the first step in creating intellectual order out of natural diversity, depends at the outset on the provision of field-specimens, it is unavoidably founded in geographic exploration. Arising directly from the complex relationships revealed by taxonomic discoveries, a series of new fields closely related to taxonomy have developed insights for interpreting particular aspects of diversity; thus from a beginning in biogeography, we have moved through ecology and biosystematics to modern studies of ethology. Evolutionary theory, at once, for all these fields, comprehensively explains and relates diversity itself, its distribution, and its behaviour.

Whether or not they agree to use such explanatory material in their work, taxonomists seldom disagree that their primary tasks are description and classification. While a good case can be made for the advantages of stability in classification, an equivalent conservatism in description has often been a severe handicap. The traditional verbal taxonomic description, being linear and thus essentially *unidimensional*, is a poor vehicle for conveying information about the *multidimensional* patterns of organic diversity set in a complex ecogeographic context.

A great need exists for the development of improved techniques in taxonomic description, not only for expressing these complexities graphically or by other analytical means, but also for collecting in the first place the data that will reveal these often neglected characteristics of diversity. To this end it may be necessary to specifically train a new class of taxonomic field-technician.

A growing demand for reliable predictive information about the variation and behaviour of economically important organisms in various conditions and places, is another challenge for the descriptive skills of the taxonomist. Even though he may defer monographic studies until utilitarian requests for knowledge of particular groups are made, he must be ready with up-to-date techniques for rapidly obtaining and expressing the desired information in most useful form. If full advantage is to be taken of modern tools of communication, then electronic data-processing and video-graphic readout methods employed in data retrieval by industry and engineering must also be harnessed for this purpose in taxonomy.

ECOGEOGRAPHIC EXPLORATION AND TAXONOMY

In recent years there has been a spate of papers reviewing the interactions of ecogeographic and taxonomic studies (e.g. Constance, 1951, 1953, 1957, 1964; Davis and Heywood, 1963; Heslop-Harrison, 1952; Keck, 1957*a*, *b*; Kruckeberg, 1967; McMillan, 1954; Rollins, 1957; Walters, 1961, etc.). In the most recent of these, about to be published, Kruckeberg has comprehensively reviewed pertinent current literature linking ecology and systematics and I shall therefore not attempt to cover the same ground.

Since Linnean times at least, field and garden have been linked with herbarium and library as the traditional accompaniments of taxonomic studies (Constance, 1964). However, though specimens must of course be sought in the field, and the place of geographic exploration can therefore be taken for granted, according to Simpson (1961) the place of ecological data has sometimes been questioned. Nevertheless, the world's great herbaria already contain a vast amount of unstudied material awaiting attention as a result of broad geographic exploratory work, and according to Holttum (1961) the days of such general uncritical collecting are effectively over. Instead, field studies these days have taken a much more specific and basically ecological direction. The change in taxonomic viewpoint brought about by the elaboration of evolutionary concepts has meant of course the shift in field work from a study of individuals to a study of populations: from a study of form to a study of process and pattern.

But even today the necessity of such population studies for revealing the particular patterns of normal geographic variation is apparently not always fully recognized. Thus while some writers may mention the existence of geographic variation, they are certainly not always explicit in suggesting that field studies should be set up specifically to uncover such patterns (e.g. Benson, 1962). Even though large samples may be recommended they seem to be expected rather to reveal irregularities due to such phenomena as "hybridization" or something called "incomplete divergence", as though these were unfortunate abnormalities. Such limitations are however becoming rarer. Instead, population studies directed at establishing variation patterns are much more likely to be regarded as basic to proper definitions of taxa (Rollins, 1957). Moreover, the sequence of studies through such xplorers as Bonnier, Clements, Hall, Turesson and others, to the still continuing work of such as Clausen and his colleagues at Stanford, has been accompanied by a gradual change in sampling methods matched to the variation patterns being discerned.

Of course, as Rollins (1957) has said, it would take a millenium of hard work by many more botanists than are now active, to produce for all plants the kind of model study we have had from the Carnegie group at Stanford. Their work nevertheless highlights the tremendous complexity we can expect from such studies, genetical, physiological and taxonomical. And it also points to the need for assessing the relative stability of any characteristic by means of experiment.

Although their writings extend over more than a decade and a half of intensive work, the reviewers mentioned at the beginning of this paper are remarkably single-minded regarding the primary ecogeographic roles in taxonomy. What specifically are these contributions the taxonomist can expect from the modern field man? Let us begin with a minimal, and entirely descriptive, basic contribution. I hardly need to remind readers that no taxonomist worthy of the name would normally bother with a specimen unless he knew its source, even in the crudest terms. A specimen is in fact virtually useless without an origin. Thus purely descriptive notes about distribution, collectively providing a knowledge of broad geographic range, are the first kind of field data that occur to anyone. These must be complemented by samples of any observed variations, a need that extends of course to any fossils or microfossils.

But these basic facts are not enough for long. Attempts to explain, interpret, and predict, cannot be avoided. As a result new kinds of data suddenly become available, and the taxonomist in the herbarium surely cannot ignore them. During the last 50 years the beginnings of an understanding of evolutionary genetics and causal ecology has led to demands for increasingly intensive ecological information from the field (Kruckeberg, 1967). Awareness of ecotypic variation has provided an awareness of the multidimensional nature of ecotypic response, which has effected the interpretation of variation and discontinuity for taxonomist and ecologist alike. Wilkins shows elsewhere in this volume, that correlations continue to be sought between environmental and morphological variation with increasing sophistication. From the generalized climatic correlations of Clausen's group we have now moved to much more precise ecophysiological correlations such as those uncovered by Hiesey and Milner, Björkman, Mooney, or Harper, to name only a few (e.g. Hiesey and Milner, 1965; Björkman and Holmgren, 1966; Björkman, 1966; Harper and Clatworthy, 1963; Mooney, 1966).

Besides concern for such patterns in space, the ecogeographer may also be concerned with patterns in time. These are of course the changing patterns called evolution. Is it really true, as Keck (1957*a*) has said, that evolution i rarely fast enough for taxonomists to witness substantial changes directly The very diversity of ecogeographic patterns is a necessary accompanimen of continuing evolutionary activity, not all of which is slow. Surely failur to interpret data correctly is more often the reason for certain fast evolutionary changes being overlooked. The changes in distribution and population structure of animals after their human introduction to new territory described for example by Wodzicki (1965), may be faster than similar patteri changes for plants, but the latter seem no more difficult to detect (e.g Barber, 1955, 1956; Barber and Jackson, 1957; Lewis, 1962).

A point strongly emphasized in Kruckeberg's as yet unpublished review is the need for a new and more precise kind of environmental taxonomy. If ecological boundaries are to become more sharply defined, he says, and if full predictability of ecophysiological modes and tolerances is to be achieved, ecological language must be made more precise. Though every ecological pattern may be unique as Langlet (1963) has warned us, we must still seek means of concisely describing each one. Other difficulties are inherent in commonly used distributional terms, such as "geographical" versus "ecological" distinctness, and the related words "sympatric" and "allopatric", as has been critically discussed by Cain (1953). Longstanding philosophical pitfalls in words like "adaptation" and "environment" have been analyzed recently by Ghiselin (1966). Kruckeberg hopes, perhaps rather optimistically, that ecologists might uncover some further generalizing insights concerning the ecological components of diversification of taxonomic categories well above the species level. He notes the special ecological correlations of particular families. He is also hopeful of a truly 20th century natural history arising out of the superintegration of ecology, systematics, and evolution. Even if this is unlikely ever to be possible for the whole biological world, he hopes it may some day be achieved in certain critical plant or animal groups.

Thus the full circle, from the field man's notes accompanying specimens about "distribution and variation", has arisen the separate field of biogeography; from notes on "habitat", have come autecology and synecology; from "behaviour", the modern studies of ethology or floral ecology. Each of these has its own generalizing insights and each contributes its share to evolutionary theory. And as J. Z. Young (1960) has put it, it is "evolutionary theory which provides a general science compounding the activities of all other sciences".

TAXONOMIC COMMUNICATION

Almost every paper in this symposium has drawn attention to the complexity of natural variation patterns uncovered by ecogeographic studies. I think the taxonomist needs reminding of his dual role (*cf.* Mason, 1950; Heslop-Harrison, 1952): of describing as well as classifying this complexity. Like Constance (1953) I believe "the role of taxonomy has by no means been completed when the initial organizational impulse exhausts itself". We have had a lot of definitions already—here is another. *Taxonomy is a system of communicable generalizations about patterns discernible in natural diversity, and the primary acts of the discipline are concerned with preliminary generalizations about objects themselves.* These are sooner or later expressed as taxonomic descriptions. Am I naïve to think this just as important as the classifying process itself? Doesn't classification really entail a taxonomist making relational generalizations about properties of objects for the purpose of communicating with others?

But this is really not the place to get involved in a discussion about the philosophic foundations of taxonomy. In their exhaustive discussion of this subject I think Gilmour and Walters (1963) summed up my position by stating the purpose of taxonomy to be the making of a broad "map" of the diversity of living things. My criticism concerns the fact, mentioned by Wilkins, that we are now in a position to make much larger scale "maps" than ever before, yet we have rarely made it an important task to do so! The usefulness of such new "maps" will be judged by their increased convenience in permitting the scientist to talk usefully about vegetation.

I do not want to dwell on this point, but I do find it difficult myself to see how inference can be avoided in the making of such "maps". Like Mayr (1966) I deplore any tendency to reduce the taxonomist to the role of computer clerk. Surely Constance (1964) is right in saying that studies of any and all relationships among organisms, including ecological ones, are pertinent branches of taxonomy. Interpretative generalizations from autecology, for example, are essential in studies of key taxonomic problems such as convergence, polyploidy, or hybridization.

But while variation and integradation may certainly raise problems of interpretation, above all they raise special problems of phytographic description. I would commend the strong emphasis given by Davis and Heywood (1963) to the question of adequate description, including the need for sound methods for expressing observed variation patterns. Theirs is a useful review of methods and concepts involved in graphically presenting variation data. Nevertheless, as a surprisingly small number of taxonomists seem to have emphasized (e.g. Heslop-Harrison, 1952; Davis and Heywood, 1963), the greater part of descriptive taxonomy today is unable to deal adequately with such variation problems. Wherever it has been discussed this weakness is attributed in the first place to the inadequacy of the samples available. For many problems much more detailed field studies would be required than are usually possible with the time and money available.

—	*R. insignis*	*R. lobulatus*	*R. monroi*	var. *dentatus*	var. *sericeus*
Leaves					
size	4–9 in.	6–8 in.	1–4 in.	1–4 in.	1–4 in.
shape	round-cordate or reniform	orbicular-reniform or subpeltate	round-reniform, ovate or cordate	broad-ovate to ovate-lanceolate	round-reniform, ovate or cordate
upper vesture	villous	glabrous or nearly so	±silky-villous or near glabrous	strigose-ferruginous	±silky-villous or near glabrous
lower vesture	villous	±clothed in long white hairs	±silky-villous or near glabrous	strigose-ferruginous	±silky-villous or near glabrous
margins	crenate and shortly lobed	shallow-lobed and crenate	coarsely crenate or crenate-lobulate	coarsely toothed	coarsely crenate or crenate-lobulate
Scape					
height	12–36 in.	12–18 in.	6–15 in.	6–15 in.	6–15 in.
branching	often very numerous	5–15	1–3	1–3	1–3
bracts	large, cut, lobed, or linear-oblong		entire or deeply lobed	entire or deeply lobed	entire or deeply lobed
vesture	villous	glabrous or with few hairs in axils	±silky-villous or nearly glabrous	±silky-villous or nearly glabrous	±silky-villous or nearly glabrous
Flowers					
diameter	1–2 in.	1–1¼ in.	½–1 in.	½–1 in.	½–1 in.
petals	5–6+	5–6	5–8	5–8	5–8
petal shape	obcordate	obovate, round or emarginate	narrow-obovate-cuneate	narrow-obovate-cuneate	narrow-obovate-cuneate
petal glands	1–2		1	1	1
Achenes					
shape	turgid, obliquely obovate, winged below	turgid, obovate keeled at back	turgid, keeled	turgid, keeled	turgid, keeled
vesture	villous		hairy to nearly glabrous	hairy to nearly glabrous	clothed in silky hairs
Distribution	Lat. to 38° S to a little south of 42° S North Island, Nelson, Wairau, Lake Tennyson, Kaikoura Mts.	Upper Awatere Valley, upper Clarence Valley, Kaikoura Ranges and foothills, Mt Fyffe and Woodside Creek (Seaward Kaikoura Range)	Lat 41°–44° S Tararua Mts, Wairau, Spenser, Kaikoura, Clarence, and Torlesse Ranges	Marlborough and Canterbury from Clarence southwards	Mt Peel and Kaikoura Ranges

TABLE I. Previous taxonomic criteria for the subdivision of *R. insignis*

The main features of the members of the *Ranunculus insignis* complex as given by Cheeseman (1926). None of these features can be used for unequivocal identification; the descriptions reveal only a part of the great variability of the members (from Fisher, 1965).

But there is also another reason. Taxonomic descriptions are highly conventionalized and conservative. As I have pointed out elsewhere (Fisher, 1965), while taxonomists seem quite willing to use drawings or photographs to show complex morphological or anatomic details, they much more rarely choose to illustrate distributional patterns in this way. In complex relationships the advantages of pictures over words and the amount of expressible information, are so obviously greater for a given effort, that one might be forgiven for suspecting taxonomists of playing a kind of secret verbal game.

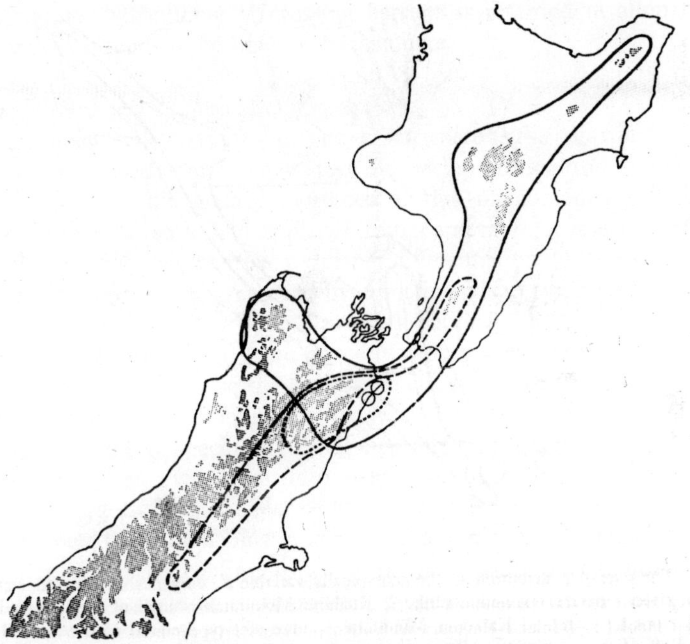

FIG. 1. The geographic ranges of the members of the *Ranunculus insignis complex* given by Cheeseman (1926) showing the large area of overlap in the region of the Kaikoura Mountains of the South Island of New Zealand. . . . . . , Distribution of *R. lobulatus* (Kirk) Cockayne; – – – – , distribution of *R. monroi* Hook.f.; ———— , distribution of *R. insignis* Hook.f.; shading indicates mountainous areas above 4000 feet in altitude).

Such games are almost necessary in poetry, but seem hardly practical for anything so utilitarian as taxonomy (*cf.* Walters, 1961). It is not as though the mind were incapable of recognizing more than the simple linear sequences of verbal description. Try verbally describing people's faces! Yet we are able to recognize hundreds of them in even quite crude photographic pictures. Geographic patterns are like faces: too complex to be easily described in words; with too many simultaneous variables for a linear sequence to convey.

As a result of this dependence upon conventional verbal descriptions, essentially quite simple situations may be made, paradoxically, to appear much more complex than they really are. Take for example the case of the *Ranunculus insignis* complex of New Zealand (Table I). As described by Cheeseman (1926) and more recently by Allan (1961), there are three species and a number of varieties, the ranges of which overlap in the Clarence headwaters among the Kaikoura Mountains of the South Island (see Fig. 1). This species complex is a conspicuously handsome member of the mountain

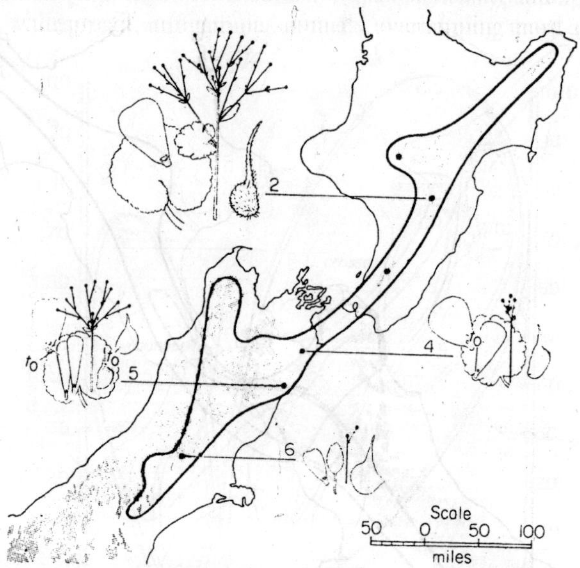

FIG. 2. The overall distribution of the ecotypically variable *R. insignis* showing four states previously given separate taxonomic rank: 2. Ruahine Mountains, source of type material of *R. insignis* Hook.f.; 4. Inland Kaikoura Mountains, source of type material of *R. monroi* Hook.f.; 5. Mt. Fyffe, Seaward Kaikoura Range, source of type material of *R. lobulatus* (Kirk) Cockayne; 6. Broken River, Craigieburn Mountains, source of type material of *R. monroi* var. *dentatus* Kirk (after Fisher, 1965).

flora. Accordingly it is very well represented in herbaria, so that it was possible for me when embarking upon a comprehensive treatment, to study much herbarium material in advance. Before any additional field work was undertaken, comparison of several hundred specimens available in the collections revealed no discontinuities within the complex by which to separate the previously described taxa. Thus the picture left by Allan, with five more or less indistinguishable taxa, was most unsatisfactory until field studies in the critical area of the Kaikoura Mountains revealed "homogeneous" populations belonging more or less centrally within an overall clinal pattern, not

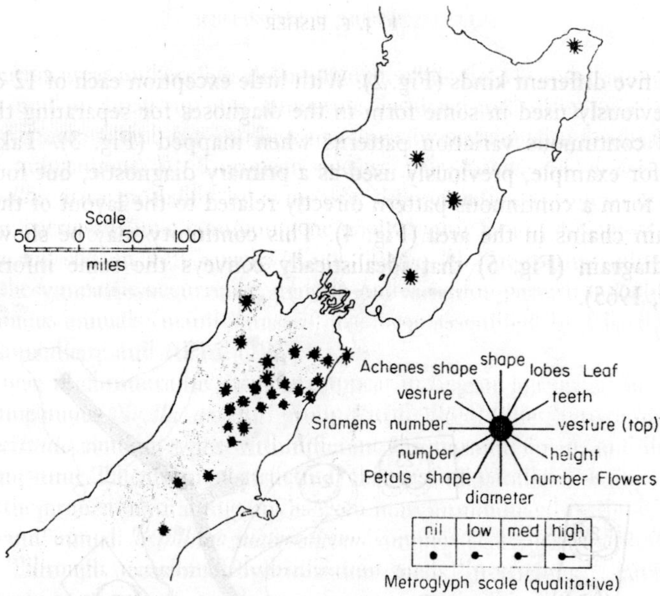

FIG. 3. The continuous variation patterns in twelve characteristics of *Ranunculus insignis* shown by radiate indices representing samples from various parts of the geographic range. Low values tend to occur towards the south, except for petal number in which high values correlate with very broad leaves (from Fisher, 1965).

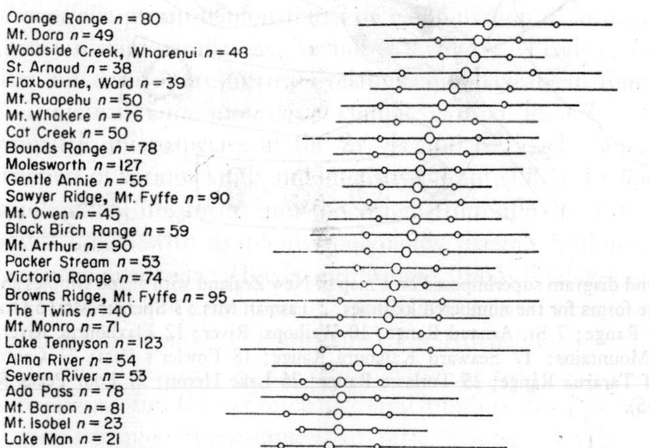

FIG. 4. A chart of leaf shape variation for 36 samples of *R. insignis*. The gradient in average shape closely follows north–south trend along the main mountainous ranges (from Fisher, 1965). O, Mean width/length ratio; o, standard deviation; —, range; *n*, number of leaves in sample.

a set of five different kinds (Fig. 2). With little exception each of 12 characters previously used in some form in the diagnoses for separating the taxa showed continuous variation patterns when mapped (Fig. 3). Take leaf-shape, for example, previously used as a primary diagnostic, but found, in fact, to form a continuous pattern directly related to the layout of the main mountain chains in the area (Fig. 4). This continuity may be shown in a trend diagram (Fig. 5) that idealistically conveys the same information (Fisher, 1965).

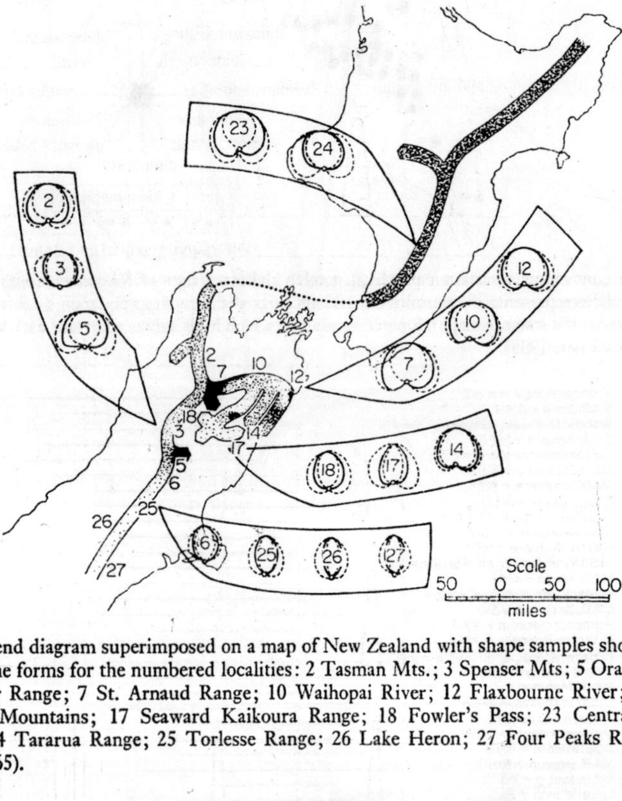

FIG. 5. Trend diagram superimposed on a map of New Zealand with shape samples showing mean and extreme forms for the numbered localities: 2 Tasman Mts.; 3 Spenser Mts; 5 Orange Range; 6 Dampier Range; 7 St. Arnaud Range; 10 Waihopai River; 12 Flaxbourne River; 14 Inland Kaikoura Mountains; 17 Seaward Kaikoura Range; 18 Fowler's Pass; 23 Central Volcanic Plateau; 24 Tararua Range; 25 Torlesse Range; 26 Lake Heron; 27 Four Peaks Range (from Fisher, 1965).

NEW METHODS OF EXPRESSING VARIATION

About 15 years ago, Constance (1951) tried to reassure us that the taxonomist really *was* aware of the great diversity of natural patterns even though, as he put it, "he does go to extraordinary lengths to conceal this recognition behind the formal facade of his taxonomic writings". Also about 15 years ago, in the first issues of *Taxon*, Heslop-Harrison (1952) warned us of this failure of taxonomists to adopt field methods directed at revealing population

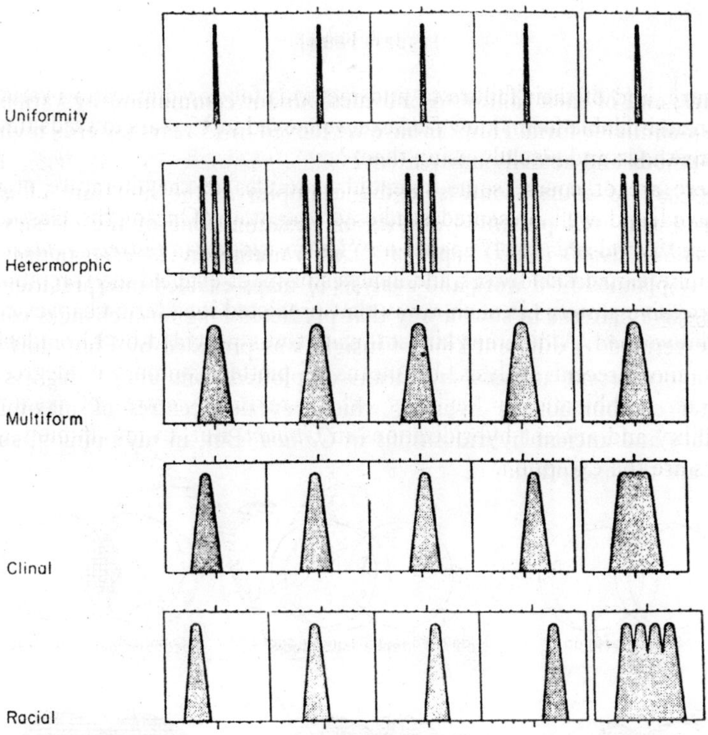

FIG. 6. A comparative scale of divergent patterns of geographic variation. In each line four samples are compared with the total variation. Line 1. *Uniformity* (any sample fully represents the total narrow variation); Line 2. *Heteromorphic* (two or more distinct forms occur in every population); Line 3. *Multiform* (a wide range occurs in every population, but regional differences are lacking); Line 4. *Clinal* (a gradual regional shift is expressed as a cline); Line 5. *Racial* (the regional shifts become more marked but overlaps are still present); Line 6. *Subspecific* (the point where slight but unequivocal distinctions become taxonomically recognizable); Line 7. *Species Complex* (regional differences justifying full specific status when features indicating close relationship are present); Line 8. *Coenospecies to Comparium* (evolutionary maturity, when species overlap in range without loss of individuality; between members of a coenospecies, some fertility, no matter how slight, reveals potential ability for gene exchange; hybrids between different coenospecies of a comparium are sterile) (from Fisher, 1965).

structure, and of their failure to find methods for communicating variation data in assimilable form. How far have we moved in 15 years towards finding such methods and actually using them?

There are of course some splendid examples in the literature of well conceived and well presented studies of variation. One of the classics of course is Woodson's (1947) paper on "Leaf variation in *Asclepias tuberosa*". Carefully planned field work and analysis of maps enabled interpretation of introgressive patterns in such a way that predicted long term changes could later be verified. A different kind of insight was provided by Ehrendorfer's (1958) more recent analysis, by means of plotted contours of degrees of character combination in hybrids, which revealed centres of maximum variability and ancient hybridizations in *Galium*. But, in my opinion, such papers are the exception.

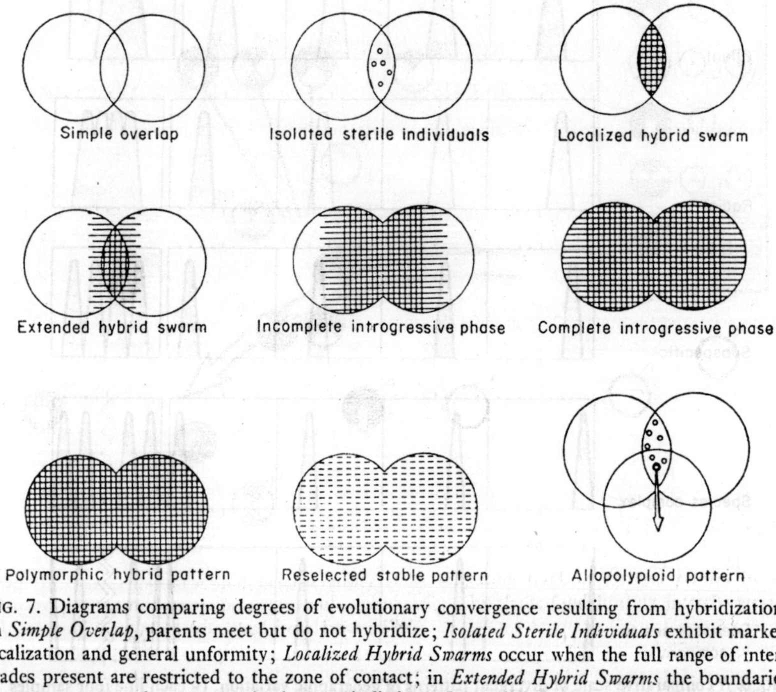

FIG. 7. Diagrams comparing degrees of evolutionary convergence resulting from hybridization. In *Simple Overlap*, parents meet but do not hybridize; *Isolated Sterile Individuals* exhibit marked localization and general unformity; *Localized Hybrid Swarms* occur when the full range of intergrades present are restricted to the zone of contact; in *Extended Hybrid Swarms* the boundaries between swarm and parent species become indistinct; *Incomplete Introgression* is the stage when only the marginal populations most distant from the area of contact remain unaffected by introgression; *Complete Introgression* has taken place when no marginal population remains unaffected by introgression; *Polymorphic Hybrid Patterns* have been reached when none of the parental forms survive; *Reselected Stability* is the final product of hybridization, where the residual variation is indistinguishable from divergence due to selection and mutation. The *Allopolyploid Pattern* derives of course from an *Isolated Sterile Individual* and combines convergence and divergence in a new stable product (after Fisher, 1965).

On the grounds that every pattern is unique, we have been warned from time to time against attempting a classification of variation patterns (e.g. Langlet, 1963; Nooney, 1965). Since the processes of divergence are more or less continuous, a case can be made, however, for seeking one or more continuous scales of some kind that might reflect degrees of divergence. A rather idealized version of one such scale of divergent variations has been attempted for the New Zealand *Ranunculi* referred to previously (Fig. 6). A scale of convergence has also been made (Fig. 7). A numerical quantitative equivalent for each of these should be easy to establish in terms of phenetic distance, or trends towards separation etc., using methods already established by the taximetrists. In suggesting such a scale I am quite conscious of the view that "quantification in science must follow, not precede, adequate qualification. Measurement can only be helpful when the proper things have been found to measure" (Gerard, 1961).

NUMERICAL EXPRESSION

Some time ago Ehrlich and Holm (1963) pointed out that it is in the area of description particularly, that Numerical Taxonomy is superior to classical methods. Generally, as we know, taximetrists have so far mainly directed their activities towards classification. What is more, sad to say, they may even be guilty of evading the question of variability, even recommending single individuals rather than population samples (e.g. "exemplar method" of Sokal and Sneath, 1963) on the rather shaky grounds that the introduction of variants adds prohibitively to labour and expense! Is it because description is much more difficult?

Taximetrists may even have to be reminded of sources of error long avoided by classical taxonomists, such as the relative unreliability of absolute measurements compared with ratios. Torre (1965) showed the misleading results of multivariate analyses based upon the comparison of grossly different linear measurements. Such weaknesses are again a function of sample inadequacies. How much more useful would the monumental computerized efforts of Irwin and Rogers (1967) have been for our understanding of variation patterns in *Cassia*, had they been able to provide themselves with less sparsely distributed samples and less limited graphic methods of presentation!

Perhaps these criticisms are a bit unfair, for I really am quite well aware of excellent descriptive work by taximetrists both sides of the Atlantic. Recently Sneath (1966), and also Marcus and Vandermeer (1966), for example, have been concerned in particular with mathematical strategies for comparing biogeographic trend-surfaces. These are greatly improved techniques for expressing complex variation patterns that have been developed by geologists (Good, 1964). The series of papers on geographic patterns in the gall-aphids (Sokal and Rinkel, 1963; Sokal and Thomas, 1965, etc.) provide not only trend maps for 18 separate characters but also integrative maps illustrating the results of multiple factor analysis of concurrent geographic variation and covariation.

When Cain (1958) predicted such improved numerical methods a few years ago, he was pessimistic of their being developed or used in the larger taxonomic institutions where their need was really greatest. His fears, however, have not been entirely justified, the New York Botanic Garden for example, is certainly one "larger taxonomic institution" where taximetric research and the development of electronic data processing methods for descriptive purposes has been active, at least until recently (Rogers, 1963). Dr Stearn's chapter in this volume shows us how some of his current work at the British Museum (Natural History) London, also bears on descriptive problems.

Before we rush headlong into the use of modern data processing equipment, however, there is room for developing caution and critical insight. We are probably all familiar with the unfortunate case analyzed by Woods *et al.* (1963) where the confusion of nomenclatural and taxonomic questions, combined with the rejection of internationally accepted practices, and an inadequate processing system, led to the relative failure of a very elaborate and expensive attempt to develop a comprehensive taxonomic index of plant family names. Wood's criticism of this case was directed at showing *how* computer methods should not be used, rather than *that* they should not be used.

The argument between cladists and pheneticists seems to have died down somewhat now that the compatibility of their views has been clearly demonstrated (Jancey, 1965; Kendrick, 1965; Sokal and Camin, 1965). If we can go on some of the comments made in this volume, there may justifiably be doubts about the worthwhileness of the very slight increase in accuracy of character correlations revealed by a numerical analysis (Gilmour and Walters, 1963). And again, justifiable doubts have certainly been expressed about the tremendous loss of information involved in reducing relationships down to a single measure of phenetic distance. But about the value of electronic data processing for other more detailed descriptive purposes there can be no doubt whatever.

PRACTICAL DIFFICULTIES

Detailed mapping of the kind we have been discussing, to be really reliable, obviously requires a phenomenal outlay of time and human effort. Dr Perring's impressive work with the Atlas of the British Flora (see

Perring, 1963) demonstrates very clearly that even with the enormous amount of voluntary help he has received from the Botanical Society of the British Isles, for example, there has been little opportunity for going very far beyond a relatively low resolution map showing only the broadest trends of variation or ecological pattern. But the problem is no less urgent for being gigantic. Alarms are being repeatedly sounded by many writers. Constance (1957) said it was a moot question whether one group of men would succeed in recording the remaining unclassified organisms before other men succeeded in inadvertently destroying them. Sharp (1962) suggested a *crash* programme of biological exploration and conservation of presently unknown genetic materials in natural areas and botanic gardens. Holttum has despaired at the alarming rate of disappearance of native vegetation of the tropics (Holttum, 1961).

But the real problem is the insufficiency of botanists or biologists to undertake even the exploration, let alone the monographic work to follow (Smith, 1964). In what form should the recruitment take place of large numbers of new workers to help keep ahead of increases in human population and the destruction of scientifically valuable and potentially economic plants (Keck, 1957)? Perhaps a new class of parascientific assistants is required to help meet the task. Just as in medicine, dentistry, engineering, or architecture, perhaps what we need is a new class of semi-professional assistants, developed to enable proliferating routine to be handled, leaving to the professional the higher levels of organization, evaluation, and research.

HUMAN ECOSYSTEMS

As already mentioned, predictive insights stem from sound ecogeographic studies. These insights may have utilitarian as well as scientific value. How many can recall the elegant studies started a few years ago by Benedict (see Benedict and Breen, 1955) in which air pollution at levels far below those normally detectable by human responses or chemical means could be precisely determined by morphogenetic responses of common weed species such as dandelion, chickweed, or bluegrass?

This brings to mind some other very serious questions. Pollution is only one of the utilitarian problems with taxonomic implications that face mankind. Conservation is not the only integrative task that ecologist and taxonomist might tackle together. In discussing man's ecological crisis, Vogt (1965) has suggested the need for a comprehensive and integrated approach to the whole human organism in its environment. Through our ignorance of ecological consequences we blindly destroy sector after sector of our human environment and resources.

Counterpart to the theory of evolution itself, such holocoenotic views of a global ecosystem in which human activities are a critical component, promise very practical comprehensive insights urgently needed for human survival (Fuller, 1963). Such a global ecosystem is no less biological for having human biological components. Conversely, its study will require not only the contributions of the ecological biologist as we have known him, but also of professionals in human ecology including the engineer, the architect, and the sociologist.

Even within biology as commonly understood in the narrower sense, consideration of biotic communities as organized systems is no longer the prerogative of the ecologist. Almost every biological discipline appears to have shifted towards such an outlook. To quote Baker (1966) "meaningful predictive insights concerning natural variation derive at once from such a view, contributing mutually to ecologists and evolutionists alike". The application of such biological insights to human affairs is itself becoming an important segment of operations research (Miller, 1965*a*, *b*), and the particular contribution of taxonomic insights is certainly not to be overlooked (Rohdendorf, 1965).

The complexity of these tasks might have alarmed us a few years ago, when the capacities for data integration were humanly frail. Even today the difficulties are great, electronics notwithstanding. Just what is to be presented to the machine and the form it should take before we can hope to get anything taxonomically significant out of it, will require much research (Constance, 1964). Computerized systems are taking the place of the traditional library in engineering and industry. The enormous searching capacities make comprehensive keyword indexing possible to replace the older elaborate classifications where redundancy had to be avoided. Thousands of references can now be searched in seconds and a bibliography automatically printed in minutes (Thiesmeyer, 1966).

This is just a beginning. Clearly, new ways will have to be developed for automatically integrating the ideas themselves if the mass of bibliographic material is to be assimilated. Most of us are already half-buried under the avalanche of potentially relevant publications. What is needed is a radically different approach to the information itself in the first place. We have all read about the substantial steps of this kind which are being made in integrative treatment of ideas by the use of video-graphic designing operations in industry. Elsewhere in this volume Dr Dale has indicated similar possibilities for taxonomic problems.

Reading a recent statement by Philip H. Coombs about educational institutions, I was struck by its relevance to taxonomic institutions. Generally, Coombs says, "the most fundamental obstacle to change is the absence of sufficient institutional mechanisms within the system whose prime business it is to change and improve the system—to change and evaluate old practices critically, to develop new and better ones, and to persuade and assist the far-flung operators of the system to innovate as is done in the more dynamic industries as a matter of course. Priority should be given to creating such mechanisms, to changing negative attitudes and to modernizing clumsy and outmoded rules and arrangements" (Coombs, 1967). Unless we forever question the basic imaginative constructs of our predecessors, according to Gerard, we condemn ourselves to working at progressively more detailed and trivial levels, merely to filling in further digits past the decimal point (Gerard, 1961).

SUMMARY

(1) Taxonomists have two tasks: description as well as classification.

(2) Intensive geographic and ecological studies reveal complex variation patterns, not adequately expressible by traditional verbal descriptive methods.

(3) Numerical taxonomists (taximetrists) are making substantial advances in taxonomic methodology of data presentation using graphic techniques.

(4) The improved descriptions, in setting out new kinds of data, will also lead to changes in classification methodology.

(5) Taxonomists should seek institutional mechanisms for continually updating techniques in the light of the needs and technologies of society on a global basis.

REFERENCES

ALLAN, H. H. 1961. *Flora of New Zealand*. Vol. I. Wellington, Government Printer.

BAKER, H. G. 1966. Reasoning about adaptations in ecosystems. *Bio-Science*, **16**: 35–37.

BARBER, H. N. 1955. Adaptive gene substitutions in Tasmanian Eucalypts. I. Genes controlling the development of glaucousness. *Evolution*, **9**: 1–14.

BARBER, H. N. 1956. The natural history of natural selection. *Aust. J. Sci.*, **18**: 148–159.

BARBER, H. N. and JACKSON, W. D. 1957. Natural selection in action in Eucalyptus. *Nature, Lond.*, **179**: 1267–1269.

BENEDICT, H. M. and BREEN, W. H. 1955. The use of weeds as a means of evaluating vegetation damage caused by air pollution. *Proc. 3rd National Air Pollution Symp.*, 177–190.

BENSON, L. D. 1962. *Plant taxonomy, methods and principles*. Ronald Press, New York.

BJÖRKMAN, O. 1966. Comparative studies of photosynthesis and respiration in ecological races. *Brittonia*, **18**: 214–224.

BJÖRKMAN, O. and HOLMGREN, P. 1966. Photosynthetic adaptation to light intensity in plants native to shaded and exposed habitats. *Physiol. Plant.*, **19**: 854–859.

CAIN, A. J. 1953. Geography, ecology and co-existence in relation to the biological definition of the species. *Evolution*, **7**: 76–83.

CAIN, A. J. 1958. The post-Linnean development of taxonomy. *Proc. Linn. Soc. Lond.*, **170**: 234–244.

CHEESEMAN, T. F. 1925. *Manual of the New Zealand Flora* (second edition). Wellington, Government Printer.

CONSTANCE, L. 1951. The versatile taxonomist. *Brittonia*, **7**: 225–231.

CONSTANCE, L. 1953. The role of plant ecology in biosystematics. *Ecology*, **34**: 642–649.

CONSTANCE, L. 1957. Plant taxonomy in an age of experiment. *Am. J. Bot.*, 44: 88–92.
CONSTANCE, L. 1964. Systematic botany—an unending synthesis. *Taxon*, 13: 257–273.
COOMBS, P. M. 1967. Education around the world. *Saturday Rev.*, 50(33): 49–50.
DAVIS, P. H. and HEYWOOD, V. H. 1963. *Principles of Angiosperm Taxonomy*. Oliver and Boyd, Edinburgh and London.
EHRENDORFER, F. 1958. Zur Phylogenic der Gattung *Galium* VI. Ein Variabilitätszentrum als "fossiler" Hybrid-Komplex: Der ostmediterrane *Galium graecum* L.—*G. canum* Req.-Formenkreis. *Österr. Bot. Z.*, 105: 229–279.
EHRLICH, P. R. and HOLM, R. W. 1963. *The Process of Evolution*. McGraw-Hill, New York.
FISHER, F. J. F. 1965. The alpine Ranunculi of New Zealand. *N.Z. Dept. Sci. Industrial Res. Bull.* 165. Government Printer, Wellington.
FULLER, R. B. 1963. *Ideas and Integrities*. Prentice-Hall. N. Jersey.
GERARD, R. W. 1961. Quantification in biology. *Isis*, 52: 334–352.
GHISELIN, M. T. 1966. On semantic pitfalls of biological adaptation. *Phil. Sci.*, 33: 147–153.
GILMOUR, J. S. L. and WALTERS, S. M. 1964. Philosophy and classification. *Vistas in Botany*, 4: 1–22, Pergamon Press, New York and London.
GOOD, D. I. 1964. Fortran II trend-surface program for the IBM 1620. *Kansas Geol. Survey sp. Dist. Publ.*, 14: 1–54.
HARPER, J. and CLATWORTHY, J. N. 1963. The comparative biology of closely related species. VI. Analysis of the growth of *Trifolium repens* and *Trifolium fragiferum* in pure and mixed populations. *J. exp. Bot.*, 14: 172–190.
HESLOP-HARRISON, J. 1952. Statistical methods in plant taxonomy. *Taxon*, 1: 53–59, 73–78.
HIESEY, W. M. and MILNER, H. W. 1965. Physiology of ecological races and species. *A. Rev. Plant Physiol.*, 16: 203–216.
HOLTTUM, R. E. 1961. Plant taxonomy as a scientific discipline. *Adv. Sci.*, 18: 234–242.
IRWIN, H. S. and ROGERS, D. J. 1967. Monographic studies in Cassia. (Leguminosae–Caesalpinioideae). II. A taximetric study of section Apoucouita. *Mem. N.Y. Bot. Gdn.*, 16: 71–118.
JANCEY, R. C. 1965. Numerical methods in taxonomy. *Proc. Linn. Soc. New South Wales*, 90: 335–340.
KECK, D. D. 1957*a*. The future of systematic botany. *Syst. Zool.*, 8: 76–82.
KECK, D. D. 1957*b*. Trends in systematic botany. *Survey of Biological Progress*, 3: 47–107. Academic Press, New York.
KENDRICK, W. B. 1965. Complexity and dependence in computer taxonomy. *Taxon*, 14: 141–154.
KRUCKEBERG, A. 1967. Ecological aspects of the systematics of plants. *Proc. International Conference on Systematic Ecology*, Univ. Michigan (in press).
LANGLET, O. 1963. Patterns and terms of intra-specific ecological variability. *Nature*, 200: 347–348.
LEWIS, H. 1962. Catastrophic selection as a factor in speciation. *Evolution*, 16: 257–271.
MARCUS, L. F. and VANDERMEER, J. H. 1966. Regional trends in geographic variation. *Syst. Zool.*, 15: 1–13.
MASON, H. L. 1950. Taxonomy, systematic botany, and biosystematics. *Madroño*, 10: 193–208.
MAYR, E. 1966. Numerical phenetics and taxonomic theory. *Syst. Zool.*, 15: 73–97.
MCMILLAN, C. 1954. Parallelisms between ecology and taxonomy. *Ecology*, 35: 92–94.
MILLER, J. G. 1965*a*. Living systems: basic concepts. *Beh. Sci.*, 10: 193–237.
MILLER, J. G. 1965*b*. Living systems: structure and process. *Beh. Sci.*, 10: 337–410.
MOONEY, H. A. 1966. Influence of soil type on the distribution of two closely related species of *Erigeron*. *Ecology*, 47: 950–958.

NOONEY, G. C. 1965. Mathematic models, reality and results. *J. theoret. Biol.*, **9**: 239–252.
PERRING, F. H. 1963. Data-processing for the Atlas of the British Flora. *Taxon*, **12**: 183–190.
ROGERS, D. J. 1963. Taximetrics—new name, old concept. *Brittonia*, **15**: 285–290.
ROHDENDORF, B. B. 1965. Diversity of the organic world and its historical development. (English summary) *Zhurnal Obshchei Biologii* (A. N. SSSR), **26**: 405–414.
ROLLINS, R. C. 1957. Taxonomy of the higher plants. *Am. J. Bot.*, **44**: 188–196.
SHARP, A. J. 1962. Responsibilities and opportunities of the taxonomist today. *Am. Biol. Teach.*, **24**: 87–90.
SIMPSON, G. G. 1961. *Principles of Animal Taxonomy*. New York Columbia Univ. Press.
SMITH, C. E., Jr. 1964. How dead? How soon? *Bio-science*, **14**: 15–16.
SNEATH, P. H. A. 1966. Estimating concordance between geographic trends. *Syst. Zool.*, **15**: 250–252.
SOKAL, R. R. 1965. Statistical methods in systematics. *Biol. Rev.*, **40**: 337–391.
SOKAL, R. R. and CAMIN, J. H. 1965. The two taxonomies; areas of agreement and conflict. *Syst. Zool.*, **14**: 176–195.
SOKAL, R. R. and RINKEL, R. C. 1963. Geographic variation of alate *Pemphigus populi-transversus* in Eastern North America. *Univ. Kansas Sci. Bull.*, **44**: 467–507.
SOKAL, R. R. and SNEATH, P. H. A. 1963. *Principles of Numerical Taxonomy*. W. H. Freeman, San Francisco.
SOKAL, R. R. and THOMAS, P. A. 1965. Geographic variation of *Pemphigus populitransversus* in Eastern North America: stem mothers and new data on alates. *Univ. Kansas Sci. Bull.*, **46**: 201–252.
THIESMEYER, L. R. 1966. Current efforts to solve the information explosion problem. *Tappi*, **49**: 74A–76A.
TORRE, D. 1965. On the use of linear measurements in multivariate analysis applied to taxonomy. *Riv. Ital. Paleont.*, **71**: 1271–1273.
VOGT, W. 1965. Man's ecological crisis (book reviews). *Natural History*, **74**: 4–7.
WODZICKI, K. 1965. The status of some exotic vertebrates in the ecology of New Zealand. *The Genetics of Colonizing Species*. ed. H. G. Baker and G. L. Stebbins. Academic Press, New York and London.
WALTERS, S. M. 1961. The shaping of angiosperm taxonomy. *New Phytol.*, **60**: 74–84.
WOODS, C. E., Jr., COWAN, R. S. and BUCHHEIM, G. 1963. Botanical nomenclature, punched cards, and machines. *Taxon*, **12**: 2–12.
WOODSON, R. E., Jr. 1947. Some dynamics of leaf variation in *Asclepias tuberosa*. *Ann. Missouri Bot. Gard.*, **34**: 353–432.
YOUNG, J. Z. 1960. *Doubt and Certainty in Science*. Oxford Univ. Press, New York.

17 | Geographical and Ecological Aspects of Infraspecific Differentiation

F. EHRENDORFER

University of Graz, Austria

INTRODUCTION

It is generally accepted today that the geographical and ecological radiation of populations is in many instances paralleled by morphological (physiological, biochemical etc.) differentiation. Backed by extrinsic (environmental) and/or intrinsic (biological) isolation this allopatric type of differentiation becomes an important aspect of speciation, and of the space-time bound processes of evolution in general. Stimulated by Darwin, the German zoologist M. Wagner (1868) and, independently, the Austrian botanist A. Kerner v. Marilaun (1869) were the first to formulate and clarify these ideas. Kerner derived his general conclusions from taxonomic work on *Cytisus* sect. *Tubocytisus* (= *Chamaecytisus*), Fabaceae (Fig. 1). He postulated that new races originate in marginal areas relative to the main areas of basic groups, where they are more or less protected by geographical or ecological isolation from hybrid swamping and can exploit new niches. Kerner also carried out transplant experiments and realized the impossibility of transformation of races by simple environmental modification, and he pointed to the possible origin of species by hybridization. All this makes him one of the foremost pioneers of modern botanical biosystematics, a fact which often tends to be overlooked. After further important work on geographical aspects of differentiation by the Russian plant taxonomist A. v. Bunge, Kerner's ideas were elaborated by his son-in-law, R. v. Wettstein during the later nineteenth century. Wettstein worked with such critical genera as *Euphrasia* and *Gentianella*, and, in 1898 published his "Principles of the morphogeographical method in plant systematics" in which he stressed the close correlation between morphological and geographical differentiation. Consequently, adjacent or allopatric distribution of races was regarded as a sign of recent evolutionary divergence and close affinity, while sympatric occurrence was thought to indicate advanced and more remote phylogenetic connections.

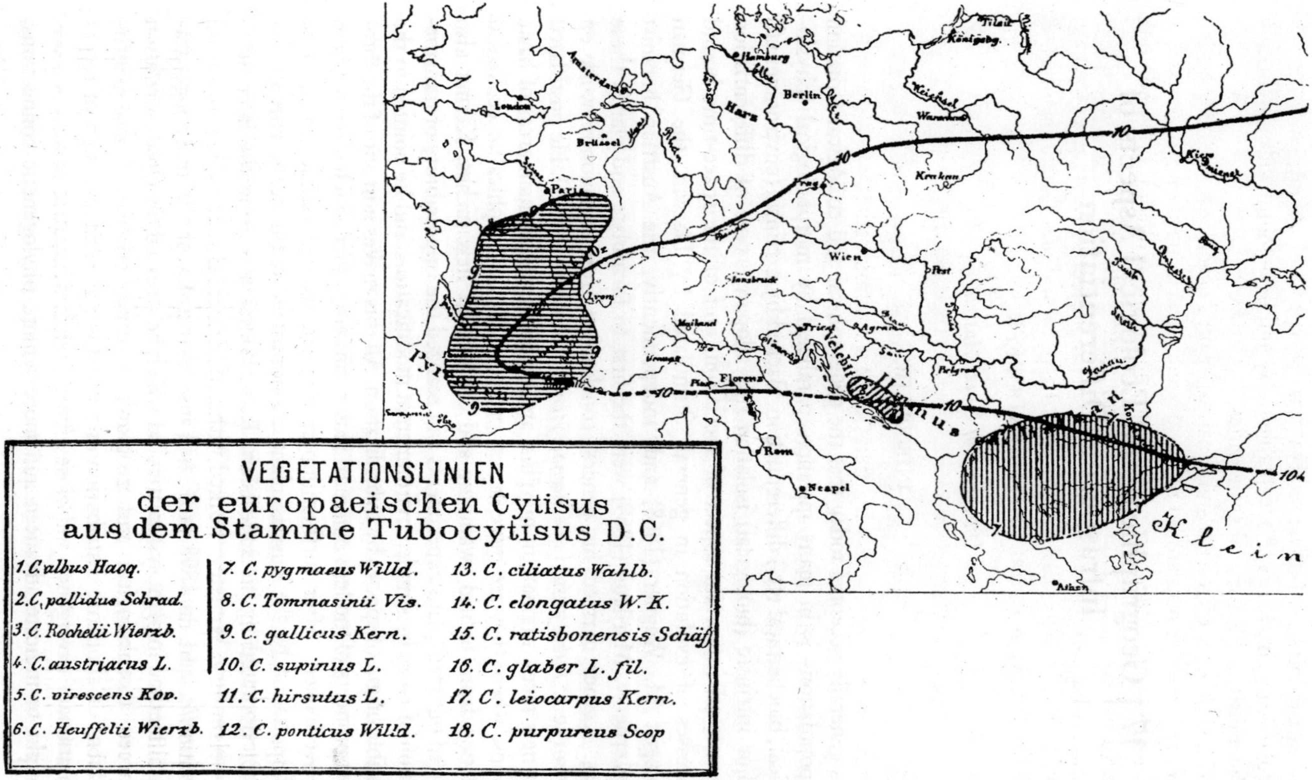

FIG. 1. One of Kerner's historical distribution maps from 1869, concerning species of *Chamaecytisus* (slightly modified).

These principles (with some reservations and keeping in mind the dangers of circular reasoning) are still acceptable today and have found wide application in botanical and zoological systematics.

The centenary of Kerner's original paper on *Tubocytisus* may be regarded as an appropriate occasion to reconsider our present position with regard to the analysis and understanding of geographical and ecological differentiation in seed plants, and its expression in taxonomic terms. As a basis for our discussion we can use the excellent review on the subject by Davis and Heywood (1963) and concentrate on some selected aspects and new developments, and cite new research as examples.

BASIC PHENOMENA AND ANALYTICAL METHODS

All evolutionary processes are regarded today as hereditary changes in populations brought about by mutation and recombination, sifted by selection, and canalized by isolation and random events. These basic phenomena are closely related to the size, position, migration, variation and reproduction of populations and their environment. Under these headings we can briefly review some of the methods available for the analysis of geographical and ecological aspects of micro-evolutionary differentiation in seed plants.

1. Size, Position and Migration of Populations

Number and spatial arrangement (density) of individuals, together with their reproductive biology, determine the range of the breeding community. Relevant data are still very meagre for seed plant populations. In Aegean cliff habitats population size of semi-shrubby *Erysimum* sect. *Cheiranthus* is usually not more than 50–100, often less; only a Phrygana population of *E. senoneri* subsp. *amorginum* consists of 2000–5000 individuals (Snogerup, 1967). Glade populations of minute annual *Leavenworthias* in Alabama range from 50–20,000 individuals (Lloyd, 1965). In contrast, populations of mediterranean annuals in California may comprise millions of individuals (Allard, 1965). For the desert annual *Linanthus parryi*, Epling *et al.* (1960) stress the enormous fluctuations of populations size and a less apparent component—the large numbers of ungerminated seeds in the ground.

Of particular importance is the geographical and ecological position of populations. Apart from the usual maps of horizontal distribution, altitudinal position can be demonstrated on transects (e.g. Casper, 1966). Occurrence on different geological substrates and soils is another important distributional feature, for instance, in western North American populations of *Ceanothus* (Nobs, 1963). Ecological position can be shown in terms of vegetation transects (e.g. Sauer, 1965), or by means of ecological gradients (such as water supply, light, nutrients, pH etc.), either by using composite indices (Ellenberg,

1963) or graphically (one- or two-dimension grids: Whittaker, 1967, and Fig. 2). Another approach is to analyse accompanying vegetation physiognomically and/or floristically (Bradshaw *et al.*, 1964 for *Alchemilla* microspecies, L. & Y. Mäkinen (1964) for populations of *Primula nutans* subsp. *finmarchica*). Different, but neighbouring plant associations have been shown by Białobrzeska (1966) to contain populations of *Carpinus betulus* which differ statistically in fruit and cupule shape. For *Aconitum*, Götz (1967) has

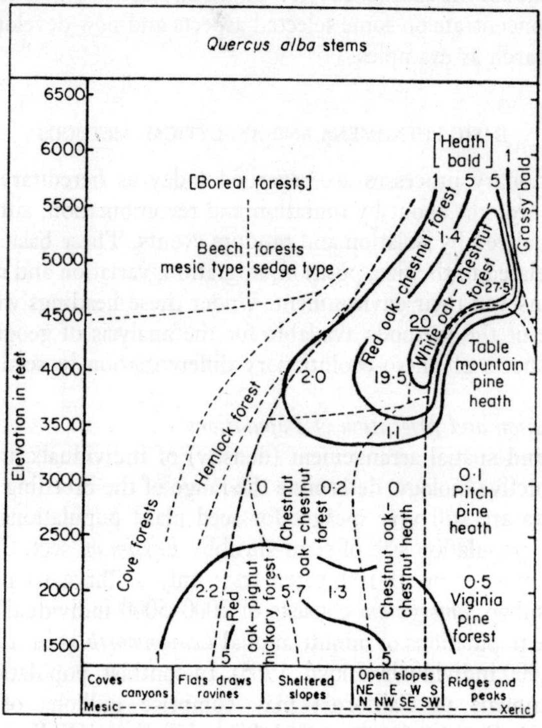

FIG. 2. Two-dimensional ecological grid showing position and density of *Quercus alba* populations in the Great Smoky Mountains, Tennessee, relative to elevation, relief, and the main forest communities (from Whittaker, 1967).

developed a combined diagram to indicate occurrence of various modifications in different vegetation types and altitudinal zones. Most comprehensive are the grid schemes (Fig. 3) which Dansereau (1952) proposes to express distribution, coverage (dominance), fidelity (ecological specialization) and dynamics (successional status).

Knowledge about the migration of populations is essential for an understanding of the direction of geographical and ecological change. Information

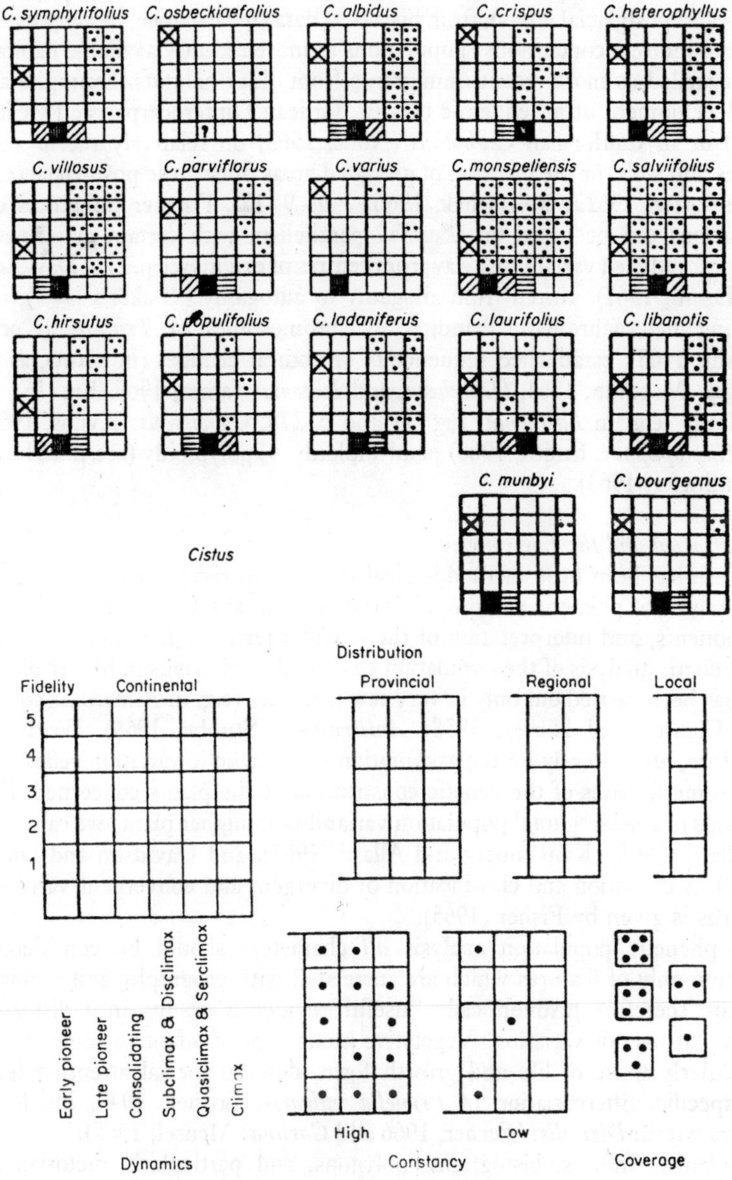

FIG. 3. Idiograms showing geographical and ecological positions of all the 17 species of *Cistus* in the Mediterranean (from Dansereau, 1952).

comes either directly from macro- or microfossils (for instance, pollen of inter- and postglacial floras), from historical data (for instance, in neophytes), or from indirect conclusions: populations in more recently available habitats are regarded as more recent immigrants from older habitats. Examples are the development of neoendemic taxa in Bornean Dipterocarpaceae (Meijer, 1963) or in Californian *Ceanothus* (Nobs, 1963) on relatively recent substrates and soils, or the invasion of glaciated areas during the postglacial (e.g. Ehrendorfer, 1962*a*; Polatschek, 1966; Figs 9, 17). Further inferences on migrations can be based on assumed parallelism with certain cytogenetic changes: reduced variability away from centres of origin (see p. 276; Davidson and Dunn, 1967), switch from allogamy to autogamy ("Baker's law"), increasing heterochromatin banding in chromosomes of *Trillium* (Kurabayashi, 1963), established sequence of structural changes (in plants, as in *Clarkia*: Mosquin, 1964, *Oenothera*, and *Isotoma*: James, 1965, Fig. 15), of dysploidy (e.g. in *Lasthenia*, Fig. 11 and p. 278, *Chaenactis*: Kyhos, 1965, and *Haplopappus*: Smith, 1966) or of diploidy → polyploidy (see p. 284 and Ehrendorfer, 1963).

2. *Variation and the Environment*

There has been little methodological progress in recent years concerning the analysis of phenetic variation, separation of modificational and genetical components, and interpretation of the latter in terms of hereditary change. The genetic analysis of the population gene pool is laborious in higher plants and has been carried out only in very few instances (e.g. in *Potentilla glandulosa*, Clausen and Hiesey, 1958; *Antirrhinum*, Stubbe, 1966). We have, therefore, often to rely on the assumption that phenetic variation reflects at least some features of the genetic constitution of the plants concerned. For methods of assessing total population variability in higher plants we can refer to Allard (1965), Kannenberg and Allard (1967), and Davidson and Dunn (1967). A definition and classification of divergent and convergent variation patterns is given by Fisher (1965).

In phenetic population analysis *all* characters should be considered; selection only of features which are correlated with geography and ecology, and are therefore taxonomically "useful" evidently results in a distorted picture of overall variation. Vegetative characters of adaptive importance, particularly those of life and growth form, deserve special attention (e.g. infraspecific differentiation of *Prunella vulgaris*: Böcher, 1949, life form spectra within *Digitalis*: Werner, 1966 and *Carlina*: Meusel, 1965).

Formulae, indices, histograms, polygons, and particularly pictorialized scatter diagrams are well known methods to bring out diversity and combination of characters within, and distinctness between, populations (see for example, Fig. 10). Special mention should be made of the remarkable taximetric computer analysis of 196 individual phenotypes in S. American groups of *Cassia* (Irwin and Rogers, 1967): on the basis of 71 characters with 395 attributes an objective clustering pattern appeared which clearly reflected regularities of geographical differentiation.

Attractive but still little used techniques in the field of morphogeographical analysis have been developed by Rothmaler (1955) and Schwarz (1938). They combine distribution maps with contours for corresponding phenetic conditions. This may be applied to certain character expressions (phenocontours), frequencies of one (or more) characters (isophenes = Isosemen, e.g. Woodson [1964] on flower colour forms in *Asclepias tuberosa*, and Fig. 9), overall number of characters (isochars = Isopsepheren), number of taxa (isoflors = Isoporien; e.g. Cook, 1966*a*, *b*, on *Ranunculus* subgen. *Batrachium*) etc. (for further details see Davis and Heywood, 1963, p. 313–321).

The importance of chromosomes as vehicles of the genetic system is evident. While abrupt structural and numerical changes are particularly important in partly sympatric differentiation (p. 281), geographical aspects of genic differentiation become apparent, for instance in the remarkable work on gene-geography in *Triticum* and *Aegilops* (Kihara, 1966).

The selective influence of the environment on variation and the adaptive importance of characters or rather character combinations are quite neglected and little understood problems, mainly left to genecological research. The remarkable stability of flower colour polymorphism during many years over distances of a few feet in the outbreeding *Linanthus parryi* (Epling *et al.*, 1960, Fig. 5), the maintenance of plant association-bound differences without evident adaptive value in the wind pollinated *Carpinus betulus* (Białobrzeska, 1966), and many other facts (Heslop-Harrison, 1964) suggest that the importance of selection for all sorts of evolutionary differentiation has been greatly underestimated until now. More extensive plotting of variation against environment should produce insights which could then be checked by various experimental approaches. The contributions by Harper (1965, etc.) and Barber (1965, etc.) on the critical and crucial seedling and juvenile stages (as compared with adult plants) demonstrate how much still can be done along these lines.

3. *Reproduction and Isolation*

As antagonistic regulators of coherence and segregation in breeding communities these two factors have vital influence on the recombination rate and the outcome of geographical and ecological differentiation. The production of gametes and diaspores is linked with the speed of generation cycles: their genetic make-up is influenced by the genetic system (chromosome number, chiasma frequency); and their mode of origin (sexual [out- or inbreeding] or asexual), and their efficiency will depend considerably on the range and accuracy of their distribution (flower and fruit biological conditions etc.). As this subject, and its relationship to problems of differentiation and taxonomy, has recently been covered by a symposium (Hawkes, 1966), it will suffice here to mention interesting recent contributions which assess relative frequencies of out- and in-breeding (Allard, 1965; Vasek, 1967*a*, *b*). For the importance and analysis of extrinsic and intrinsic, pre- and postzygotic isolation mechanisms we can refer to the contribution of Solbrig in this symposium.

PRINCIPLES OF ALLOPATRIC DIFFERENTIATION

The detailed study of the chasmophytic, semishrubby and self-fertile members of *Erysimum* sect. *Cheiranthus* in the Aegean by Snogerup (1967) can serve to demonstrate some salient features of allopatric differentiation (Fig. 4). Geological data suggest that the geographical segregation of this

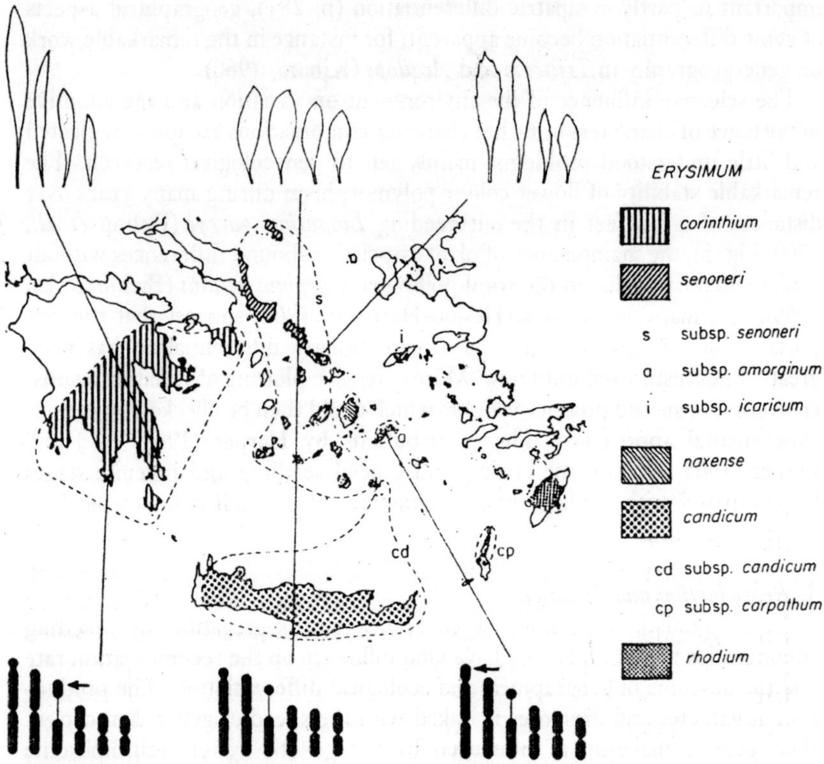

FIG. 4. *Erysimum* sect. *Cheiranthus* in the Aegean: total distribution, leaf variation, and haploid chromosome sets from selected populations (modified from Snogerup, 1967).

group began during the middle Pliocene, about 5 million years ago. The present situation has been produced during 200,000–500,000 generations by uneven distribution of genes (affecting morphological and physiological characters) and chromosome structure changes under the influence of selection and random events (genetic drift mostly in very small populations). Different levels of morphological differentiation—species, subspecies, local populations (e.g. in *E. senoneri* subsp. *senoneri*) and individuals—are apparent, and different degrees of reduced fertility and vigour (due to chromosomal and genic change) can be followed from within populations to experimental

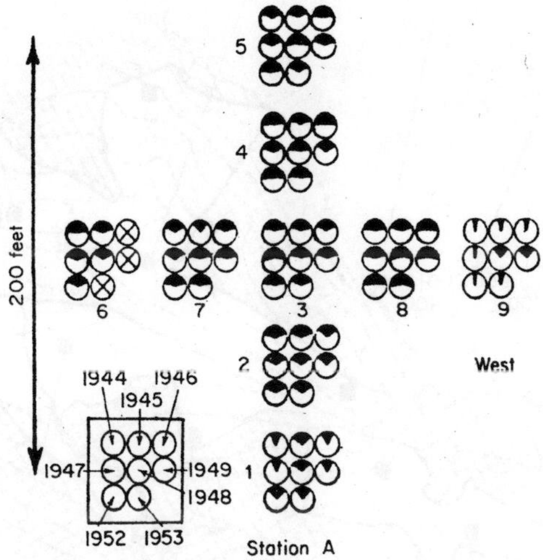

FIG. 5. Stability of flower colour polymorphism in a south Californian population of *Linanthus parryi*: relative frequencies of blue (solid) and white flowered plants in 9 closely adjacent spots of 100 square feet over a period from 1944 to 1953 (slightly modified from Epling *et al.*, 1960).

interracial and interspecific hybrids. There are only loose correlations between length of isolation, morphological divergence and intensity of crossing barriers (*cf.* the relative recently isolated but strongly diverging *E. naxense*). Differentiation is predominantly divergent, and natural hybrid introgression has only been observed on Tinos between native *E. senoneri* subsp. *senoneri* and cultivated *E. cheiri*. A basically similar pattern of gradual allopatric differentiation, more or less paralleled by the gradual building of crossing barriers has been established for other perennial groups with little chromosomal divergence like New Zealand alpine *Ranunculi* (Fisher, 1965), *Haloragis erecta* (Forde, 1964*a*, *b*), the Australian *Senecio lautus* group (Ali, 1964, 1966), and south-eastern N. American *Ruellia* species (Long, 1966).

Primary intrapopulation variability is either oligogenic or polygenic, and therefore polymorphisms tend towards the heteromorphic or the multiform pattern (Fisher, 1965). Both types may be combined with clinal, stepped or distinct interpopulation differentiation. Among allogamous annuals with

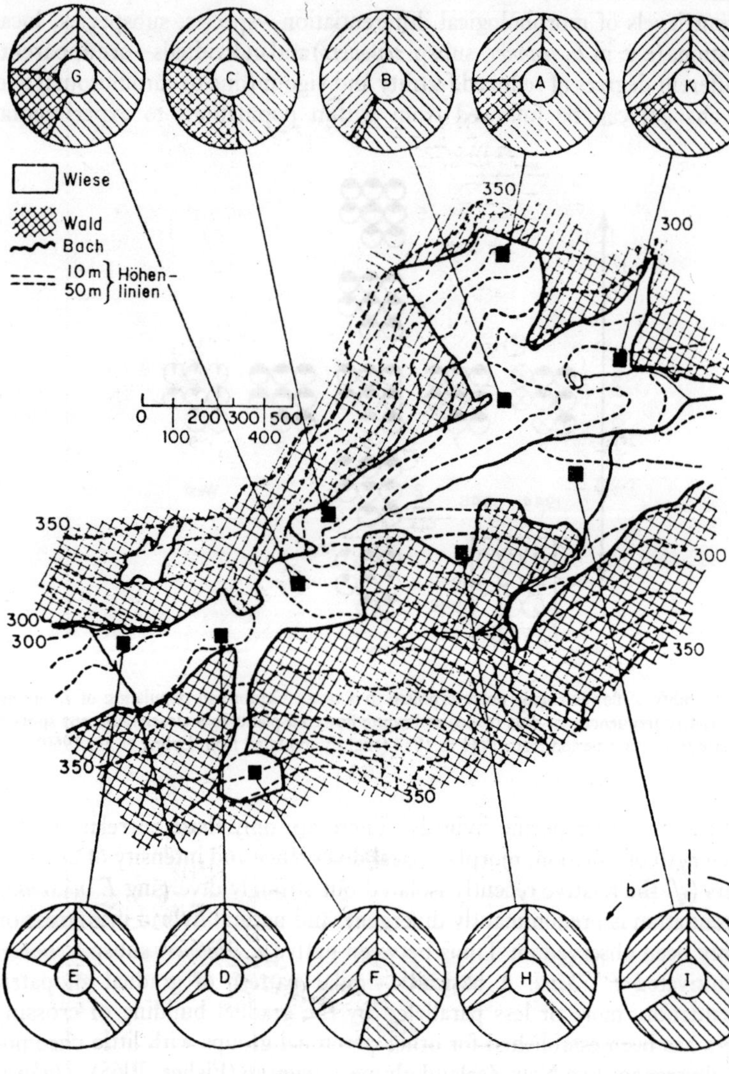

FIG. 6. Eco-geographical micro-differentiation of a population of *Galium pumilum* near Vienna. Percentage of hairy (b) and early flowering (f) individuals indicated for subpopulations on S and N exposed slopes (circles in upper and lower row); scale is in metres (from Ehrendorfer, 1953).

populations demonstrating oligogenic flower colour polymorphism *Linanthus parryi* (Epling *et al.*, 1960) is an example of great stability (Fig. 5), while *Justicia simplex* (Joshi and Jain, 1964) shows great genetic fluctuations from year to year. Examples of stepped clines of oligogenic polymorphism corre-

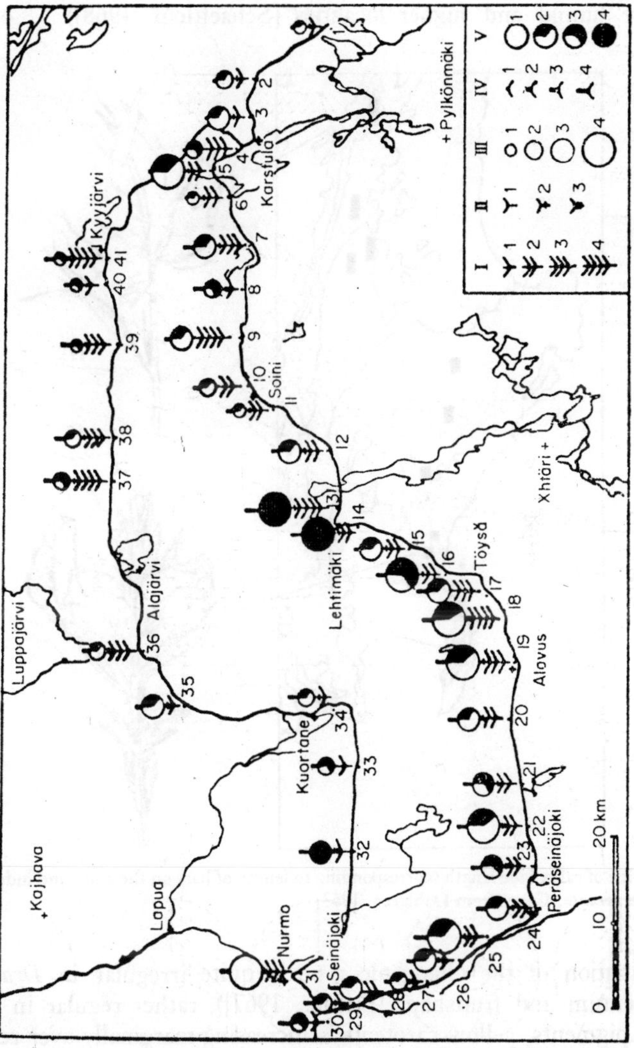

FIG. 7. Variation of *Melampyrum pratense* along two transects in central Finland. Symbols stand for branching (I), number of intercalary leaves (II), flower size (III), length of anthers (IV), and flower colour (V) (from Jalas and Rikkinen, 1962).

lated with ecology are *Galium pumilum* (hairy forms more on south-facing slopes and at lower altitudes, glabrous forms more on north-facing slopes and at higher altitudes [Ehrendorfer, 1953, Fig. 6]), *Trifolium repens* (decrease of genes for cyanogenic substances with temperature [Daday, 1965]) and the *Euphrasia rostkoviana* group (non-glandular forms replace glandular ones in more humid and higher localities [Schaeftlein, 1968]). Geographical

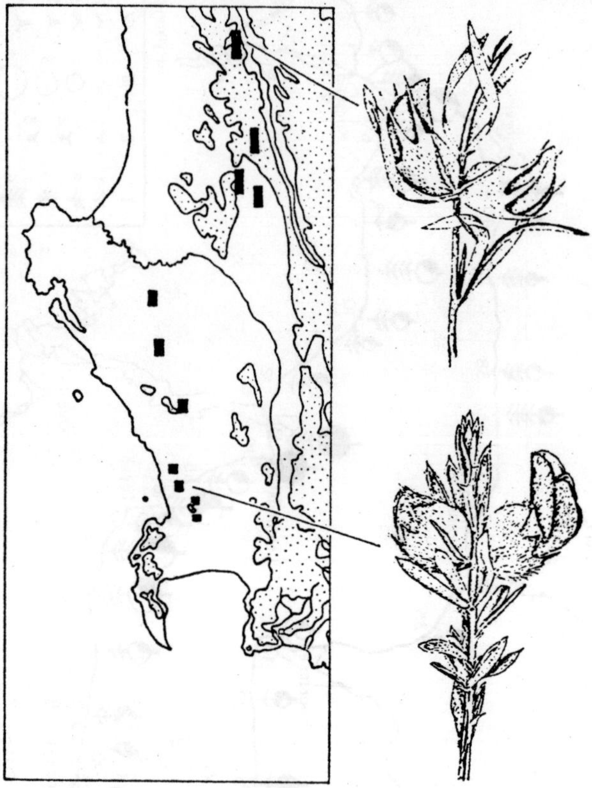

FIG. 8. Cline of calyx lobe length (corresponding to length of bars on the map) and indumentum in *Aspalathus ternata*, Cape (from Dahlgren, 1963).

differentiation of the oligogenic type is quite irregular in *Draba dubia* (indumentum and fruitshape [Buttler, 1967]), rather regular in *Asclepias* (flower pigments, yellow carotenoids increasing marginally over red anthocyanins [Woodson, 1964]). Examples of polygenic patterns are found in *Melampyrum pratense* (branching-pattern, flower-size [Jalas and Rikkinen, 1962, Fig. 7]) and *Pinus sabiniana* (fruit characters [Griffin, 1964]), both corresponding to rather irregular stepped clines, while *Pinus sylvestris* (dry-substance, growth [Langlet, 1959]) and *Aspalathus ternata* (calyx-lobe-length and indumentum [Dahlgren, 1963, Fig. 8]) show almost smooth clines. On the basis of the above mentioned and other examples one can make the generalization that smooth clines reflect smooth gradients, while stepped clines or distinct population groups mostly correspond to similar steps or breaks in the environment (or to secondary racial contacts, p. 276).

An important feature of geographical variation is that characters may vary quite independently from each other. This was demonstrated for example for turpentines and phenolic constituents as compared with morphological characters in *Pinus muricata* (Forde and Blight, 1964) and in *Baptisia leucophaea* var. *laevicaulis* (Brehm and Alston, 1964).

As a result of divergent geographical and ecological differentiation we can often observe a hierarchy of older, comprehensive and more widely distributed groups, which consist of younger, restricted and more localized races or populations. Good examples are *Primula nutans* subsp. *finmarchica* with the Arctic var. *finmarchica* and with one population-group of var. *jokelae* on the Gulf of Bothnia and another on the White Sea (L. & Y. Mäkinen, 1964), *Digitalis obscura* in Spain and the Rif with subsp. *obscura*, and subsp. *laciniata* var. *laciniata* and var. *riphaea* (Werner, 1964), and the Central Mediterranean groups of *Campanula fragilis* and *C. garganica*, each further segregated into allopatric species, subspecies, etc. (Damboldt, 1965). In all these cases the groups at the various levels of differentiation are disjunct. The following examples are more complex and show contiguous or even overlapping infraspecific races: the Californian *Lomatium dasycarpum-mohavense* (Theobald, 1966), the world-wide group of *Potentilla anserina* (Rousi, 1965), and the paleotropical *Dichrostachys cinerea* (Brenan and Brummitt, 1965). Racial contacts in such groups often seem to be secondary: *Potentilla anserina* subsp. *anserina* and subsp. *pacifica*, for example, are clearly separated in California but hybridize in eastern N. America (Rousi, 1965).

The hybrid origin of geographical and ecological variation patterns and even of new racial entities can be demonstrated, using examples from *Galium* as guide lines. Variation within the tetraploid *Galium anisophyllum* subsp. *alpino-balcanicum* in the Eastern Alps (Fig. 9) seems to be clinal and might be classified as initial and divergent. A statistical analysis of thousands of plants has shown however that the minute differential characters are specific and concentrated in three less glaciated refugial areas, around Brenner Pass, and further on in the southern and the north-eastern Alps. From there, frequencies decrease towards the more heavily glaciated areas where much intermingling is encountered. This pattern has very likely originated by fusion and amalgamation of formerly isolated and more clearly marked

Würm ice age populations. Separation of divergent and convergent evolutionary processes becomes nearly impossible in such cases, a fact which emerged also from the study of New Zealand *Ranunculi* (Fisher, 1965) or European *Knautia* (Ehrendorfer, unpubl.).

On the southern slopes of the Alpine system there is a more or less broad altitudinal belt where hybrid populations connect (and separate) the northern *Galium pumilum* and the Submediterranean *G. rubrum* (Ehrendorfer, 1955). Similar transitional belts have recently been found, for example, between *Centaurea raphanina* subsp. *raphanina* and subsp. *mixta* in the Cyclades

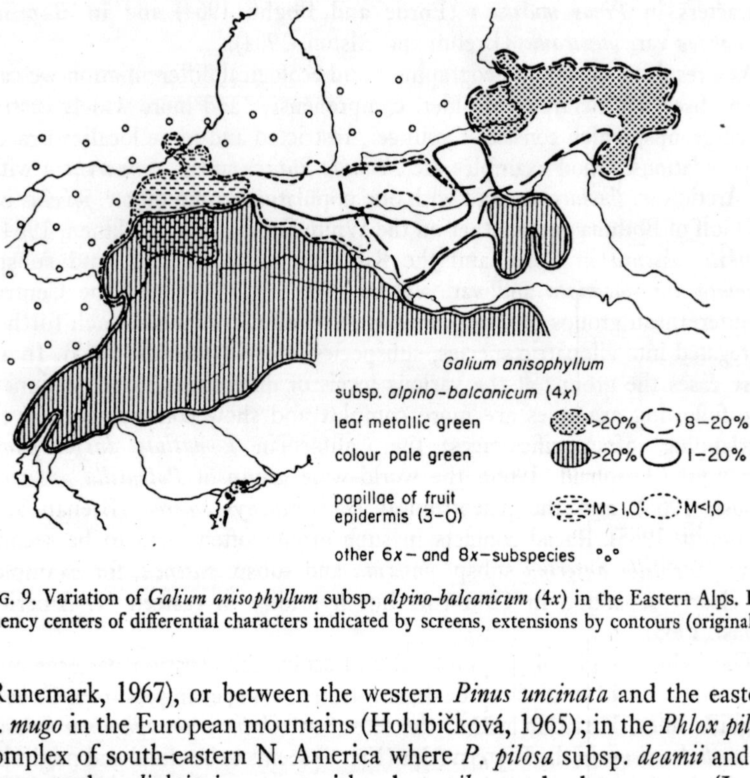

FIG. 9. Variation of *Galium anisophyllum* subsp. *alpino-balcanicum* (4*x*) in the Eastern Alps. Frequency centers of differential characters indicated by screens, extensions by contours (original).

(Runemark, 1967), or between the western *Pinus uncinata* and the eastern *P. mugo* in the European mountains (Holubičková, 1965); in the *Phlox pilosa* complex of south-eastern N. America where *P. pilosa* subsp. *deamii* and *P. amoena* subsp. *lightipei* connect with subsp. *pilosa* and subsp. *amoena* (Levin and Smith, 1966).

Hybrid contact zones may finally serve as centres of origin for new populations which migrate into areas unoccupied by the parental types (hybridization-differentiation cycles, without change of ploidy level). This is the way in which *Galium dumosum* has evidently originated in S.W. Anatolia from a contact zone between the S. Aegean *G. graecum* and the interior *G. canum*

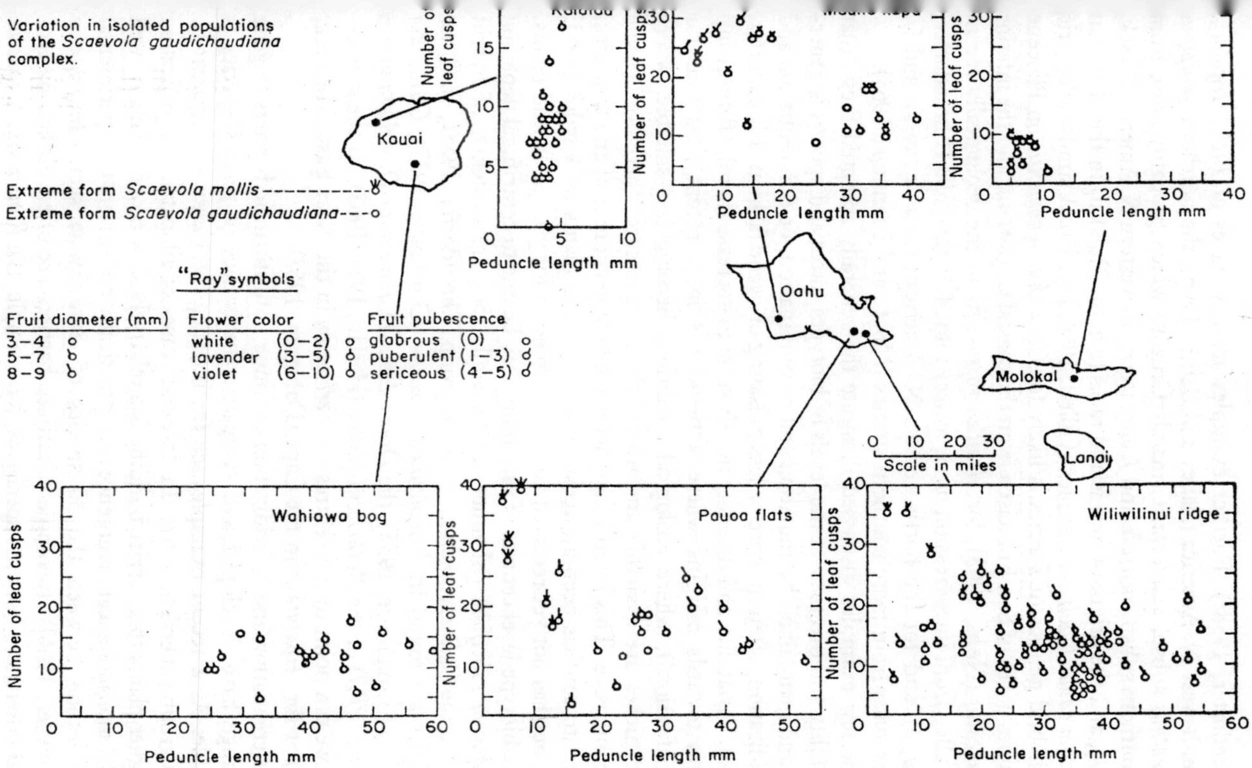

FIG. 10. Hybridization between *Scaevola gaudichaudiana* and *S. mollis* in the Hawaiian Islands; scatter diagrams of selected populations; on Molokai (Hanalilolilo) a new and characteristic recombination population (from Gillett, 1966).

(Ehrendorfer, 1958*c*). Further examples are new races of *Ptelea trifoliata* in south-west N. America (Bailey and Bailey, 1965), the northern *Juniperus occidentalis* subsp. *occidentalis* (mainly Oregon) which probably arose from the southern subsp. *australis* and *J. osteosperma* in western N. America (Vasek, 1967*b*), and the origin of new island races (e.g. on Molakai) in the Hawaiian *Scaevola gaudichaudiana* complex (Gillett, 1966, Fig. 10). A similar allopatric and hybrid genesis of species is likely for *Ceanothus sonomensis* on Pliocene volcanics north of San Francisco from the coastal *C. gloriosus* and the interior *C. cuneatus* (Nobs, 1963), for *Achillea roseo-alba* in the geologically recent Po-Valley from introgression of *A. setacea* into *A. asplenifolia* (Ehrendorfer, 1959*a*, *b*), and for the north-western N. American *Cirsium tweedii* and *C. hookerianum* from more southern species (Moore and Frankton, 1965).

The few examples discussed illustrate the generally accepted thesis that variability and diversity at the levels of biotypes, races and species is generated and maintained by mutation and recombination, or differentiation and hybridization, often in more or less pulsating or cyclic fashion (Ehrendorfer, 1959*a*). Variability and diversity mostly seem to decrease thereby from group centres towards margins where settlement is more recent (less time for diversification!), where ecological conditions become less suitable, where fewer niches are available and where selective pressures therefore become more intensive. This interesting phenomenon of geographical and ecological differentiation has been appreciated for cultivated plants by Vavilov (1951, etc.), but has not yet received sufficient attention for groups of wild plants. At the biotype level centres of variability may be due to centrifugal depletion in *Agauria salicifolia* (Sleumer, 1938), while they are evidently of hybrid origin in the *Galium canum-G. graecum* group (Ehrendorfer, 1958*b*, *c*). Other examples are found in *Eriophyllum lanatum* (Constance, 1937), *Gaillardia pulchella* (Stoutamire, 1955), the *Draba dubia-D. tomentosa-D. stellata* group (Buttler, 1967) or the *Bothriochloininae* (Harlan, 1951, 1963). At the species-level we can point to the centres for *Verbascum* in the Near East (Murbeck, 1939) or for *Aspalathus* in the Cape (Dahlgren, 1963).

Differentiation and hybridization continue until intrinsic barriers to gene exchange have developed and sympatric coexistence of related species is possible. As a recent example, see for instance the *Carex sempervirens-C. firma* group (Dietrich, 1967). In "normal" cases of allopatric diversification in higher plants this barrier building is a relatively slow process, lagging well behind morphological divergence (e.g. practically no barriers to experimental hybridization between the 30 species of *Antirrhinum* sect. *Antirrhinum* [Rothmaler, 1956]). Incomplete intrinsic barriers are often backed by ecological differences between species as, for example, the *Ranunculus lappaceus* group (Briggs, 1962), *Primula vulgaris* and *P. veris* (Woodell, 1965) and *Cercocarpus* (Brayton and Mooney, 1966), and/or by selective pressure on hybrids as, for example, in N. American *Lespedeza* (Clewell, 1964) where hybrids originate in forest openings but are soon suppressed by succession. Prezygotic incompatibility barriers may be selected in sympatric races in order to avoid waste of gametes and zygotes, and maintain an optimum seed-set. Indications of the origin of such intrinsic and prezygotic barriers have been found in some annuals, as in the leafy-stemmed group of *Gilia* (Grant, 1966) where there is much less incompatibility in allopatric than in sympatric members. Whether the standard types of intrinsic and postzygotic barriers in higher plants (as in *Laya*, Clausen, 1951, and Stebbins, 1966) are also more or less dependent on selection is still open to question. It seems at least likely that a loosening of selective pressure favours not only morphological diversification but also the origin of chromosomal change and corresponding barrier building (Lindsay and Vickery, 1967, for *Mimulus*).

It is relatively easy to transpose "normal" patterns of allopatric differentiation into terms of the taxonomical hierarchy. The use of "subspecies" for geographical and/or ecological races which are more or less morphologically differentiated, but connected with other races by transitional forms, and therefore insufficiently distinct, has become more or less universally accepted. In contrast, "varietas" is nowadays mostly relegated to the status of a container for specimens of uncertain status. "Forma" is not generally considered as a term for populations but rather for certain biotypes with striking oligogenic characters. They should mostly remain without formal names and can be marked by formulae, indices or numbers (e.g., Ehrendorfer, 1958*c*; Buttler, 1967). Different frequencies of certain forms may some times serve to characterize subspecies. Difficulties and uncertainties arise in regard to the use of "species" or "subspecies" where taxa are widely disjunct or where they are connected by hybridization. Personally, I favour rather narrowly circumscribed species which can be grouped conveniently into informal species aggregates. Other problems are encountered where there is clinal or extensive infraspecific "box-in-box" differentiation, or where parallel mutations or similar environmental conditions produce parallel or convergent variation. Possible solutions are to recognize (unnamed) "minor variants" and to recognize—if necessary—only one taxon, even in cases where polyphyletic origins are likely.

POSSIBILITIES FOR SYMPATRIC DIFFERENTIATION

Sympatric differentiation implies the origin of intrinsic isolating barriers and new races within the dispersal range of the parental population. The problem has been extensively discussed by Mayr (1963) who minimizes its importance and even questions its occurrence in animal speciation. Less attention has been focused on this question by botanists, but a number of findings suggest that one should not rashly extend such conclusions to higher plants.

As a recently analysed example we can point to some annual members of the mostly outcrossing Compositae genus *Lasthenia* from western N. America (Ornduff, 1966, and Fig. 11). In *L. chrysostoma* we find, in addition to strong

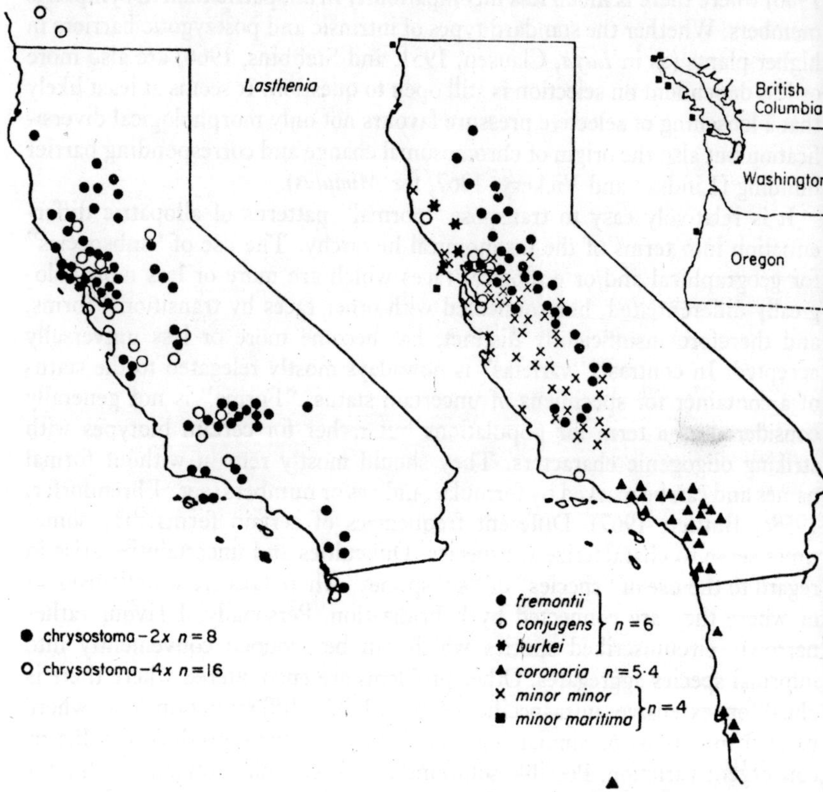

FIG. 11. Distribution and abrupt cytogenetic differentiation n *Lasthenia*, western N. America: infraspecific polyploidy in *L. chrysostoma* (sect. Baeria), dysploidy in sect. *Ptilomeris* (slightly modified from Ornduff).

and not clearly eco-geographically correlated morphological variation, "cryptic" diploid and tetraploid cytotypes, often in closely adjacent populations. Within *L. coronaria* chromosomal differentiation has proceeded to dysploidy ($n = 5 \rightarrow 4$), creating corresponding crossing barriers, but here again there is no clear parallelism between this, the remarkable morphological diversity, and distribution. *L. minor* subsp. *maritima* has switched from allogamy to autogamy, backing the coexistence of different chromosomal types (with influence on crossing fertility) within single localities. Similar crossing barriers evidently make possible the common sympatric occurrence even of closely related *Lasthenia* species (e.g. *L. fremontii* and *L. conjungeus*).

The *Lasthenia* example indicates various avenues towards the prerequisite of sympatric differentiation, i.e. restriction of intrapopulation gene flow. As the following discussion is intended to demonstrate, this is promoted by strongly divergent ecological differentiation, reduction of outbreeding or sexuality, and accomplished by the more or less abrupt origin of intrinsic barriers. While the building of crossing barriers is retarded in allopatric differentiation, it tends to become precocious here.

1. Strongly Divergent Ecological Differentiation

Brief comments only on this topic are necessary here as the contribution by Wilkins in this volume and the review by Heslop-Harrison (1964) have adequately stressed the great importance of this phenomenon for allo- and sympatric evolution in higher plants. Initial and more or less sympatric ecological divergence will be basic to further differentiation when part of the ecocline or ecotype series is removed by climatic and vegetational changes. Numerous ecological races, in open coastal associations or in open serpentine habitats have become isolated from formerly widespread parental groups in tundras or grasslands because the latter were eliminated or replaced by the invasion of closed forest communities; today, such relict ecotypes have often achieved subspecific or specific status. If environmental gradients are steep, crossing barriers may even originate *in situ*: cytogenetic changes producing a pre- or postzygotic barrier effect would have a selective advantage under such conditions, for example in a local serpentine ecotype; individuals without barriers would be continuously contaminated with pollen from the surrounding non-serpentine ecotype; their progeny would be bad competitors on serpentine soil against well adapted progeny from individuals protected by such a barrier from unwanted amalgamation. Evidently much experimental work has still to be done in this fascinating field.

2. Reduction of Outbreeding and Sexuality (Prezygotic Mechanism)

Differences in flower-structure and -colour, coupled with flower constancy of pollinators, differences in flowering-time, and incompatibility reactions may contribute to sympatric differentiation but their overall importance is probably quite small. As examples, we may point to the sympatric selection of new flower types by pollinators from among hybrid swarms (e.g. in *Ophrys* [Stebbins, 1959]) or to the intrapopulation differentiation of *Galium pumilum* (Ehrendorfer, 1955, Fig. 6), which is backed by somewhat different flowering times on N. and S. slopes. In contrast, facultative or obligate autogamy and apomixis evidently greatly enhance the possibilities of sympatric divergence.

The Mediterranean sect. *Jubogalium* is an appropriate (large-scale) example of the different variation pattern 1) in primitive, allogamous perennials, reflecting a former allopatric differentiation (e.g. the disjunct species pair *G. cappadocicum*-*G. jungermannioides* [E. Anatolia-Amanus to Antilibanon] or the S.E. Mediterranean group of *G. graecum*-*G. thiebautii*-*G. canum* [all 2*x*], secondarily fused by hybridization); 2) in intermediate, localized allogamous annuals (e.g. the S.W. Anatolian *G. pamphylicum*); and 3) in strongly advanced, widespread, autogamous annuals, overlapping most of the other

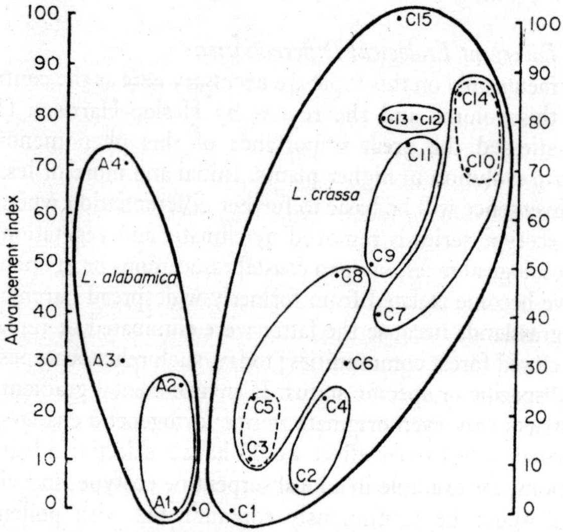

FIG. 12. Origin of autogamous from allogamous races in *Leavenworthia alabamica* and *L. crassa*. Vertical axis (advancement index) indicates increase of characters linked with autogamy, horizontal axis stands for other morphological divergence (from Lloyd, 1965).

species and showing extensive mosaic variability but little correlation with geographical distribution (Mediterranean + Irano-Turanian *G. setaceum*, with "cryptic" polyploidy: 2*x* and 4*x*) (Ehrendorfer, 1958*b*, *c*, 1965*a* and unpubl.).

On a smaller scale, the detailed analysis of the annual Cruciferae *Leavenworthia alabamica* and *L. crassa* (Lloyd, 1965, Fig. 12) is representative of the break-down of self-incompatibility and outbreeding, and its replacement by inbreeding. This is paralleled by progressive reduction of numbers, size, colour, pollen-quantity, filament and style length of the flowers (expressed in an advancement index 0–100). The primitive allogamous races have wider distribution areas and are less differentiated, while the advanced autogamous groups tend to break up into numerous localized and often more or less sympatric races which can produce new lines by occasional hybridization.

The autogamous W. European species pair *Ranunculus hederaceus-R. omiophyllus* arose probably by sympatric differentiation through the origin of genic barriers from a common gene pool; cryptic 2*x* and 4*x* cytotypes are *in statu nascendi* in both species (Cook, 1966*a*, *b*). Interesting information about the sympatric occurrence, genesis, and variation pattern in Californian autogamous annuals (mainly grasses), has been assembled by Allard (1965), and Kannenberg and Allard (1967).

Changes of chromosome structure appear in Aegean species of the annual and autogamous *Nigella arvensis* group (Strid, 1965). The Syros population of *N. aristata* contains types with different chromosome forms and different flowering-time. The origin of structural changes and parallel crossing barriers under the protection of autogamy is even more pronounced in the Mediterranean and annual *Trifolium subterraneum* complex (Katznelson and Morley, 1965). Through occasional hybridization such "precocious" barriers in selfers can segregate to produce numerous new and isolated races, as in perennial grasses of *Elymus glaucus* agg. (Stebbins, 1959). The compensation of inbreeding by increasing structural heterozygosity is exemplified by S.W. Australian populations of *Isotoma petraea* (James, 1965), while the annual *Veronica hederifolia* group demonstrates the combination of autogamy with wide sympatry of very similar 2*x*, 4*x* and 6*x* cytoypes (Fischer, 1967, Fig. 16). Similar, but even more intricate relationships have been found in the western N. American *Gilia inconspicua* complex with at least 23 sibling and mostly sympatric microspecies at the 2*x*, 4*x* and 8*x* levels, separated by intrinsic barriers of various kinds and intensity (Grant, 1964). In *Boisduvalia* (Onagraceae) annual life forms and the switch from allo- to auto-, and to cleistogamy is linked with dysploidy, polyploidy, barrier building and extensive sympatric occurrence (Raven and Moore, 1965). Few comments are necessary on the well known pattern of apomixis or agamospermy in various (mostly perennial or woody) Angiosperms. The balance between predominant asexual and occasional sexual reproduction or even hybridization in such groups allows for the maintenance and production of most diverse polyploid and aneuploid races and their often sympatric occurrence. As recent contributions we may cite work on the *Ranunculus cassubicus* group, *Lithophragma*, *Taraxacum* and various grasses: *Bothriochloininae*, *Poa* etc. (for references see Ehrendorfer, 1965*b*, 1967).

3. *Abrupt Origin of Intrinsic Barriers* (*Postzygotic Mechanisms*)

The examples mentioned above have already shown how intrinsic barriers, mainly through structural changes, dysploidy and polyploidy, can originate

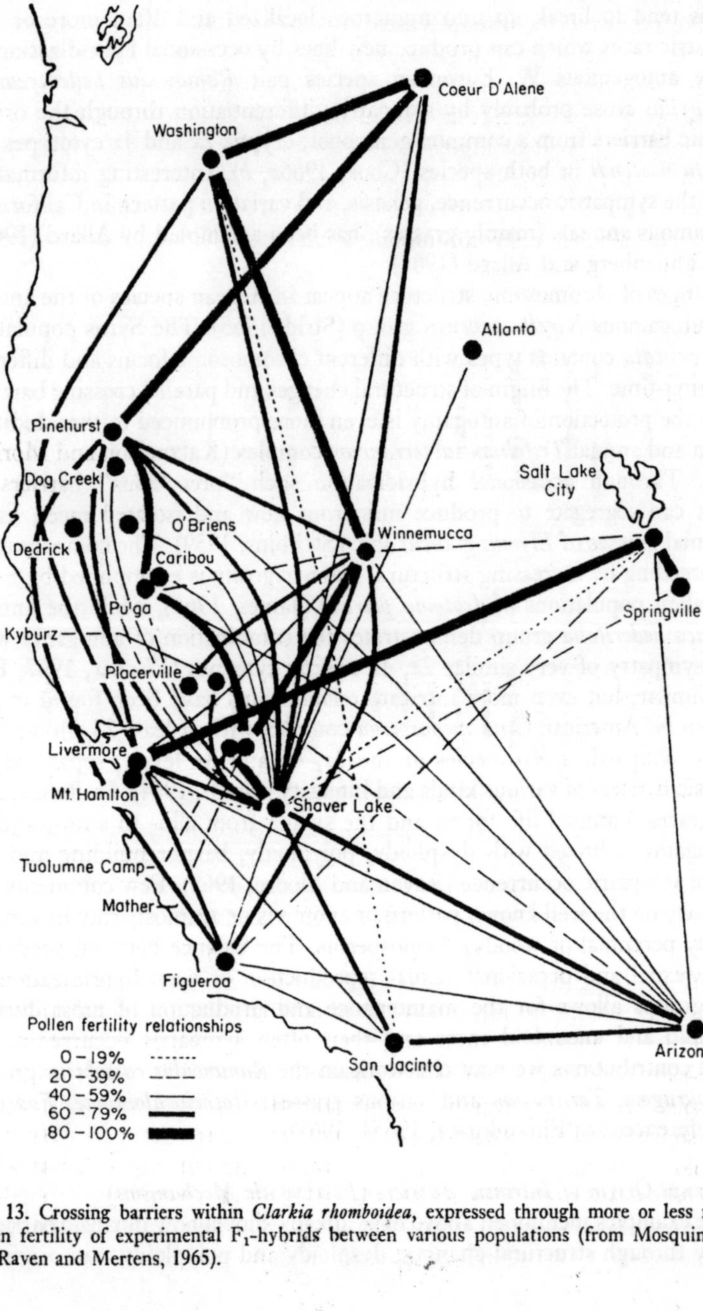

FIG. 13. Crossing barriers within *Clarkia rhomboidea*, expressed through more or less reduced pollen fertility of experimental F_1-hybrids between various populations (from Mosquin, 1964; and Raven and Mertens, 1965).

even under sympatric conditions when gene flow is restricted. A similar result can be expected when selective pressures in relation to such cytogenetic changes are only slightly negative or positive, as in newly opened habitats or in cases of favourable effects of heterosis, supergene construction, recombination rate, protection from hybrid amalgamation, etc.

In many rapidly expanding groups structural changes and even dysploidy, together with corresponding barrier building, have been observed, often without visible morphological effects. Apart from the examples from *Lasthenia* we can refer to initial population differentiation in *Anemone*

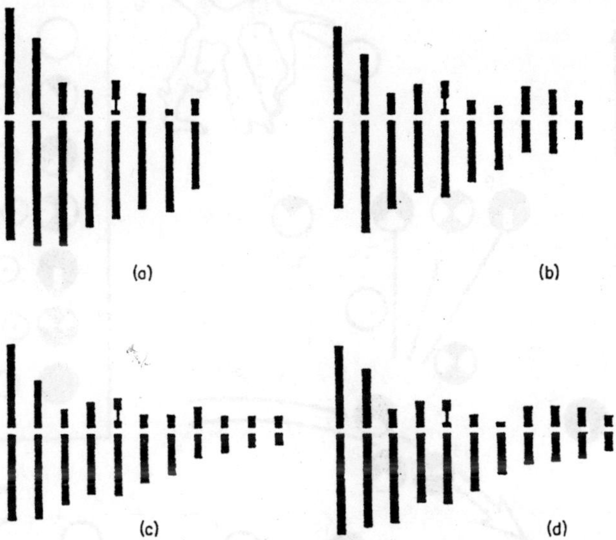

FIG. 14. Chromosome structural and numerical differentiation within *Ornithogalum lanceolatum* in the Near East (a) Lebanon, $2n = 16$, (b) Anamur, $2n = 20$, (c) Muğla, $2n = 22$, (d) Seyhan, $2n = 22$ (and 20!) (slightly modified from Cullen and Ratter, 1967).

cylindrica (Heimburger and Kamitakahara, 1963) and in *Trillium kamtschaticum* (Fukuda, 1967), crossing barriers within *Clarkia rhomboidea* (Mosquin, 1964, Fig. 13), intra- and interpopulation divergence in *Ornithogalum lanceolatum* (Cullen and Ratter, 1967, Fig. 14) and several cases of near-sympatric species origin by saltation (Lewis, 1966). Even short periods of isolation and catastrophic selection affecting small population fragments will sometimes be sufficient to set in motion an avalanche of structural and dysploid change, as for example in various annual species like *Clarkia lingulata* and *C. franciscana* (Lewis, 1966), *Cruciata articulata* (Ehrendorfer, 1965*a*), *Knautia orientalis* and *K. degenii* (Ehrendorfer, 1962*b*, and unpubl.), *Haplopappus* (Smith, 1966), *Chaenactis* (Kyhos, 1965) etc.

Polyploidy is regarded as the most typical cause of abrupt and sympatric origin of intrinsic barriers. Nevertheless, much recent information indicates that ploidy barriers can often be bridged rather easily by hybridization as in *Betula* (Dugle, 1966) or in *Salix glauca* (Argus, 1965), where continuous clines traverse different ploidy levels. Referring to recent work by way of illustration, we encounter a continuous series from occasional autopolyploids

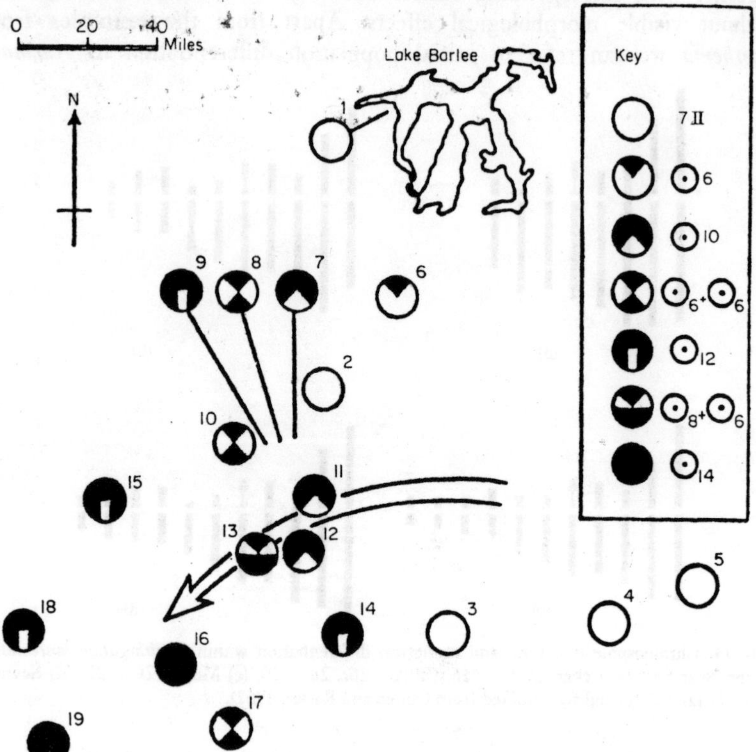

FIG. 15. Increasing structural heterozygosity along migration track of *Isotoma petraea* in SW Australia. Change from populations with seven meiotic bivalents (and floating interchanges) to those with a ring of 14 (modified from James, 1965).

with similar morphology and irregular distribution (e.g. *Lasthenia chrysostoma* [Fig. 11]; *Cardaminopsis petraea* [Polatschek, 1966]; *Ranunculus* subgen. *Batrachium* [Cook, 1966*a*, *b*]; *Zinnia juniperifolia* [Torres, 1965]), to groups where diploids and polyploids show no or very weak morphological divergence but a more regular vicarious distribution of diploids and polyploids (e.g. *Thlaspi alpinum* [Fig. 17] and the *Leucanthemum maximum* group

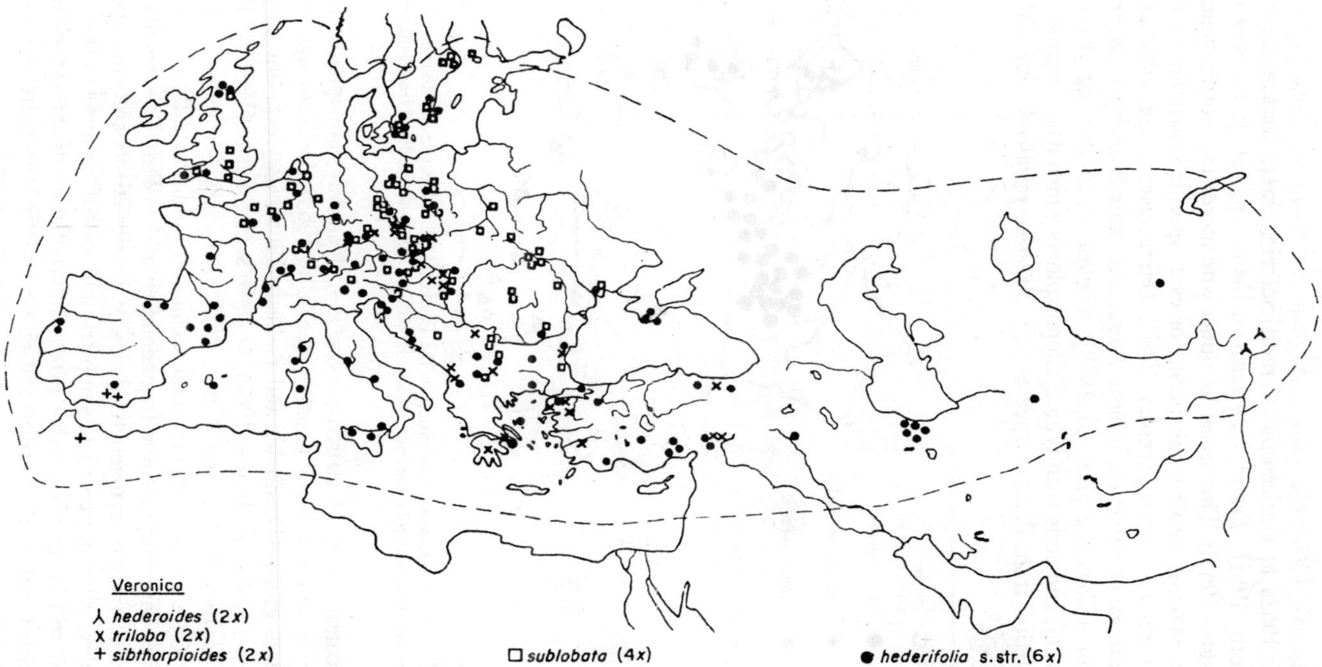

FIG. 16. Widely sympatric distribution of (micro-)species on different ploidy levels in the autogamous *Veronica hederifolia*-group (modified from Fischer, 1967).

[Polatschek, 1966]; *Galium anisophyllum* and *Knautia arvensis* [Ehrendorfer, 1958*a*, 1962*a*, *b*]; *Gutierrezia sarothrae* [Solbrig, 1964]; *Campanula rotundifolia* [Podlech, 1965]; *Eriophyllum lanatum* [Mooring, 1966]; *Urginea maritima* [Battaglia, 1965]). This series continues with increasing morphological divergence, making possible subspecific or even specific separation of cytotypes (e.g. *Caltha palustris* [Wcisło, 1967]; *Draba cinerea* agg. [Böcher, 1966]; *Lamiastrum* [Polatschek, 1966]; *Pinguicula vulgaris* group [Casper, 1966]; *Galium mollugo* agg. [Krendl, 1968]; *Achillea millefolium* agg. [Ehrendorfer, 1959*a*, *b*]) and ends with typical allopolyploids which have originated by hybridization from distinct species (e.g. *Spartina townsendii* agg. [Marchant, 1967]).

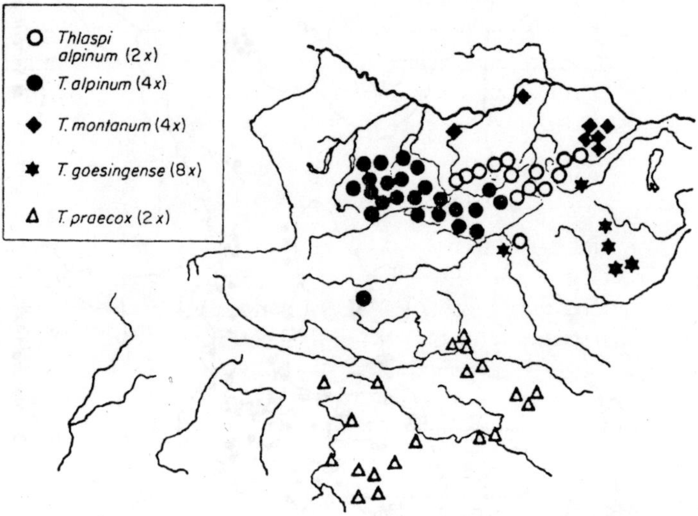

FIG. 17. The *Thlaspi montanum* group in the Eastern Alps: allopatric distribution of species on different ploidy levels and cytotypes: 2*x* and 4*x* within *T. alpinum* (modified from Polatschek, 1966).

A combination of structural changes, dysploidy, aneuploidy and polyploidy is found in many Crassulaceae (Uhl, 1963), the *Saxifraga exarata-S. moschata* group (Damboldt and Podlech, 1965), *Claytonia* and *Hedyotis* + *Oldenlandia* (W. H. Lewis, 1967), *Mimulus* sect. *Simiolus* (Vickery, 1966) and *Ornithogalum* (Cullen and Ratter, 1967); *Eleocharis* (Strandhede, 1965) and other Cyperaceae and Juncaceae exhibit in addition diffuse centromeres, together with chromosome fragmentation and fusion. This results in a remarkable array of cytogenetic diversity, barrier effects and correspondingly complex, often more or less sympatric and mosaic patterns in these groups.

It is still uncertain to what extent differentiation of the cytoplasm, often combined with male sterility and incompatibility reactions may contribute to (partly) sympatric divergence as, for example, in *Solanum* (Grun and Aubertin, 1966; Caspari *et al.*, 1966).

The taxonomic treatment of groups with a tendency to restriction of gene flow and precocious crossing barriers, and therefore with mosaic and partially sympatric differentiation, evidently presents numerous problems. The application of informal "species aggregates", "species", "subspecies", and "varieties" in such cases should be guided mainly by practical considerations. The best procedure in my opinion is to follow the general standards set in related outbreeding taxa with allopatric differentiation. For example, classification of the polyploids mentioned above might therefore vary from unnamed forms to varieties, subspecies or species. Here, as in other instances, the rank of "subspecies" will sometimes have to be accorded to taxa which are genetically isolated but otherwise still insufficiently differentiated. The use of hard and fast rules, "biological definitions" etc. is bound rather to increase than to solve our taxonomic difficulties in such groups (for an excellent and detailed discussion see Davis and Heywood, 1963).

CONCLUSIONS AND OUTLOOK

In the third and fourth sections of this review I have tried to characterize two types of ecological and geographical differentiation in seed plants and to document them by recent examples. Some of the relevant conclusions are summarized schematically in Fig. 18; individuals are indicated by circles, their distance signifies ecogeographical position, their size and colouring stands for morphological and genetical constitution; broken to continuous lines around and between groups of individuals characterize incomplete to complete barriers to crossing and gene flow. Two groups (I and II) can be followed in their evolutionary history along four levels of the time scale from bottom upwards. These two groups stand for the types of differentiation mentioned: I follows an allopatric pattern, II tends towards a sympatric pattern. In I, gene flow is relatively unrestricted (outbreeding), and we can follow the development of more or less clinal and large scale differentiation, geographical isolation of a subgroup and retarded, allopatric origin of only weak barriers, well correlated with morphological divergence. The next phases indicate hybrid fusion and partial amalgamation, and the origin of a new recombination subgroup with some novel characteristics. Differentiation in I is more by "segregation", relatively slow and "conservative". In II, gene flow is somewhat reduced (e.g. by predominant inbreeding), differentiation is rather irregular and follows a mosaic pattern; no or little correlation with precocious, often abrupt and more or less sympatric origin of barriers is discernible. The barriers may reinforce a certain sorting of variation and further divergence. Initial races appear, develop in a centrifugal direction and become isolated or sympatric again. There is no hybrid merging but occasional origin of new hybrid races, more or less surrounded by recombined or abruptly established barriers. Differentiation in II is more by "budding", relatively fast and progressive. This scheme is clearly simplified and idealized, and numerous cases exist which fall more or less between the two extremes emphasized. Its usefulness as a working hypothesis will have to be tested.

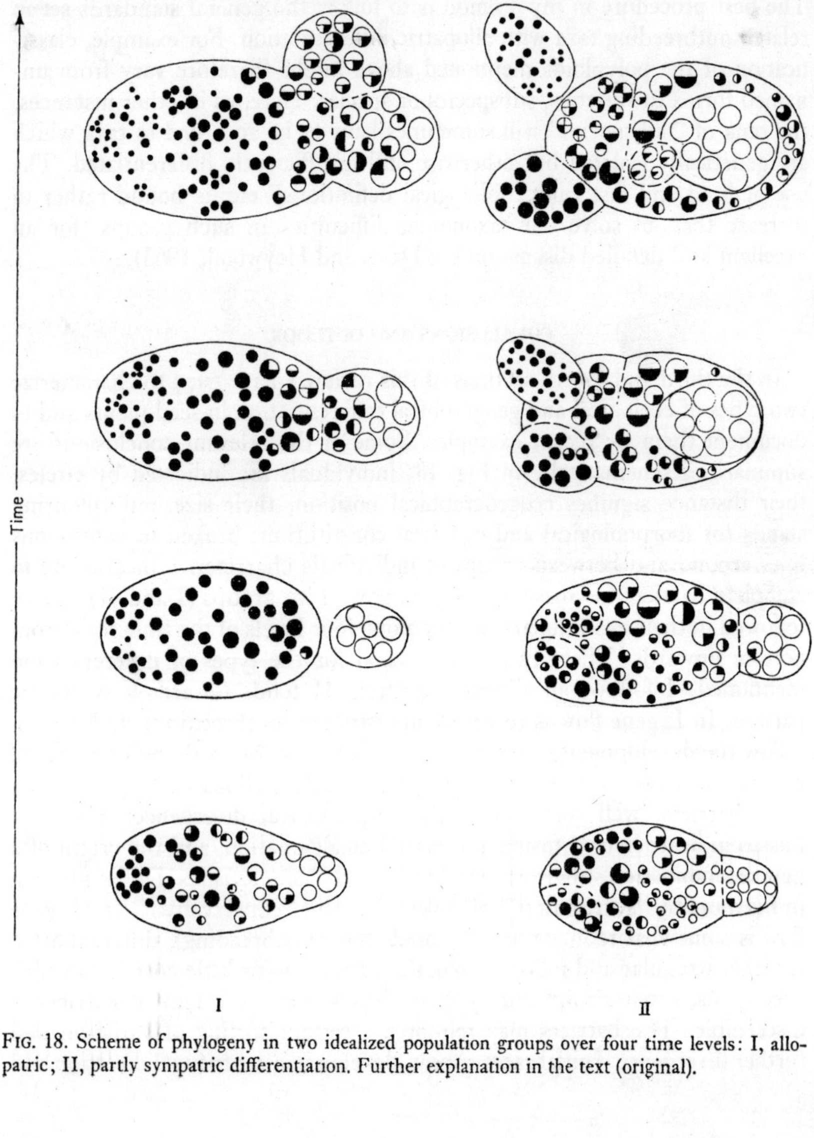

FIG. 18. Scheme of phylogeny in two idealized population groups over four time levels: I, allopatric; II, partly sympatric differentiation. Further explanation in the text (original).

Further progress in our understanding of geographical and ecological differentiation should mainly come from detailed analysis of relatively uncomplicated populations and their cytogenetic constitution, from broad surveys of groups based on a multitude of methods, and from a comparative and synthetic approach digesting the enormous amount of scattered data already available. There are a few questions which seem particularly important and fascinating. Differentiation *in situ* or during migration? Regularities of character and gene distribution (primitive—derived, dominant—recessive, accumulation, etc.) relative to centre and margin of area? Initial differentiation—geographical or ecological and local? Environmental influence on speed and patterns of differentiation? For the last question one should refer to general remarks by Levins (1962, 1963), as well as Stebbins (1952) on arid habitats and Carlquist (1966) on insular floras. I have suggested elsewhere (Ehrendorfer, 1959*a*, 1962*a*), that in relation to the ecosystem there is a cooperation of centrifugal differentiation and centripetal hybridization, and that stable communities tend to harbour old relict types, while successional and climax biotopes often allow for expansion of recent elements. A comparison of depauperate, localized and disjunct with actively evolving and more or less continuous groups (e.g. *Galium baldense* agg.—*G. anisophyllum*, *G. pumilum*: $2x$ and $4x$ ancestors—$8x$, *Knautia velutina* groups and *K. arvensis* group: Ehrendorfer, 1958*a*, 1962*a*; *Cistus varius* or *C. bourgeanus* and *C. monspeliensis* or *C. salviifolius*: Dansereau, 1952, and Fig. 3) suggests an intimate causal and "cybernetic" connexion between various geographical, ecological and cytogenetic aspects, which might be expressed somewhat like this:

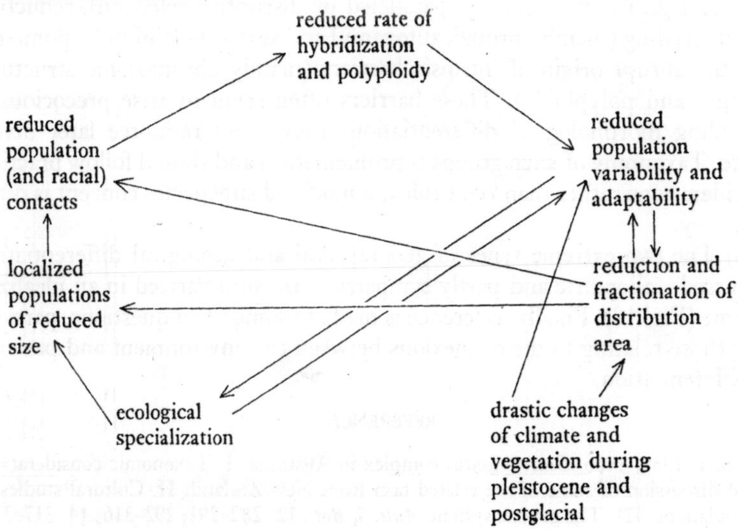

Since Kerner's publication on *Tubocytisus* and the principles of geographical divergence in 1869 much new information has come to hand, but at the same time many new questions have been raised. Keen observations and experiments, analysis and synthesis, so well balanced in the contributions of Kerner, set a high standard for the immense work still to be done in the field of geographical and ecological aspects of infraspecific differentiation.

SUMMARY

(1) The classical pattern of correlated morphological and geographical (or ecological), allopatric differentiation was established by Kerner-Marilaun, elaborated by Wettstein and soon widely adopted, particularly in zoological taxonomy.

(2) The methods available for the analysis of some of the basic phenomena involved in ecogeographical differentiation (size, environmental position and migration of populations, variation and selection, reproduction and isolation) are illustrated from recent contributions.

(3) The allopatric pattern is found most typically in allogamous groups without restriction on intrapopulation gene flow. Polymorphism, clinal and stepped initial differentiation, development of "box-in-box" population systems, and secondary hybrid fusion are discussed. Application of the subspecies concept is often very suitable but difficulties arise with clinal or hierarchical differentiation, hybridization etc.

(4) More complex and partly sympatric patterns are found in groups where intrapopulation gene flow is restricted. This may be due to strongly divergent ecological differentiation (paralleled by disruptive selection), reduction of outbreeding (mainly through autogamy) and sexuality (through apomixis), and the abrupt origin of intrinsic barriers (mainly chromosome structural changes and polyploidy). These barriers often seem to arise precociously, preceding morphological differentiation; they often reinforce later divergence. Taxonomy of such groups is problematical and should follow practical considerations rather than rigid rules; a modified subspecies concept is often necessary.

(5) The two extreme types of geographical and ecological differentiation discussed—allopatric and partly sympatric—are summarized in an idealized scheme (Fig. 18). Finally, reference is made to some open questions, particularly those relating to the connexions between the environment and patterns of differentiation.

REFERENCES

ALI, S. I. 1964, 1966. *Senecio lautus* complex in Australia. I. Taxonomic considerations and discussion of some of the related taxa from New Zealand. II. Cultural studies of populations. III. The genetic system. *Aust. J. Bot.*, 12: 282–291, 292–316; 14: 317–327.

ALLARD, R. W. 1965. Genetic systems associated with colonizing ability in predominantly self-pollinated species. *In*: H. G. Baker and G. L. Stebbins (ed.). *The Genetics of Colonizing Species*, 49–75. Academic Press Inc., New York.

ARGUS, G. W. 1965. The taxonomy of the *Salix glauca* complex in North America. *Contr. Gray Herb. Harvard Univ.*, **196**.

BAILEY, H. E. and BAILEY, V. L. 1965. Phylogenetic studies on the genus *Ptelea* (Rutaceae). Chromatography as an aid to clarification of taxonomic relationships between varieties of *Ptelea trifoliata* subsp. *angustifolia* and subsp. *pallida*. *Lloydia*, **28**: 27–43.

BARBER, H. N. 1965. Selection in natural populations. *Heredity*, **20**: 551–572.

BATTAGLIA, E. 1965. Nuovi reperti di biotipi cariologici 2*n*, 4*n*, 6*n*. *Caryologia*, **17**: 557–565.

BIAŁOBRZESKA, M. 1966. Variability of the leaves and fruits of *Carpinus betulus* in analogous associations of the primeval forests of Puszcza Bialowieska, Puszcza Borecka, Puszcza Niepołomicka and of the Carpathian Mountains. I. Białowieza National Park. *Acta Soc. Bot. Polon.*, **35**: 401–424.

BÖCHER, T. W. 1949. Racial divergencies in *Prunella vulgaris* in relation to habitat and climate. *New Phytol.*, **48**: 285–324.

BÖCHER, T. W. 1966. Experimental and cytological studies on plant species. IX. Some arctic and montane crucifers. *Biol. Skr.*, **14**(7).

BRADSHAW, M. E., DANSEREAU, P. and VALENTINE, D. H. 1964. Notes on the genus *Alchemilla* in southeastern Canada. *Can. J. Bot.*, **42**: 89–104.

BRAYTON, R. and MOONEY, H. A. 1966. Population variability of *Cercocarpus* in the White Mountains of California as related to habitat. *Evolution*, **20**: 383–391.

BREHM, B. G. and ALSTON, R. E. 1964. A chemotaxonomical study of *Baptisia leucophaea* var. *laevicaulis* (Leguminosae). *Am. J. Bot.*, **51**: 644–650.

BRENAN, J. P. M. and BRUMMITT, R. K. 1965. The variation of *Dichrostachys cinerea* (L.) Wight & Arn. *Bol. Soc. Brot.*, 2ª ser. **39**: 61–115.

BRIGGS, B. G. 1962. Interspecific hybridization in the *Ranunculus lappaceus* group. *Evolution*, **16**: 372–390.

BUTTLER, K. P. 1967. Zytotaxonomische Untersuchungen an mittel- und südeuropäischen *Draba*-Arten. *Mitt. bot. St Samml., Münch.*, **6**: 275–362.

CARLQUIST, S. 1966. The biota of long-distance dispersal. IV. Genetic systems in the floras of oceanic islands. *Evolution*, **20**: 433–455.

CASPARI, E., WATSON, G. S. and SMITH, W. 1966. The influence of the cytoplasmic pollen sterility on gene exchange between populations. *Genetics*, **53**: 741–746.

CASPER, S. J. 1966. Monographie der Gattung *Pinguicula* L. *Bibl. Bot.*, **127/128**.

CLAUSEN, J. 1951. *Stages in the Evolution of Plant Species*. Cornell Univ. Press Ithaca, New York.

CLAUSEN, J. and HIESEY, W. H. 1958. Experimental studies on the nature of species. IV. Genetic structure of ecological races. *Carnegie Inst. Wash. Publs*, **615**.

CLEWELL, A. F. 1964. The biology of the common native *Lespedeza* in southern Indiana. *Brittonia*, **16**: 208–219.

CONSTANCE, L. 1937. A systematic study of the genus *Eriophyllum* Lag. *Univ. Calif. Publs Bot.*, **18**(5): 69–136.

COOK, C. D. K. 1966*a*. A monographic study of *Ranunculus* subgenus *Batrachium* (DC.) A. Gray. *Mitt. bot. Staatssammlg., Münch.*, **6**: 47–237.

COOK, C. D. K. 1966*b*. Studies in *Ranunculus* subgenus *Batrachium* (DC.) A. Gray. III. *Ranunculus hederaceus* L. and *R. omiophyllus* Ten. *Watsonia*, **6**: 246–259.

CULLEN, J. and RATTER, J. A. 1967. Taxonomic and cytological notes on Turkish *Ornithogalum*. *Notes Roy. Bot. Gard. Edinburgh*, **27**: 293–339.

DADAY, H. 1965. Gene frequencies in wild populations of *Trifolium repens* L. IV. Mechanism of natural selection. *Heredity*, 20: 355–365.

DAHLGREN, R. 1963. Studies on *Aspalathus*. Phytogeographical Aspects. *Bot. Notis.* 116: 431–472.

DAMBOLDT, J. 1965. Zytotaxonomische Revision der isophyllen *Campanulae* in Europa. *Bot. Jb.*, 84: 302–358.

DAMBOLDT, J. and PODLECH, D. 1965. Zytotaxonomische Untersuchungen an *Saxifraga*-Sippen der grex *Exarato-moschatae* Engl. et Irmsch. (Zur Zytotaxonomie der Gattung *Saxifraga* L. II). *Ber. Dtsch. bot. Ges.*, 77: 332–339.

DANSEREAU, P. 1952. The varieties of evolutionary opportunity. *Rev. Canad. Biol.*, 11: 305–388.

DAVIDSON, R. A. and DUNN, R. A. 1967. A correlation approach to certain problems of population-environment relations. *Am. J. Bot.*, 54: 529–538.

DAVIS, P. H. and HEYWOOD, V. H. 1963. *Principles of Angiosperm Taxonomy*. Oliver and Boyd, Edinburgh and London.

DIETRICH, W. 1967. Die Zytotaxonomie der *Carex*-Sektion *Frigidae* in Europa. *Feddes Repert.*, 75: 1–42.

DUGLE, J. R. 1966. A taxonomic study of western Canadian species in the genus *Betula*. *Can. J. Bot.*, 44: 929–1007.

EHRENDORFER, F. 1953. Ökologisch-geographische Mikro-Differenzierung einer Population von *Galium pumilum* Murr. (Sect. *Leptogalium* Lange). *Österr. Bot. Z.*, 100: 670–672.

EHRENDORFER, F. 1955. Hybridogene Merkmalsintrogression zwischen *Galium rubrum* L. s.str. und *G. pumilum* Murr. s.str. *Österr. Bot. Z.*, 102: 195–234.

EHRENDORFER, F. 1958*a*. Die geographische und ökologische Entfaltung des europäisch-alpinen Polyploid-Komplexes *Galium anisophyllum* Vill. seit Beginn des Quartärs. *Uppsala Univ. Årsskr.*, 1958(6): 176–181.

EHRENDORFER, F. 1958*b*. Sect. nova *Jubo-Galium* (Rubiaceae: *Galium*), ein alter aufgesplitterter Sippenkomplex mit Zentrum im südöstlichen Mittelmeergebiet. *Österr. Bot. Z.*, 105: 212–228.

EHRENDORFER, F. 1958*c*. Ein Variabilitätszentrum als "fossiler" Hybrid-Komplex: Der ostmediterrane *Galium gracecum* L.-G. *canum* Req.-Formenkreis. *Österr. Bot. Z.*, 105: 229–279.

EHRENDORFER, F. 1959*a*. Differentiation-hybridization cycles and polyploidy in *Achillea*. *Cold Spring Harbor Symp. quant. Biol.*, 24: 141–152.

EHRENDORFER, F. 1959*b*. *Achillea roseo-alba* Ehrendf., spec. nov., eine hybridogene di- und tetraploide Sippe des *Achillea millefolium*-Komplexes. *Österr. Bot. Z.*, 106: 363–368.

EHRENDORFER, F. 1962*a*. Cytotaxonomische Beiträge zur Genese der mitteleuropaischen Flora und Vegetation. *Ber. Dtsch. Bot. Ges.*, 75: 137–152.

EHRENDORFER, F. 1962*b*. Beiträge zur Phylogenie der Gattung *Knautia* (Dipsacaceae), I. Cytologische Grundlagen und allgemeine Hinweise. *Österr. Bot. Z.*, 109: 276–343.

EHRENDORFER, F. 1963. Cytologie, Taxonomie und Evolution bei Samenpflanzen. *In*: *Vistas in Botany*, 4: 99–186. Pergamon Press, Oxford.

EHRENDORFER, F. 1965*a*. Dispersal mechanisms, genetic systems, and colonizing abilities in some flowering plant families. *In*: H. G. Baker and G. L. Stebbins (eds). *The Genetics of Colonizing Species*, 331–351. Academic Press Inc., New York.

EHRENDORFER, F. 1965*b*, 1967: Systematik und Evolution der Samenpflanzen. Bericht über die Jahre 1963/64 und 1965/66. In: *Fortschritte der Botanik*, 27: 348–418; 29: 237–303. Springer-Verlag, Berlin, Heidelberg, New York.

ELLENBERG, H. 1963. Vegetation Mitteleuropas mit den Alpen in kausaler, dynamischer und historischer Sicht. *In*: H. Walter, (ed.). *Einführung in die Phytologie*, 4(2). Eugen Ulmer, Stuttgart.

EPLING, C., LEWIS, H. and BALL, F. M. 1960. The breeding group and seed storage: a study in population dynamics. *Evolution*, 14: 238–255.

FISCHER, M. 1967: Beiträge zur Cytotaxonomie der *Veronica hederifolia*-Gruppe (Scrophulariaceae). *Österr. Bot. Z.* 114: 189–233.

FISHER, F. J. F. 1965. The alpine Ranunculi of New Zealand. *N.Z. Dept. Sci. Industr. Res. Bull.*, 165: 192 p.

FORDE, M. B. 1964*a*. *Haloragis erecta*: A species complex in evolution. *N.Z. J. Bot.*, 2: 425–453.

FORDE, M. B. 1964*b*. Variation in natural populations of *Pinus radiata* in California. Part 3 and 4. *N.Z. J. Bot.*, 2: 459–485, 486–501.

FORDE, M. B. and BLIGHT, M. M. 1964. Geographical variation in the turpentine of Bishop Pine. *N.Z. J. Bot.*, 2: 44–52.

FUKUDA, I. 1967. The formation of subgroups by the development of inbreeding systems in a *Trillium* population. *Evolution*, 21: 141–147.

GILLETT, G. W. 1966. Hybridization and its taxonomic implications in the *Scaevola gaudichaudiana* complex of the Hawaiian Islands. *Evolution*, 20: 506–516.

GÖTZ, E. 1967. Die *Aconitum variegatum*-Gruppe und ihre Bastarde in Europa. *Fedd. Repert.*, 76: 1–62.

GRANT, V. 1964. The biological composition of a taxonomic species in Gilia. *Adv. Genet.*, 12: 281–328.

GRANT, V. 1966. The selective origin of incompatibility barriers in the plant genus *Gilia*. *Am. Nat.*, 100: 99–118.

GRIFFIN, J. R. 1964. Cone morphology in *Pinus sabiniana*. *J. Arn. Arb.*, 45: 260–273.

GRUN, P. and AUBERTIN, M. 1966. The inheritance and expression of unilateral incompatibility in *Solanum*. *Heredity*, 20: 131–138.

HARLAN, J. R. 1951. Anatomy of gene centers. *Am. Nat.*, 85: 97–103.

HARLAN, J. R. 1963. Two kinds of gene centers in *Bothriochloininae*. *Am. Nat.*, 97: 91–98.

HARPER, J. L. 1965. Establishment, aggression, and cohabitation in weedy species. *In*: H. G. Baker and G. L. Stebbins (eds). *The Genetics of Colonizing Species*, 243–265. Academic Press Inc., New York.

HAWKES, J. G. (ed.) 1966. *Reproductive Biology and Taxonomy of Vascular Plants*. Bot. Soc. Brit. Is. Conf. Report: Pergamon Press, Oxford.

HEIMBURGER, M. and KAMITAKAHARA, I. E. 1963. A multiple interchange variant in a natural population of *Anemone cylindrica*. *Evolution*, 17: 358–367.

HESLOP-HARRISON, J. 1964. Forty years of genecology. *Adv. Ecol. Res.*, 2: 159–247.

HOLUBIČKOVÁ, B. 1965. A study of the *Pinus mugo* complex (variability and diagnostic value of characters in some Bohemian populations). *Preslia*, 37: 276–288.

IRWIN, H. S. and ROGERS, D. J. 1967. Monographic studies in *Cassia* (Leguminosae-Caesalpinioideae). II. A taximetric study of section Apoucouita. *Mem. N.Y. Bot. Gard.*, 16: 71–118.

JALAS, J. and RIKKINEN, K. 1962. Populationsstudien in *Melampyrum pratense* L. in Finnland. I. Über die Variation im Wasserscheidegebiet zwischen Pohjanmaa und Nord-Häme. *Ann. Bot. Soc. Zool. Bot. Fenn.* "*Vanamo*", 32: (4) 1–24.

JAMES, S. H. 1965. Complex hybridity in *Isotoma petraea*. I. The occurrence of interchange heterozygosity, autogamy and a balanced lethal system. *Heredity*, 20: 341–353.

JOSHI, B. C. and JAIN, S. K. 1964. Clinal variation in natural populations of *Justica simplex*. *Am. Nat.*, 98: 123–125.

KANNENBERG, L. W. and ALLARD, R. W. 1967. Population studies in predominantly self-pollinated species. VIII. Genetic variability in the *Festuca microstachys* complex. *Evolution*, 21: 227–240.

KATZNELSON, U. and MORLEY, F. H. W. 1965. Speciation processes in *Trifolium subterraneum* L. *Israel J. Bot.*, 14: 15–35.

KERNER, A. VON MARILAUN. 1869. *Die Abhängigkeit der Pflanzengestalt von Klima und Boden. Ein Beitrag zur Lehre von der Entstehung und Verbreitung der Arten, gestützt auf die Verwandtschaftsverhältnisse, geographische Verbreitung und Geschichte der Cytisusarten aus dem Stamme Tubocytisus D.C.* Innsbruck.

KIHARA, H. 1966. Factors affecting the evolution of common wheat. *Ind. J. Genet.*, 26(A): 14–28.

KRENDL, F. 1968. Cytotaxonomie der *Galium mollugo*-Gruppe in Mitteleuropa. *Österr. Bot. Z.*, 114: 508–549.

KURABAYASHI, M. 1963. Karyotype differentiation in *Trillium sessile* and *T. ovatum* in the Western United States. *Evolution*, 17: 296–306.

KYHOS, D. W. 1965. The independent aneuploid origin of two species of *Chaenactis* (Compositae) from a common ancestor. *Evolution*, 19: 26–43.

LANGLET, O. 1959. A cline or not a cline—a question of Scots Pine. *Silvae Genetica*, 8: 13–22.

LEVIN, D. A. and SMITH, D. M. 1966. Hybridization and evolution in the *Phlox pilosa* complex. *Am. Nat.*, 100: 289–302.

LEVINS, R. 1962. 1963. Theory of fitness in a heterozygous environment. I. The fitness set and adaptive function. II. Developmental flexibility and niche selection. *Am. Nat.*, 96: 361–378, 97: 75–90.

LEWIS, H. 1966. Speciation in flowering plants. *Science, N.Y.*, 152(3719): 167–172.

LEWIS, W. H. 1967. Cytocatalytic evolution in plants. *Bot. Rev.*, 33: 105–115.

LINDSAY, D. W. and VICKERY, R. K., Jr. 1967. Comparative evolution in *Mimulus guttatus* of the Bonneville Basin. *Evolution*, 21: 439–456.

LLOYD, D. G. 1965. Evolution of self-compatibility and racial differentiation in *Leavenworthia* (Cruciferae). *Contr. Gray. Herb. Harv.*, 195: 3–134.

LONG, R. W. 1966. Artificial interspecific hybridization in *Ruellia* (Acanthaceae). *Am. J. Bot.*, 53: 917–927.

MÄKINEN, L. and Y. 1964. The distribution, ecology, morphology and taxonomy of *Primula nutans* Georgi ssp. *finmarchica* (Jacq.) Löve & Löve. *Ann. Bot. Fenn.*, 1: 273–291.

MARCHANT, C. J. 1967. Evolution in *Spartina* (Gramineae). I. The history and morphology of the genus in Britain. *J. Linn. Soc.* (*Bot.*), 60(381): 1–24.

MAYR, E. 1963. *Animal Species and Evolution.* Harvard Univ. Press, Cambridge; Oxford Univ. Press, London.

MEIJER, W. 1963. Notes on Borneo Dipterocarpaceae. *Acta bot. neerl.*, 12: 319–353.

MEUSEL, H. 1965. Die Evolution der Pflanzen in pflanzengeographisch-ökologischer Sicht. *In*: *Beitr. Abstammungsl.*, 2: 7–39. Berlin.

MOORE, R. J. and FRANKTON, C. 1965. Cytotaxonomy of *Cirsium hookerianum* and related species. *Can. J. Bot.*, 43: 597–613.

MOORING, J. S. 1966. Chromosome numbers in the *Eriophyllum lanatum* (Compositae, Helenieae) complex. *Madroño*, 18: 236–239.

MOSQUIN, T. 1964. Chromosomal repatterning in *Clarkia rhomboidea* as evidence for post-pleistocene changes in distribution. *Evolution*, 18: 12–25.

MURBECK, S. 1939. Weitere Studien über die Gattungen *Verbascum* und *Celsia*. *Acta Univ. lund.* Ser. 2, 35 (1).

NOBS, M. A. 1963. Experimental studies on species relationships in *Ceanothus. Carnegie Inst. Wash. Publs*, 623.

ORNDUFF, R. 1966. A biosystematic survey of the goldfield genus *Lasthenia* (Compositae: Helenieae). *Univ. Calif. Publ. Bot.*, 40.

PODLECH, D. 1965. Revision der europäischen und nordafrikanischen Vertreter der Subsect. *Heterophylla* (Wit.) Fed. der Gattung *Campanula* L. *Feddes Repert.*, 71: 50–187.

POLATSCHEK, A. 1966. Cytotaxonomische Beiträge zur Flora der Ostalpenländer, I + II. *Österr. Bot. Z.*, 113: 1–46, 101–147.

RAVEN, P. H. and MERTENS, T. R. 1965. *Plant Systematics: Theory and Practice.* BSCS Pamphlets, 23.

RAVEN, P. H. and MOORE, D. M. 1965. A revision of *Boisduvalia* (Onagraceae). *Brittonia*, 17: 238–254.

ROTHMALER, W. 1955. *Allgemeine Taxonomie und Chorologie der Pflanzen. Grundzüge der speziellen Botanik.* 2. Aufl. Gronau Verlag, Jena.

ROTHMALER, W. 1956. Taxonomische Monographie der Gattung *Antirrhinum. Feddes Repert. Beih.*, 136.

ROUSI, A. 1965. Biosystematic studies on the species aggregate *Potentilla anserina.* L. *Ann. Bot. Fenn.*, 2: 47–112.

RUNEMARK, H. 1967. Studies in the Aegean flora. XII. Cytologic and morphologic investigations in *Centaurea. Bot. Notis.*, 120: 161–176.

SAUER, J. 1965. Geographic reconnaissance of western Australian seashore vegetation. *Aust. J. Bot.*, 13: 39–69.

SCHAEFTLEIN, H. 1968. Beiträge zur Kenntnis einiger mitteleuropäischer Euphrasien. I. Der taxonomische Wert der dr–sigen Behaarung. *Phyton* (*Austria*), 12: 48–90.

SCHWARZ, O. 1938. Phytochorologie als Wissenschaft, am Beispiele der vorderasiatischen Flora. *Feddes Repert. Beih.*, 100: 178–228.

SLEUMER, H. 1938. Die Gattung *Agauria* (DC.) Hook. f. *Englers Bot. Jb.*, 69: 374–394.

SMITH, E. B. 1966. Cytogenetics and phylogeny of *Haplopappus* section *Isopappus* (Compositae). *Can. J. gen. Cyt.*, 8: 14–36.

SNOGERUP, S. 1967. Studies in the Aegean flora. IX. *Erysimum* sect. *Cheiranthus*, B. Variation and evolution in the small-population system. *Opera Bot.*, 14: 86 p.

SOLBRIG, O. T. 1964. Infraspecific variation in the *Gutierrezia sarothrae* complex (Compositae–Astereae). *Contr. Gray Herb.*, 193: 67–115.

STEBBINS, G. L. 1952. Aridity as a stimulus to plant evolution. *Amer. Nat.*, 86: 33–44.

STEBBINS, G. L. 1959. The role of hybridization in evolution. *Proc. Am. Phil. Soc.*, 103: 231–251.

STEBBINS, G. L. 1966. *Processes of organic evolution.* XII + 191 p. Prentice Hall Inc., Englewoods Cliffs, N.J.

STOUTAMIRE, W. P. 1955. Cytological differentiation in *Gaillardia pulchella. Am. J. Bot.*, 42: 912–916.

STRANDHEDE, S. O. 1965. Chromosome studies in *Eleocharis*, subser. *Palustres.* III. Observations on Western European taxa. *Opera Bot.*, 9(2): 86 p.

STRID, A. 1965. Studies in the Aegean flora. VII. Chromosome morphology in the *Nigella arvensis* complex. *Bot. Notis.*, 118: 139–165.

STUBBE, H. 1966. *Genetik und Zytologie von Antirrhinum L. sect. Antirrhinum.* G. Fischer Verlag, Jena.

THEOBALD, W. L. 1966. The *Lomatium dasycarpum-mohavense-foeniculaceum* complex (Umbelliferae). *Brittonia*, 18: 1–18.

TORRES, A. M. 1965. The chromosome races of *Zinnia juniperifolia. Am. J. Bot.*, 52: 760–765.

UHL, C. H. 1963. Chromosomes and phylogeny of the Crassulaceae. *Cact. Succ. J.*, 35: 80–84.

VASEK, F. C. 1967*a*. Outcrossing in natural populations. III. The Deer Creek population of *Clarkia exilis*. *Evolution*, 21: 241–248.

VASEK, F. C. 1967*b*. The distribution and taxonomy of three western Junipers. *Brittonia*, 18 (1966): 350–372.

VAVILOV, N. I. 1951. The origin, variation, immunity and breeding of cultivated plants. *Chronica Bot.*, 13 (1/6).

VICKERY, R. K. Jr. 1966. Speciation and isolation in section *Simiolus* of the genus *Mimulus*. *Taxon*, 15: 55–63.

WAGNER, M. 1868. *Die Darwin'sche Theorie und das Migrationsgesetz der Organismen*. Duncker u. Humblot, Leipzig.

WCISŁO, H. 1967. Karyological studies in *Caltha palustris* L. s.l. *Acta biol. cracov.*, 10: 1–23.

WERNER, K. 1964. Die Verbreitung der *Digitalis*-Arten. *Wiss. Z. Martin-Luther-Univ. Halle-Wittenb.*, 13(6): 453–486.

WERNER, K. 1966. Die Wuchsformen der Gattungen *Isoplexis* (Lindl.) Benth. und *Digitalis* L. *Bot. Jb.*, 85: 88–149.

WETTSTEIN, R. v. 1898. *Grundzüge der geographisch-morphologischen Methode der Pflanzensystematik*. Fischer Verlag, Jena.

WHITTAKER, R. H. 1967. Gradient analysis of vegetation. *Biol. Rev.*, 49: 207–264.

WOODELL, S. R. J. 1965. Natural hybridization between the cowslip (*Primula veris* L.) and the primrose (*P. vulgaris* Huds.) in Britain. *Watsonia*, 6: 190–202.

WOODSON, R. E. 1964. The geography of flower color in Butterflyweed. *Evolution* 18: 143–163.

Author Index

The numbers in *italics* indicate the pages on which names are mentioned in the reference lists.

Subject Index

L

M

N